HIGH ANGLE RESCUE TECHNIQUES

REVISED EDITION

by

Tom Vines
Steve Hudson

EDITED By:
Donna Carter

ILLUSTRATION By:
Stella Carpenter Twilley

DESIGN By:
Innovative Projects, Inc.

PHOTOGRAPHY By:
William L. Renaker
Steve Hudson

Additional Photographs Provided By:
Pigeon Mountain Industries, Inc.
Seattle Manufacturing Corp.
Petzl
Mountain Tools
Lirakis Safety Harness
US Office of Aircraft Services
US NAVY
Bruce J. Waltz

A PUBLICATION OF

The National Association for Search and Rescue

KENDALL/HUNT PUBLISHING COMPANY
2460 Kerper Boulevard P.O. Box 539 Dubuque, Iowa 52004-0539

ISBN 0–8403–7363–5

Printed in the United States of America
10 9 8 7 6 5 4 3

ACKNOWLEDGEMENTS

As with most projects of the size of *High Angle Rescue Techniques,* this text represents the work of many more people than just the two authors. We would like to thank all those individuals and organizations who helped out with the creation of this textbook. It is impossible to list all the individuals who have taught us techniques, made suggestions, listened to ideas, told us of screw-ups (ours and theirs), introduced us to others with special skills and all the other things that have helped shape our own knowledge. Many of these people were providing such help years before we came up with the idea for this book. Now we can say "thank you!"

Several individuals and organizations do deserve special acknowledgement for their direct contributions to our efforts. Photographic assistance, equipment and models were provided by: Chattanooga-Hamilton County Rescue Service, Chattanooga Fire Department, Walker County Georgia Rescue, Pigeon Mountain Industries, Inc., Ken Fagan, Bill Renaker, Allen Padgett, Karen Padgett, Kenneth Huffines, Lisa Smith, Diane Cousineau, John Reid, Beth Elliott, Reggie Ferguson, Mark Wolinsky, Don Black, Sammy Manning, Bill Lord, Melinda Lord, Randolph Lane, Hank Moon, Buddy Lane and Doranne Lane. Also Robert Canan, Linda Wilske, Michael J. Casey, Robin Pinkstaff, Steve Dewell, Joy Eden, and Merle E. Froslie. Special thanks to Peggy McDonald and Sharon Pepper of the NASAR office for the hands-on work of production. All of these people could not have donated their time unless their spouses, families and employers were willing to spare a little of their time also. Once again . . . THANK YOU ALL.

TABLE OF CONTENTS

Introduction

How This Book Came to Be Written

The past several years have seen a great revolution in all aspects of the high angle environment, in particular, rope rescue. There have been great changes not only in rope and hardware, but also in the concepts of how things should be done. Many questionable beliefs about techniques and training have been swept aside, while the truly sound ones have survived these changes.

One of the most significant changes has been the sweeping aside of the self-anointed "experts". We have all met these and, most likely, been subjected to them in training. You know how the "expert" first appears: as if he were Moses, just down from Mount Sinai, allowing himself to deliver to us his law engraved on the stone tablets. He is often very self-assured and intolerant of other people's opinions. When you question him, you are questioning the very word of God. Yet, these "experts," who claim to be the highest authority on the subject, often create misleading and, sometimes, dangerous concepts.

Our feeling is that the world of rope rescue is too complex and changing for anyone to consider himself an expert. We believe that the best service that we can make to those people in the field is provide them with the best possible information available, collected from many reliable sources.

HIGH ANGLE RESCUE TECHNIQUES was written over many years. Neither one of the authors considers himself to be an expert. We do feel, however, that we have been fortunate to be in the position of having access to a great deal of information from many different sources in North America and in Europe. Nevertheless, *HIGH ANGLE RESCUE TECHNIQUES* cannot be claimed to be the absolute and last word on all kinds of rescue in any place. No one publication can be. Fortunately, in recent years, *a body of knowledge* about rope rescue has developed that is made up of many publications, meetings, and viewpoints. We see *HIGH ANGLE RESCUE TECHNIQUES* as a contribution to this body of knowledge.

We believe that no one should ever stop learning. Each time we work with someone new, teach a class, work a rescue, or just sit around discussing rope work, we learn more about the high angle environment. In today's world of rescue, we must continue to learn or we will soon fall behind.

Because the technology of rope work is constantly advancing, a book such as this one will always be out of date. The constant advance means that a better/faster/safer way will always be found. It may take several years for that new way to gain acceptance by most, but if it stands the test of time, it will eventually be accepted. As someone interested in the high angle rescue field, you must continually keep yourself informed of these improvements, but you must also be wary of the "overnight sensations" in techniques and equipment that may not prove to be so great in the light of day.

We welcome any comments on this manual with regard to its relevancy to your own work and to its comprehensibility. We also welcome any suggestions on technique or equipment that you have found useful.

We believe that the greatest benefit for all of us in the rescue community is not from those "experts" who consider themselves God's official spokesman on rope rescue, but from all of us working together to develop a common source of knowledge.

How To Use This Book

This book is designed specifically as a training manual to be used under the guidance of a qualified instructor. **It is not written as a self-instructional text.** High angle ropework is a dangerous activity that requires good basic understanding of the high angle environment combined with the precise use of well-developed skills, all learned with qualified guidance.

Listed below are examples of the sections and special features that we have used in each chapter of the book. Each section or feature is explained and an example is shown of how your attention is drawn to it.

OBJECTIVES–

At the completion of this chapter, you should be able to:

This manual is written with a number of objectives in mind that relate to knowledge or skills that you should be able to display after completion of a particular chapter. Mastery of this information or these skills indicates that you have taken steps towards competency as a high angle rope technician. At the beginning of each chapter, the particular objectives are defined. To help measure your progress towards mastery of the information and skills in each chapter, review "Questions for Review" (see below) at the end of each chapter.

TERMS– *relating to (subject) that a High Angle Rope Technician should know:*

The terminology used in modern high angle rope activities is taken from a number of different languages and from such diverse activities as mountaineering, caving, sailing, firefighting, and textile manufacturing. A command of the language of the high angle environment is essential for good communication in that environment. At the beginning of each chapter is a listing of new words introduced in that particular chapter. For your convenience, all of these words are also reproduced in the Glossary at the end of the manual.

QUESTIONS *for* REVIEW

To assist you in the learning process, the end of each chapter has a "Questions for Review" section on the major elements in that chapter.

★ ACTIVITIES ★

To assist in reviewing specific skills, many of the chapters have "Activities" section at the end of each chapter.

WARNING NOTE

One of the most important features of this manual is the use of the **WARNING NOTE.** As will be said numerous times in this manual, the high angle environment is a particularly dangerous one, and there are a number of important warnings to be kept in mind when operating in it.

When you come across **WARNING NOTE** in the text, it is essential that you regard it as a sign warning of danger. Stop, carefully read the note, and make certain that you understand it before proceeding further.

Advantages/Disadvantages

There are a number of activities in which people tend to make hard and fast rules. High angle activity is one of them. We believe strongly that there must be firm, never-to-be broken rules on safety. However, in terms of equipment and techniques, there may be choices. It may be dangerous to tell a person that there is only one way to do something. If that one way does not work in a situation, the person—lacking alternatives—will be in trouble.

Our philosophy is that the ideal high angle rope technician is one who is well-trained and who has the ability to deal with any kind of situation he may encounter by using a variety of intelligent options. Consequently, in most situations involving equipment and technique, we have listed various options so that you can make intelligent choices. We are certain that some people will be disappointed by not being told that they absolutely have to do things a certain way. But we believe that the approach in this manual is not only more honest, but also more effective in the long run since it leads to a better informed person who can make intelligent decisions whatever the situation.

As you examine the alternatives, it is not necessarily the number of advantages or disadvantages that should guide your decision, but how relevant they are to your situation.

ALTERNATIVE APPROACH

This manual will usually describe what many consider the preferable approach for certain techniques. However, we recognize that there are differences of opinion on many techniques and equipment, and some of these alternatives, though not what we prefer, are legitimate for some situations. One of the primary examples in this manual is our preference for the Figure 8 family of knots over the Bowline. While we believe that the Figure 8 does have certain distinct advantages, the Bowline is commonly used by many people. In many situations such as this, we will list the ***ALTERNATIVE APPROACH.***

SUGGESTION:

In many aspects of high angle activities there are small hints, often learned after years of experience, that may not be necessary for doing the job, but may make it easier. Throughout the manual these will be listed under **SUGGESTION**.

In the same way, there are numerous bits of knowledge that will assist you in achieving excellence in your work. Throughout the manual these will be included in the **SIDEBARS**.

PREREQUISITES

To remind users of the manual of this important principal, starting with Chapter 9, "Rappeling," a list of prerequisites is stated before each chapter. You will note that this list of prerequisites gets longer in the more advanced chapters. **It is very important that the student does not move on to these more advanced skills without first *THOROUGHLY* accomplishing the prerequisite skills.** If the student finds he has lost some of the knowledge of or ability to perform these basic skills, then he should go back and review them.

HIGH ANGLE RESCUE TECHNIQUES is divided into two sections. The first nine chapters are concerned with personal rope skills, the kind that every *individual* must have to become a high angle rope technician. The second half of the book, Chapters 10 through 17, focuses on rescue skills that every rescue team must have.

HIGH ANGLE RESCUE TECHNIQUES is a step-by-step manual, creating in effect a series of building blocks in high angle training. In this manner, the most critical skills, such as knowledge of rope and equipment, are taught initially, followed by skills such as belaying and rappeling, until they are all put together for team rescue skills.

It is essential that the book be used in this manner with the basic skills used as the foundation for the more advanced, team skills. We have observed that when things go wrong, the single most important cause is inattention to basic skills.

Chapter 1

The High Angle Environment

OBJECTIVES–

At the completion of this chapter, you should be able to:

1. Explain the differences between rock climbing and and Single Rope Techniques (SRT).
2. Discuss the importance of being prepared for self-rescue.
3. Explain the importance of continual practice of high angle skills.

TERMS– *pertaining to the high angle environment that a high angle rope technician should know:*

Anchors—The means of attaching the high angle system to secure points so that the system will not fail.

High Angle—The high angle environment in which one must be secured with rope and other equipment to keep from falling.

Mountaineering—The use of skills such as climbing, snow and ice travel, and camping to ascend a mountain.

Rock Climbing—Ascending while making direct contact with the rock and commonly using rope and other equipment for safety in the event of a fall.

Rope Rescue—The performing of rescue in a high angle environment where the use of rope and related equipment is necessary.

Single Rope Techniques (SRT)—Ascending and descending directly on the rope without direct aid by contact with the rock.

System—The combination of all the various elements, including rope, hardware, anchors, etc., used in the high angle environment.

Technical Rescue—*See Rope Rescue*

Vertical Caving—Travelling through caves that have vertical, or near vertical, sections that require the use of rope and ascending and descending equipment.

The High Angle Environment

Historically, mankind has been able to conquer much of the world around him by devising means to operate comfortably in environments, such as the ocean, that nature had originally created hostile to him. One of the last environments that mankind has conquered is the high angle environment. Here, gravity is the great adversary, and though mankind has not vanquished it, he has been able to devise the means to temporarily overcome it and in some cases to use it to his advantage.

There are actually a number of different ways of operating in the high angle environment, depending on what the needs are.

Mountaineering

Men have been engaged in the vertical sport of mountaineering for hundreds of years. It is actually a combination of a number of activities including climbing, camping, snow and ice travel and often the test of will to survive against the very worst of natural forces. While the most obvious goal is to reach the top of a mountain, there are often less obvious goals such as surviving the worst that nature can throw at you: including cold, high wind, and high altitude sickness.

Since the World War II, there have been tremendous strides in mountaineering, among the more significant being the development of rope using synthetic fibers.

Climbing

Sometimes further defined as **rock climbing,** this is a much more specialized activity than mountaineering, and often may not involve much change in altitude or a walk much farther away than the parking lot. Often it may involve the ascent of a short piece of rock that is so difficult that the climber may spend weeks working out the right "moves" so that he can make the climb efficiently. Although "ropeless-free solo climbing," ascent alone and without safety gear such as rope, is in vogue among some highly skilled climbers, most climbing is still done while roped. The two essential elements of roped climbing are: **1)** a rope that has a great deal of stretch so that it will catch the climber when he falls and **2) protection** or anchoring hardware so that the rope can be attached to the rock and "protect" the climber.

Vertical Caving

Vertical caving is one of the youngest of the vertical sports, having its beginnings only since the 1950s. It differs significantly from climbing in that instead of climbing the rock itself, one descends and ascends on the rope and does not use the rope for protection in case of a fall from the rock. This is known as **Single Rope Techniques** or **SRT.**

Although vertical caving is not a generally widespread sport, it has already had significant influence on the equipment and techniques used in other vertical activities, particularly rescue. One of the more significant of these has been the development of a rope that is very durable and stretches very little (***static rope***).

Rope Rescue

Depending on who is involved in it and the conditions, rope rescue may also be called **vertical rescue** or **technical rescue.** In these situations, a rope and other associated gear are necessary so that the subject of the rescue is kept stabilized, kept safe, or kept from falling.

Originally, most of the equipment and techniques used in rope rescue was the same as that used in mountaineering. But during the past decade, there has been a tremendous shift from these types of equipment and techniques to ones that are more specialized for rescue. One significant example is rescue rope, which has shifted from a stretchy line, as used in mountaineering, to a line with very little stretch and a more durable construction, similar to that used in vertical caving.

Fire Service Rescue

The use of life support rope and associated equipment in the fire service has undergone tremendous changes during the past several years. Part of this has been due to well- publicized tragic occurrences, such as the incident in New York City in June of 1980 when a rope broke during a rescue attempt and two firefighters fell to their deaths. Further impetus for these changes have come from such fire service organizations as the International Association of Fire Fighters (IAFF) and the International Society of Fire Service Instructors (ISFSI).

Perhaps the most significant of these changes has been the realization that the continued use of natural fiber rope for life safety is a dangerous and irresponsible practice, with the resulting massive changeover by most fire departments to synthetic fiber ropes.

Tactical Operations

The increase in the number of terrorist incidents and other forms of violence has led to the greater preparation by the law enforcement and the military to employ high angle operations. Many of these innovations have been in operational procedures, such as with helicopters, but the tactical practitioners have contributed to the high angle technology that is used by many other disciplines. One significant example of this is the Figure Eight with Ears descender. An early form of this device was developed by the Special Air Services (SAS) in England, but was then later adapted initially in North America through the joint effort of California Law Enforcement personnel and mountaineer/inventor, Russ Anderson.

Industrial Rescue

Industrial rescue incidents occur in high angle areas where there is the potential of falls, entrapment, accidents or medical emergencies. Some examples would be refineries, chemical plants, and both open pit and underground mining. The principles for high rescue are the same in industrial environments as in natural environments, but environmental circumstances may either help or hinder industrial rescuers. In industrial rescue situations, there is usually an abundance of structural anchors available, there often are many rescuers available, and the response time is generally short. However, often there will be hazardous materials and atmospheres that mean the rescuers must protect themselves with protective clothing and self-contained breathing apparatus. In addition, the presence of hazardous materials can mean that the rescuers have to use specialized equipment such as different rope materials to protect from damage. Industrial rescue may be further complicated by confined spaces.

The High Angle Rope Technician

While there has been a tradition of specializing in a particular discipline, such as rock climbing, many people are taking those techniques and equipment from all disciplines in the high angle environment to put it to use for their own particular needs. There are those who might be described as **high angle rope technicians,** who are competent in a number of high angle skills and who can use them for particular needs, such as rope rescue.

There are some basic principles that a high angle rope technician uses to be both safe and successful in his work.

The Technology

While most people associate the word technology with computers or space travel, the dictionary use of the word is **"the application of knowledge."** Thus the high angle rope technician uses the technology of the high angle environment to conquer gravity and move about with ease in any direction he desires to take his body. Or he uses the technology of rope rescue to safely retrieve a subject from a dangerous situation to one that is safe and where any injuries can be cared for.

The High Angle System

One ideal that should be kept in mind by anyone working in the high angle environment is: **THINK SYSTEMS.** While some elements are essential and important, it is the rare activity in the high angle environment that is completed with only one rope or only one person.

The high angle environment system consists of many elements—rope, hardware, anchors, and other things—that cannot be viewed by themselves. To operate safely and properly, they must be viewed together as a system.

Just as a chain is only as strong as its weakest link, the high angle system is only as effective as its weakest element. Let's say that in the equipment system you are using a rope that has a tensile test strength of 9,000 pounds, but it is attached to an anchor that pulls out at 500 pounds. The strength of the entire system is only 500 pounds.

The same idea applies to people systems in the high angle environment. It is human nature for a few motivated and aggressive individuals to be the ones who participate in every activity and attend every training session. It is also human nature for some persons to stand back and let others do the work.

But inevitably it is the well-trained individual who is not available during the unexpected emergency, and the burden is shifted to the poorly trained and badly motivated. In such an instance, the human chain can fail and disaster may result. And if a group has only a few well-trained members and the majority is badly trained and poorly motivated, then the team will probably be only as effective as this majority.

This first chapter examines the necessary elements of the high angle system and how their strengths and weaknesses determine the effectiveness of the system and, ultimately, the outcome of the operation.

Comfort in The High Angle Environment

The fear of heights is natural to human beings and a degree of it is necessary for survival. A person without any fear or respect for the hazards involved in high angle work is a danger to himself and others. But until a person feels at ease in operating in the high angle environment, his discomfort will prevent him from being effective in these activities. As with any unaccustomed environment, there is a whole new approach to movement, to using equipment, and to working alongside other people. The only way to become accustomed to working in the high angle environment, and to become effective in it, is to spend time in it. In other words, **practice.**

Attention to Detail

Attention to detail is necessary in the high angle environment not only to be effective, but also to prevent injury to yourself and others. This environment is unforgiving and the kind of lapse that might go unnoticed on level ground could result in severe injury or death in the high angle environment.

There is of course a need for balance. You must not be obsessed to the point of being unable to perform a task. But it takes attention to quickly examine a system and instantly know if all the necessary details are in place. And this only comes with practice.

While attention to detail is an essential trait to the high angle rope technician, the nature of high angle operations requires that a person has the ability to improvise.

Every high angle situation is different, with varying circumstances in terms of weather and terrain. So a well-trained individual with good judgment is preferable to one who is well-trained to perform only one way.

Prepare for Self-Rescue

Whatever the activity, a person who works in the high angle environment must always keep in mind that anything can go wrong and someday probably will. So, you should be ready for something to go wrong, and be ready to extricate yourself from it. This means that you should be mentally prepared and trained for an emergency. But it also means always having the gear on your person that is necessary to perform self-rescue, including a small assortment of carabiners, a couple of slings made from webbing or rope, and either a set of Prusik loops or a pair of ascenders (Chapter 10, "Ascending," explains the specific purpose of this gear).

Back up Others

In the same way, every person should be ready to extricate a partner or any other coworker who gets into a difficult situation. Whenever a team member is not performing a specific task in the high angle environment, he should be directing his attention to the activities of others. Everyone, no matter how intelligent and experienced, has an occasional lapse. All team members should be ready to deal with any unsafe condition that may develop.

Care of Equipment

One sign of a good mechanic or carpenter is his care for his tools. The same is true of a high angle rope technician, who must care for his rope and hardware. But in the case of high angle gear, good care is even more critical, for lives will depend upon these tools. Considerations for avoiding loss and damage to rescue gear include the following:

- Do not lay unsecured equipment at the edge of a drop. It has a good chance of being kicked or knocked over the edge resulting in damage or loss or, even worse, causing injury to those below.
- Do not lay equipment on the ground or on the floor. High angle hardware, such as carabiners, is easily lost in leaves, dirt, or debris. Also, dirt and grit cause it to malfunction and wear out. When first arriving on site, hang a sling from a tree or other convenient spot and clip all hardware to it when not in use.
- When working on a vertical face, keep all of your equipment attached to something. A convenient place is the equipment sling that many harnesses have.
- Inspect all gear after each operation. The time to discover that gear is damaged is during this inspection, not when it is being used.
- Ropes need special care (Chapter 4 details these concerns.) A traditional rule is **NEVER STEP ON A ROPE.** Many experienced vertical people have almost a religious fanaticism about never stepping on a rope, but there are good reasons for this rule. Stepping on a rope grinds grit into the core where it can damage the load-supporting fibers. But more than that, to many people, a person who steps on his rope displays a contempt for the well-being of the person who uses the rope.

Attention to Skills

The difference between a competent high angle rope technician and one who can only talk about doing it is that the competent technician really has the skills for the job. It is essential that the person not just have an understanding of the skills, but that use of the skills be instinctive. When there is a sudden and unexpected emergency, people react with actions that are instinctive. The only way to make high angle skills instinctive is practice.

To assist in demonstrating that every student has indeed performed the skills satisfactorily, an evaluation sheet should be used. An example of an evaluation sheet is shown in Appendix II. Note that this checklist has a space for the instructor to indicate that the student has performed the skills satisfactorily and a space for the student to confirm that he has performed the skills.

This method helps to double check the training system and may protect the instructor at a later date should the student fail to perform the skill correctly and claim the failure was due to a training lapse.

Continued Skills Maintenance

Being instructed in high angle techniques is only the first step. The reality of the situation is that it is difficult for individuals to maintain their skills without constant and regular training.

Safety

There can be no doubt that the primary objective of high angle activity is **DO IT SAFELY.**

This primary objective is achieved both through mental and physical concepts:

Mental Concepts

- If you feel "spaced out" due to fatigue, heat, or cold, stand back from potentially dangerous activity.
- If you become over-excited, back off for a bit. (One technique is to feel your pulse. If it is racing well beyond your normal rate, sit down, cool off.)
- If you feel you are getting in over your head, ask for help (a belay, if appropriate) or back off.
- Everyone makes mistakes, so everyone checks everyone else, even the most experienced, for possible lapses.
- No one goes into the high angle environment with his capacities diminished by alcohol or drugs.
- Practicing technique helps people to perform safely in the high angle environment.

Physical Concepts

- Establish safety lines; everyone at the edge must be tied in.
- Use redundant systems, for example, more than one anchor.
- Everyone wears a helmet.
- Check equipment constantly to make certain it is in safe condition.

Warning Calls

One of the most common dangers in the high angle environment is from falling objects, either hardware dropped by others or dislodged rocks. Whenever a hard object begins its fall, even if no one is thought to be below, the universal warning is, **"Rock!"** yelled ***very loudly.*** Do not say, "look out," "heads up," or anything else other than "Rock".

Chapter 2

Personal Equipment and Protection

OBJECTIVES–

At the completion of this chapter, you should be able to:

1. Discuss the kinds of clothing and personal protection required for the high angle technician.
2. Describe the characteristics of a secure and comfortable seat harness.
3. Tie a secure emergency seat harness.
4. Describe the standards that apply to equipment used for personal safety in a high angle environment.
5. Discuss the need for and the method of selecting personal equipment required in the high angle environment.
6. Discuss how national and international standards for rescue equipment apply to your work.
7. Describe the dangers posed by the careless use of knives in the high angle environment and the alternatives to the use of knives.

TERMS– *relating to personal equipment and protection that a High Angle Rope Technician should know:*

Chest Harness—A type of harness worn around the chest for upper body support. In the high angle environment, it should never be used as the only source of support, but always be used in combination with a seat harness.

Emergency Seat Harness—A temporary, tied harness to be used when a manufactured, sewn seat harness is not available.

Full Body Harness—A type of harness that offers both pelvic and upper body support as one unit.

Helmet—Head covering that protects against head injury both from falling objects and from head impact. When used in this manual, "helmet" indicates head protection specifically designed for high angle work.

NFPA—National Fire Protection Association. A national organization that sets safety standards, among them life safety equipment for firefighters.

Safety Belt—A belt-like harness worn around the waist to prevent falls from elevated positions. It should never be used as sole means of suspension.

Seat Harness—A system of nylon or polyester webbing that wraps and supports the pelvic region to attach the wearer to the rope or other protection in the high angle environment.

UIAA—The Union of International Alpine Associations. An organization that sets performance standards for ropes, harnesses, ice axes, helmets, and carabiners to be used by climbers and mountaineers.

The correct choices of clothing and personal equipment to be used in the high angle environment can provide a greater margin of safety, add to the comfort of the user, and enable the high angle rope technician to perform the job more effectively and more efficiently.

Headgear

One of the most critical pieces of personal equipment protection is the helmet. This is not only to protect from falling objects, such as rocks or climbing hardware, but also to reduce the severity of brain injury should the wearer happen to fall and hit his head.

Only helmets specifically designed for high angle work should be purchased. Other types of helmets may provide only an illusion of protection and, in the end, actually be dangerous to the wearer. For example, **construction type or motorcycle helmets are not suitable for high angle work.** They are not designed to offer protection from the forces that may be applied to the head in the high angle environment. They also may be uncomfortable or inconvenient to wear in the high angle environment. Motorcycle helmets, for example, tend to be very uncomfortable during hot weather and reduce hearing. Construction helmets offer only minimum protection at best and tend to slip off the head when protection is most needed.

It is very important the helmet have a secure chin strap. Helmets with elastic chin straps are not suitable for high angle work. As the elastic chin strap stretches, the helmet may flip off the head, leaving the wearer without head protection. This often occurs when the helmet is stressed, such as when the wearer falls or is hit by falling objects.

Any helmet used in high angle work should also have what is called a "three-point suspension." This means that in addition to support points on each side of the helmet, there is a third one at the rear. The third suspension point helps prevents the helmet from falling forward over the eyes.

The shells of high angle helmets are constructed of such substances as glass-reinforced polyamide or a fiberglass composite. This shell should have the rigidity to resist impacts delivered to the helmet and penetration by sharp objects. But at the same time, the shell should have some "give" to absorb some of the blow which otherwise would be directly transmitted to the skull and spine. The design of the helmet should protect the head against objects falling from above and hitting from the side.

The inside suspension of the helmet should hold the shell away from the skull during blows to the helmet and provide air circulation and comfort, particularly during hot weather

A brim helps prevent rain water or spray from dripping into the face. But the helmet should have a profile that is narrow enough so that the wearer has a good field of vision and so that the back of the helmet does not catch on a pack when the wearer raises his head.

Helmets that have passed the UIAA certification are usually adequate for high angle work. Helmets with the NFPA certification have good impact resistance, but many fire helmets are inadequate for high angle rope work. Those with extended rear brims are cumbersome in the high angle environment, and most of them are very uncomfortable in hot weather.

Don't try to go economy when purchasing a helmet for high angle work. A few extra bucks for a helmet with good suspension and impact absorption is cheap compared to the cost of a good bang on your brain.

Clothing

Clothing should protect the high angle rope technician against adverse environmental conditions and provide maximum comfort for any activity. Shirts and pants must be cut full so that they will not bind when arms are extended above the head or when legs are raised.

Because persons involved in high angle activity may have to be outdoors for long periods of time during inclement weather, they must be able to protect themselves against the chilling effects of rain and cold. One important piece of outerwear is the "shell" made of waterproof material such as coated nylon. This can protect both against precipitation and the cooling effects of wind. But remember that with intense physical activity, the lack of air circulation under the shell can result in the production of large amounts of perspiration. This may result in the interior of the shell being as saturated as the outside. So there must be some way for "venting" heat away from the body before perspiration begins. One way of doing this is with a zipper down the front of the garment. Some outdoor clothing has zippers along the side or under the armpits to help release the body heat before it produces perspiration.

There are now several types of fabric, such as Goretex™, designed to allow body perspiration to escape while protecting from precipitation. This concept, however, still remains controversial. Some users of this type of fabric claim that it adequately protects from precipitation, while others contend that under incessant and heavy rain the fabric allows wetness to soak through. But whatever the specific type of rain wear you use, if you can stand for half an hour under a shower and still not get soaked through, then it will probably protect you in the outdoor high angle environment.

Under the shell should be an insulating layer of clothing. This provides warmth, but it must also protect the wearer against chilling even when wet, either from outside precipitation or from perspiration. Cotton is the least desirable fabric in wet/cold environments, since it tends to lose its insulating qualities when wet. The traditional choice for a fabric that retains warmth when wet is wool. A more recent development, and a material that is more comfortable next to the skin, is polyester pile. Both fabrics, particularly pile, offer little shielding from wind, so they need to be worn in combination with an outer windbreaker shell.

For the outdoors, underwear made of synthetic materials such as polypropolene is being used more and more. Its advantages are that it dries quickly and it tends to draw moisture away from the skin and prevent it from having a cooling effect on the body.

CAUTION: Synthetic material underwear may not be appropriate for persons working in helicopters or other situations where flash fire is a possibility. Under intense heat, such as a flash fire, synthetic material underwear may melt into the skin, complicating the burn injury.

Footwear

Among the requirements for footwear worn by the high angle rope technician are comfort, protection, and adhesion.

Though boots are increasingly being partially or completely fabricated of materials such as Goretex™ or plastic, leather is still the material having the qualities most needed in a high angle multi-purpose boot.

Boots should provide support to the ankles and protect the feet from scrapes, cuts, and bruises. Yet they should be pliable enough to be comfortable after hours of standing or walking. To help the wearer maintain balance against surfaces found in the high angle environment, the soles should have adhesion, not the slickness found on the soles of street shoes. Rubber boots, such as those commonly used in the fire service, are not appropriate for high angle operations. They do not have the needed foot protection, and they encumber the feet and legs.

Lug soles, particularly those of the material Vibram™, have been very much in fashion, but they may not be necessary as long as the boot sole provides adhesion. Furthermore, in some cases a lug sole may actually be a disadvantage. Some types of lug soles may become dangerously slick once they are wet or caked with mud.

For specialized rock climbing, technical climbing boots with soles constructed of special rubber compounds may adhere to the rock better. Many of these climbing boots have little or no welt, so they may be better for certain climbing techniques such as "edging" along the rock. But they are specifically made for climbing, so they are not comfortable for walking or standing for long periods of time.

A good choice of socks is important for warmth, comfort, and prevention of injury such as blisters. A two-sock combination is often used. An inner sock, made of a synthetic such as polypropolene, wicks moisture away from the foot so that it stays dry. A thick outer sock, made of a material such as wool, increases the warmth and provides some protection for the foot.

Gloves

Gloves are worn in the high angle environment to protect hands against the weather and, more importantly, to protect them against burns and abrasions from a running rope. Gloves shield the hands and prevent discomfort that might cause the high angle rope technician to lose control of the rope.

While providing protection, hands must also retain a sense of feeling so that the fingers can manipulate equipment. For this reason, gloves constructed of soft leather, such as deerskin or goat skin, offer the best compromise. Heavily insulated gloves, such as those sometimes worn by firefighters, should not be used in ropework. They tend to prevent the proper feel of the rope when rappeling or rope handling.

Commercial versions of rope handling gloves are available that have added protection across the palm, but with thinner material on the fingers so they permit the hands to retain a sense of feeling.

Seat Harnesses

Seat harnesses are among the most important pieces of equipment for high angle ropework, and it is essential that they be carefully chosen for safety, security, and comfort.

High angle seat harnesses are constructed of nylon or polyester webbing that wraps the pelvic region to support it and attaches the rescuer to the rope or other protection (*see Figure 2.1* for an example of a sewn, manufactured seat harness).

An unsuitable seat harness, or one that is badly fitted, can result in such severe discomfort that it can inhibit the wearer from performing a task. Even worse, it can be dangerous to the wearer.

WARNING NOTE

Lifebelts, ladderbelts, pompier belts, and similar "safety belts" with support only around the waist must not be used as the single point of support in high angle activities.

These types of equipment are designed only as a safety element to help prevent falls from ladders or other elevated positions. When a person hangs free in them, the belts can constrict the waist and rib cage to impair breathing and cause possible damage to interior organs. They can also slip up under the armpits where they may cause permanent damage to nerves and result in permanent paralysis to the arms. Their use in place of a seat harness as the single support may result in injury and/or permanent disability.

The most secure and comfortable of the seat harnesses are those which are pre-sewn and manufactured. While the tied "emergency" harnesses may be used in a pinch (two types are described below), they are no substitute for a well-designed, sewn, manufactured harness.

It is extremely difficult to construct a tied seat harness that is as secure and comfortable as a well-designed and carefully manufactured product. One reason for this is that knots are not as secure as good stitching. Another reason is that the narrow webbing usually employed in tied harnesses does not give the support of the wider material used in manufactured harnesses. This narrow material causes constricted blood circulation and severe discomfort to body parts.

In a high angle situation, a person may have to be hanging in the harness for a relatively long period—20 minutes is not at all unusual, and periods of more than an hour are a possibility. While there is no harness that will be totally comfortable in these circumstances, the tied seat harness tends to cause greater discomfort and possible circulatory problems.

While hanging in an inadequate seat harness, the wearer's discomfort begins as the narrow webbing or rope compress the

Fig. 2.1 Manufactured Sewn Seat Harness

kidneys and thighs. This becomes even more painful as circulation to the legs is constricted. After only a few minutes, leg muscles are deprived of profusion and become useless. Ultimately, the person may be endangered from dangerous blood pressure changes.

SUGGESTION:

For those situations where you may be suspended in a seat harness for long periods of time, especially in a free drop (not in contact with a wall), even the best harness will become uncomfortable. The addition of foot loops, stirrups, or etrier (short ladder made of webbing) to stand in momentarily will provide relief by restoring circulation.

The foot supports can be attached with a Prusik knot or ascender on the rope above your rappel device or rope attachment point (see Chapter 10, "Ascending," for information on Prusiks and ascenders), or they can be attached directly to the seat harness main attachment point.

Some of the characteristics of a secure and comfortable seat harness are:

- The webbing should be wide (at least two inches) for comfort at critical points such as waist and thighs. The webbing can be made even more comfortable with padding.
- Stitching should be sewn securely and evenly, and be of contrasting color so that abrasion and wear can be detected.
- Leg/thigh supports, such as leg loops, add comfort and support by spreading the body weight, or the force of a fall, over other portions of the body, such as the thighs and buttocks.
- The harness should allow freedom of movement both when hanging in it and when wearing it on the ground.
- The harness should be easy to put on and to adjust.
- The harness should not slip down when you walk around.
- When you fall and are caught by the harness, it should allow you to easily return to an upright position.
- The harness must not allow you to fall out when you hang upside down.
- The harness should have a front tie-in point that is designed so that you maintain a correct center of gravity whatever activity you are performing.
- The stress points such as the tie-in should be faced with extra webbing and/or use heavy duty metal connectors.

Because of the variations in anatomy and because there are different kinds of activity in the high angle environment, no one seat harness design is suitable for everyone. Before deciding on a particular seat harness and investing the money, you should try several different designs to see which one is best for you.

WARNING NOTE

A seat harness by itself should not be used as the only tie-in point for swiftwater operations.

If you are attached only to a seat harness in swift water, the force of the current can easily force your upper body back over into the water, making it impossible for you to right yourself and resulting in your drowning. Swift water technicians usually employ a tie-in point higher on the body, such as a chest harness. Obtain instructions on this procedure from a qualified, swift water technician.

DIFFERENCES BETWEEN CLIMBING AND RESCUE HARNESSES

Most climbing harnesses are made to be lighter weight, often are cut for ease of movement, and are made for recovering from leader falls. Harnesses specifically designed for rescue are generally heavier and bulkier because they have wider webbing and padding for comfort while sitting in them for long periods. They are not designed for lead climbing. Rescue harnesses generally have a metal attachment point where carabiners are clipped directly into. Climbing harnesses have either two or three reinforced webbing loops to tie the belay rope directly into with a loop created from a Figure 8 Follow Through or a Bowline.

Climbing and rescue harnesses are also designed to meet test standards by different organizations. Climbing harnesses are generally designed to meet the UIAA standards, while rescue harnesses are designed to meet standards developed by NFPA and ANSI.

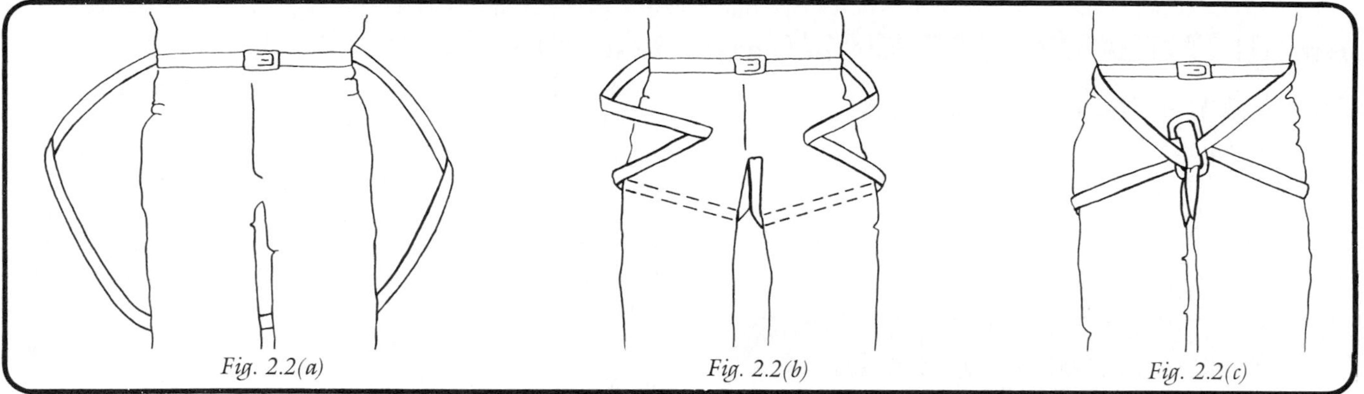
Fig. 2.2(a) Fig. 2.2(b) Fig. 2.2(c)

Tied Emergency Seat Harnesses

Tied seat harnesses may be useful in an emergency when a manufactured harness is not available. However, they cannot be used for hanging in for long periods of time, should not be used for lead climbing, and should not be used in performing intricate or difficult high angle operations.

The Diaper Harness

The Diaper Harness is one of the simplest of the tied seat harnesses. It consists of a continuous loop of webbing (two inch webbing is more comfortable than one inch webbing) that is connected by sewing, by a very substantial buckle, or a water knot backed up. The loop is brought together in front by equalizing three portions, one between the legs, and two from either side, and attached by a large locking carabiner *(see Figure 2.2).*

The Mountaineer Harness

The "Mountaineer Harness" is one harness type that is found in a number of versions throughout North America. One of the earliest descriptions was in the book, *MOUNTAINEERING, THE FREEDOM OF THE HILLS.*

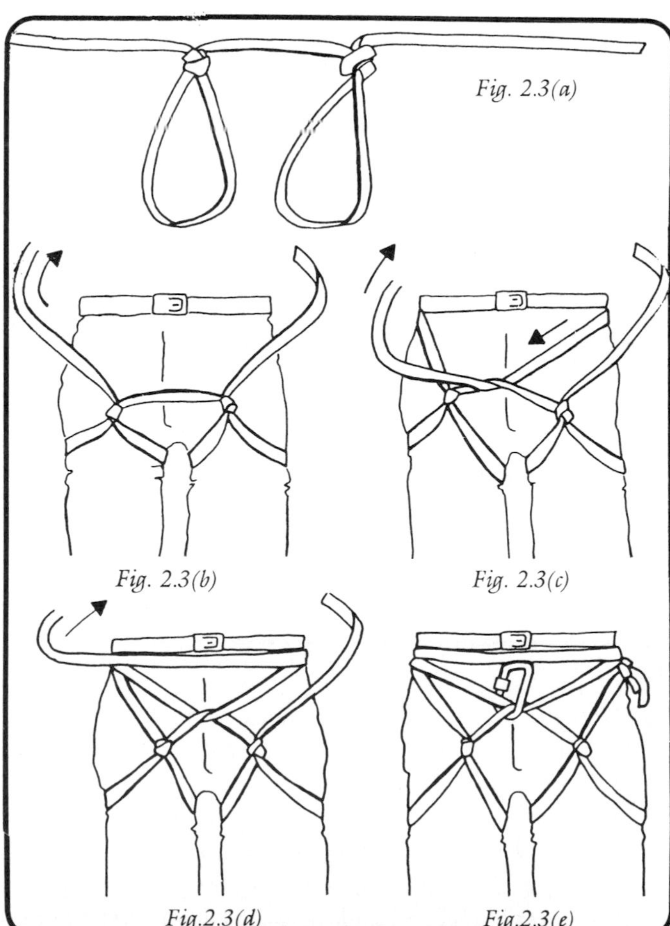
Fig. 2.3(a) Fig. 2.3(b) Fig. 2.3(c) Fig. 2.3(d) Fig. 2.3(e)

Fig. 2.3 Mountaineer Harness

PROCEDURE FOR TYING THE "MOUNTAINEER" HARNESS

(See Figure 2.3)

1. Take a piece of tubular webbing about 20-feet long.
2. Leaving about 18 inches of tail, tie a loop near one end using a Figure 8 on a Bight knot. This loop should be slightly larger than your left thigh.
3. Move over approximately 10 inches on the webbing and tie a second loop using a Figure 8 on a Bight. This loop should be slightly larger than your right thigh *(2.3a).*
4. Now step into each loop and bring it up around a thigh, with the 18-inch tail on the left, the 10-inch piece at the front and the long remaining portion on the right *(2.3b).*
5. Take the long end and run it around the back and over the buttocks to the front.
6. Run this end under and through the 10-inch section at the front and pull everything snug *(2.3c).*
7. Repeat step 5 *(2.3d).*
8. Bring the end around the back again, to the side and tie it together with the 18-inch tail, using a square knot. Back the square knot up by bringing the two ends together and tying them together with a water knot. (If enough webbing is left over, then repeat steps 5 and 6 again before tying off.)
9. Clip the tie-in carabiner across the portion where the ends have crossed the 10-inch section *(2.3e).*

ALTERNATIVE APPROACH

The Quick-Don Harness ("Swiss Seat")

One commonly used harness that can be tied quickly in emergency situations is the Quick-Don Harness (also known as the "Swiss Seat"). This harness can usually be tied quicker than the Mountaineer Harness, but it does not offer the same support or security.

WARNING NOTE

1. **One danger posed by the Quick-Don Harness is tying it incorrectly under stress so that it fails. For this reason, anyone who anticipates the use of this harness must consistently practice the tying of it so he will tie it correctly in an emergency.**
2. **If at any point the webbing is severed, the entire harness will fail.**
3. **Because it does not offer support in the proper areas of the anatomy, the webbing in this harness will constrict circulation and, possibly, impair nerve function, if the wearer hangs in the harness for more than a few minutes.**

Rescue Harnesses

A recent development is the seat harness specifically designed for rescue use. Rescue harnesses often feature wider webbing and optional padding for increased comfort and are usually heavier and bulkier than climbing harnesses. Some of the rescue harnesses also have a large "D" ring for front attachment point. These types of harnesses are not designed for absorbing shock loading, and **should not be used for lead climbing activities.**

Full Body Harness

In certain circumstances, a full body harness, or a combination of seat harness with a chest harness, may be preferable to only a seat harness. These may include the following situations:

- During technical rock climbing when it is necessary to be held upright should a fall occur. The Union of International Alpine Associations (UIAA), which sets standards for certain equipment used by climbers and mountaineers, considers a full body or seat/chest harness to be the only acceptable systems for climbing *(see Figure 2.5 for an example of a seat/chest harness combination and Figure 2.6 for an example of a full body harness).*

PROCEDURE FOR TYING THE QUICK-DON HARNESS ("Swiss Seat") *(Figure 2.4)*

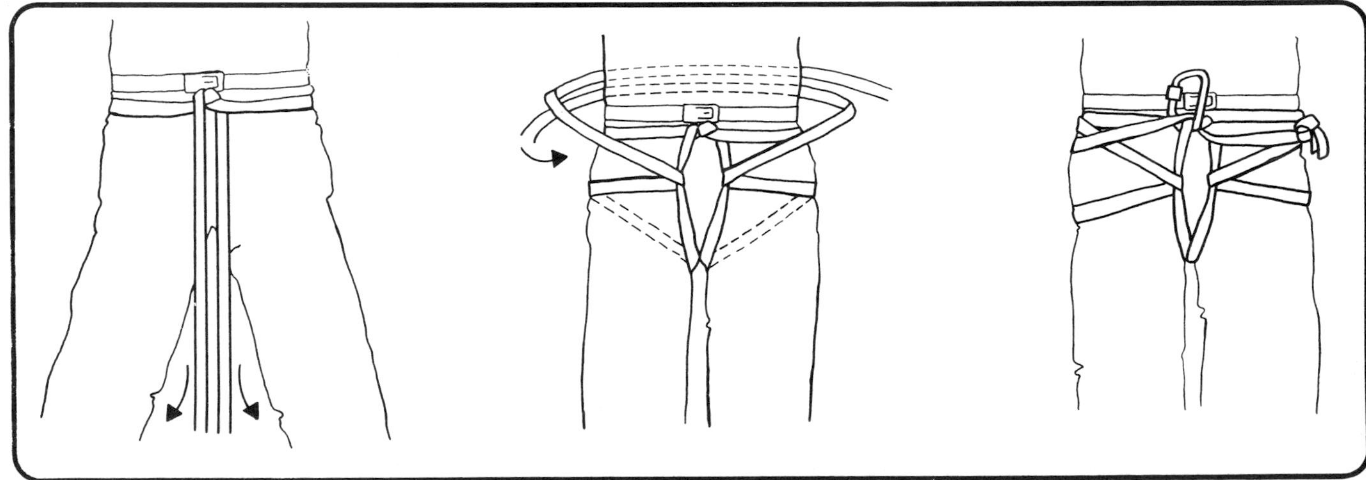

Fig. 2.4 Quick-Don Harness

1. Take a length of webbing at least 13 feet long (longer if you have large buttocks and thighs).
2. Find the center.
3. Place the center in the small of the back.
4. Pull the two ends to the front and tie an overhand knot.
5. Drop the ends in front. Reach though your legs from behind and pull both ends through to the rear.
6. Now run each end under a buttock, and around the outside of the thigh. Now run the webbing under itself in front.
7. Pull each end in the opposite direction, around the waist and bring them together on one side.
8. Tie off with a square knot and back it up by bringing the two ends together and tying them with a water knot.

Fig. 2.5
Combination Seat/Chest Harness

Fig. 2.6 Full Body Harness

- When a person is involved in an extremely dangerous activity that requires that he constantly be held upright. One example would be where a person is entering a tank containing dangerous fumes and he is equipped with a harness attached to a tether.
- To be placed on a subject in certain rescue situations. For examples of this, see Chapter 12.

In these situations a full body harness, or combination seat/chest harness, may provide needed security, but in some circumstances a full body harness may be a disadvantage. In certain rescue situations, some persons may find the full body harness too constraining, thus preventing the range of motion needed for rescue activities.

Seat/Chest Harnesses

The seat/chest harness combination may be preferable for some rescuers since it allows the wearer to choose options on the basis of the specific conditions. Worn with a seat harness, a chest harness can be quickly connected or disconnected to the system, depending on the needs of the user. Should the wearer need to be held upright without using upper body strength, then he can quickly connect it; if he finds it too constraining, then he can unclip it but still continue wearing it. An added advantage to a rescuer carrying a chest harness is that he can use it to hold a rescue subject upright in certain types of operations.

Chest Harnesses

Chest harnesses are comfortable, easily adjusted, easily combined with a seat harness, and easy to put on and take off. **A chest harness must not be used alone for high angle activities, but always in combination with a seat harness.**

OSHA/ANSI/NFPA/UIAA Standards

Standards from the Federal Occupational Safety and Health Administration (OSHA), the American National Standards Institute (ANSI), the National Fire Protection Association (NFPA), and the International Union of Alpine Associations (UIAA) apply to the construction of some harnesses.

Each of these standards applies to a specific group of users, but ***in no case does any standard apply to everyone who uses high angle equipment.*** OSHA and ANSI standards apply to certain workplace activities, while NFPA standards apply to firefighters and UIAA standards apply to mountaineers.

For specific information on any standard, contact the responsible organization. See Appendix III for addresses of standards setting organizations.

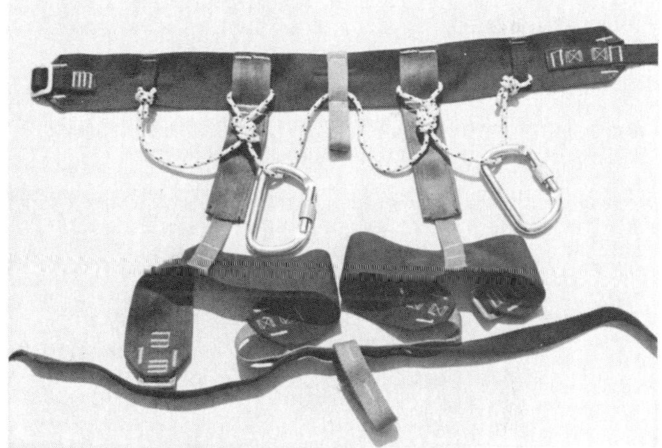

Fig. 2.7 Seat Harness Equipment Loops

Securing Hardware in the High Angle Environment

High angle hardware, such as carabiners, is easily lost when working in the high angle environment. Even worse, it can easily injure another person who happens to be in the path of its fall. When working in the high angle environment, keep all equipment attached to something secure. One convenient place is the equipment sling that many harnesses have.

If equipment slings or loops are not manufactured into the harness, they can be easily created with the use of utility cord, as shown in Figure 2.7.

Shoulder Slings

Shoulder slings for equipment are preferred by some persons. These may work well, particularly if the equipment carried is not bulky and does not interfere with high angle activities. However, some strangulation deaths have occurred when climbers have fallen and accidentally snagged the shoulder slings.

Light Sources

High angle operations, particularly in rescue, often take place at night, so all personnel should have with them a reliable source of light. Because it will be necessary for both hands to be free during high angle operations, these light sources should be in the form of head lamps.

Many traditional head lamp designs feature a belt battery pack that is attached by a cord to the head lamp. While this may provide a sufficient source of light, high angle technicians sometimes find that the cord snags and either becomes entangled or breaks. One possible solution is to wear the cord inside clothing, but should the wire short, then the wearer may discover a whole new meaning to the term "hot wired."

Some of the more recent designs of head lamps are more self-contained and can be worn as a complete unit on the helmet, with the battery either on the rear or contained near the lamp. When choosing this kind of lamp, you must be certain that it will remain stable on the head and not easily fall off.

The lamp should have an adjustable beam and have a secure switch so that the light cannot be turned on accidentally when in storage.

The battery type that gives by far the longest service and is resistant to the effects of cold is the lithium cell. However, it is also the most expensive and has been known to cause explosions by venting gas. The alkaline battery is the next most desirable in terms of long life and operating temperature, and is not as expensive as the lithium.

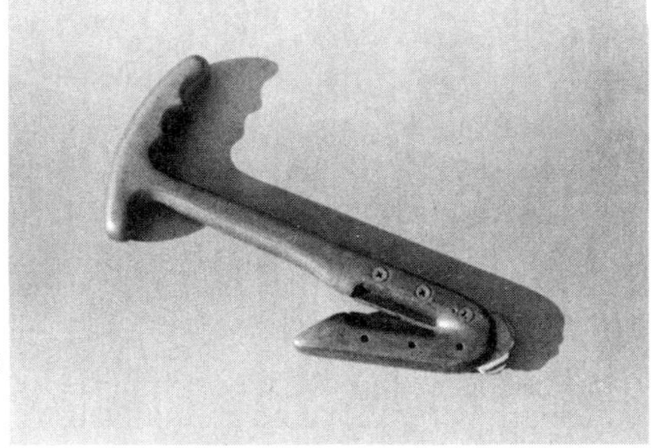

Fig. 2.8 Seat Belt Cutter

Knives in the High Angle Environment

Carrying knives in the high angle environment has almost become a custom for some. But the careless use of knives can have terrible consequences. Other than presenting the real danger of personal injury, knives in the high angle environment are a threat to life because of the ease with which they destroy life-supporting equipment.

A naked knife blade is a particular danger to ropes that are loaded with the weight of one or more persons. When stretched, as they are when supporting weight, rope yarns are very susceptible to being cut by any sharp object. Just the touch of a knife edge to such a loaded rope can cause the yarn under tension to "explode" and the rope to catastrophically separate.

Typically a person would want to reach for a knife when his tee shirt or hair has gotten sucked into his rappel device and he is stranded. However, in such a situation the person is likely to be under stress, possibly in pain, and to have limited freedom of movement. It would be very difficult for him to cut his way out without touching the knife edge to the rope.

One alternative to a knife is the instrument that is used by emergency services personnel to cut seat belts (see Figure 2.8). This instrument has a recessed blade so that it does not accidently cut a life line as easily as a naked knife blade.

But even better than cutting is the use of advanced skills and optional equipment to extricate oneself from a jammed rappel device and similar situations. One alternative is using a Prusik knot or an ascender to take the weight off the rappel device and extricating oneself from such a predicament. The skills and equipment required for this procedure are described in Chapter 10.

QUESTIONS for REVIEW

1. Name two ways that a helmet protects the wearer in the high angle environment.
2. Mark the following statements true or false:
   (a) Helmets with elastic chin straps are acceptable for high angle ropework.
   (b) Rubber boots are appropriate wear for high angle ropework.
3. Describe the two main requirements for footwear worn by the high angle technician.
4. Explain how wearing gloves helps a person maintain control of the rope in the high angle environment.
5. Explain why lifebelts, ladderbelts, and pompier belts must not be used as a single point of support in the high angle environment.
6. Describe what might happen to a person who is hanging for a long period of time in an unsuitable seat harness in the high angle environment.
7. Name three organizations, one federal, one national, and one international that have standards applying to harnesses in special functions or occupations.
8. Name three circumstances in which a full body harness, or a combination seat/chest harness, may be desirable.
9. Name a situation in which a seat harness *should not* be used.
10. What is a convenient place to secure high angle hardware in the high angle environment?
11. What is the danger in using knives in the high angle environment?
12. Perform the following tasks:
    A. Tie a quick-don harness ("Swiss Seat").
    B. Tie a "mountaineer" harness.

Chapter 3

Rope and Related Equipment

OBJECTIVES-

At the completion of this chapter, you should be able to:

1. Describe how to select rope according to the job you will be using it for in the high angle system.
2. Describe the types of fiber and rope constructions you will need to perform the job.
3. Describe the design uses, weaknesses, and limitations of the ropes you use.
4. Discuss how an appropriate safety factor is determined and how you would use it to establish rope specifications.

TERMS– *relating to rope that a High Angle Rope Technician should know:*

Abrasion—The damaging wear on rope and other gear caused by their rubbing against harder material.

Dynamic Rope—A type of rope designed for high stretch to reduce the shock on the climber and anchor system. Usually employed in rock climbing and mountaineering.

Fall Factor—A calculation used to estimate the impact of forces on a rope when it is subjected to stopping a falling person.

Kevlar™—Trade name for a type of Aramid fiber manufactured by the Dupont Corporation, which has high tensile strength, low elongation, and high resistance to heat.

Kernmantle—A rope design consisting of two elements: an interior core (kern) which supports the major portion of the load on the rope, and an outer sheath (mantle) which serves primarily to protect the core and also supports a minor portion of the load.

Laid Rope—A rope design that consists of fiber bundles twisted around one another.

Nylon 6—A type of nylon used in rope manufacturing. One trade name for this type is Perlon.

Nylon 6,6—A type of nylon used in rope manufacturing. It is manufactured by both Dupont and Monsanto in North America.

Perlon—A trade name for one type of nylon type 6.

Polyolefins—A group of fiber types used in manufacturing ropes that are often used in water applications. In this group are polypropylene and polyethylene.

Polyester—A type of fiber used in some rope manufacturing. Also known by the trade name Dacron™.

Software—A category of high angle equipment that is not hardware. In this category are rope and webbing.

Static Rope—A type of rope designed for low stretch. It is used in applications such as rescue, rappeling, and ascending where high stretch would be a disadvantage and where no falls, or very short falls, are expected before being caught by the rope.

Safety Factor—The ratio between the maximum load expected on a rope and the rope's breaking strength. The larger the ratio, the greater the safety factor.

Tensile Strength—A measurement of the greatest lengthwise stress that a rope can resist without failing.

Fig. 3.1 Typical Dynamic Rope Core

Fig. 3.2 Typical Static Rope Core

Rope is the universal link in high angle work and in technical rescue. During the past several years, there have been great strides in the technology and manufacture of rope. But rope is only as good as the use to which it is put, and will perform only as well as the care it has been handled with.

Determining the Right Rope for the Job

There are many different kinds of rope for high angle work. Each kind has a specific design and fiber that determine how it will react to natural and human forces. Before choosing a rope, the potential user must determine whether it will be for climbing rock or ice, for rappeling and ascending, for rescue, for swiftwater, or for another specialized activity. The incorrect use of a rope can result in severe problems for the user and, in some cases, tragedy.

Ropes for Technical Climbing And Mountaineering

Ropes for climbers and mountaineers are designed to catch the user if he falls while climbing. They must have stretch to absorb the energy of the fall without causing harm to the climber. This kind of rope is called **dynamic,** a term that relates to the use of force in motion.

Most of the stretch in a dynamic rope is built into it during manufacture and is achieved through various designs that elongate under load, much like a spring *(Figure 3.1).*

Another characteristic of dynamic ropes climbing and mountaineering is that they are soft and pliable. These qualities are necessary in climbing because the rope is constantly being moved, knotted and unknotted, and run through hardware. This quality of being soft and pliable is achieved in part by manufacturing climbing ropes with a sheath, or outer surface, that is loose and thin.

Standards for climbing ropes have been established by the UIAA.

Ropes for Rappeling and/Ascending

When used for rappeling or ascending, the rope does not act as a safety for the person on the rope, but as a means of travel. The user actually travels down the rope using a rappel device or travels up the rope using ascenders. One example of this kind of activity is vertical caving.

In this kind of situation, the stretch of a dynamic rope would be a disadvantage, and its thin sheath would be susceptible to abrasion.

For these kinds of activities, many prefer to use a **static** rope, which has very little stretch. In static ropes, the interior fiber bundles are constructed nearly parallel to one another (see Figure 3.2) so that most of the stretch is through the inherent stretch in the nylon. Static ropes typically stretch only about 18–20 percent before breaking.

In addition, static ropes characteristically have a thicker sheath to offer greater protection to the core. The result of this is that they do not have the ease of handling of dynamic ropes and may be slightly more difficult to tie knots in.

Ropes for Rescue Use

Ropes designed for rescue share some of the characteristics of ropes designed for rappeling and ascending: low stretch and high resistance to damage from abrasion.

In rescue, the quality of low stretch means greater control of the rope for those persons performing a rescue. One example of this would be when a stretcher is being lowered by rope. As the stretcher is eased over the edge of a vertical face, a greater load suddenly comes onto the rope. If a dynamic rope were being used, then there would be a great deal of stretch, with a significant drop in the stretcher. With a static rope there would be less stretch and therefore more control.

In addition, there is not as much “creep” in a system using a static rope. When a rope is first weighted, there is the initial stretch. But additional stretch, known as “creep”, will slowly come into the rope over a length of time as it remains loaded. This creep tends to be greater in a dynamic rope. In a rescue situation, creep would be a disadvantage. It would, for example, make it more difficult for a dynamic rope to hold a litter in a constant position on a vertical face while a patient is loaded into the stretcher.

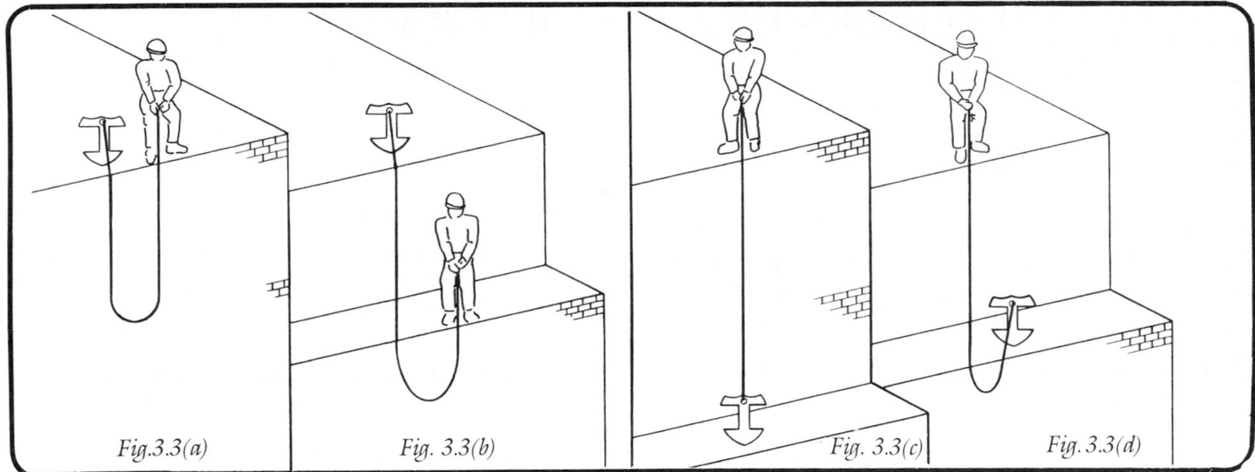

Fig. 3.3 Comparison of Fall Factors

The Role of the Fall Factor

One method of estimating the forces at work on a rope is by computing a measurement known as the **fall factor** (see Figure 3.3). As shown in the first part of the illustration, a person is attached to a rope that will catch him if he falls, but keep him from hitting the ground. The other end of the rope is directly attached to a point of "protection" that will not come loose. The fall factor is calculated by dividing the distance the person on the rope falls by the length of the rope between him and the point of protection:

$$\frac{\text{Distance person falls}}{\text{Length of rope}} = \text{Fall Factor}$$

In this first example, *(see Figure 3.3 [a])* consider that the length of the rope is 100 feet. If the person were climbing up from below and slipped just before he reached the point of protection and fell exactly a hundred feet, then the calculation would be:

$$\frac{\text{Distance fallen: 100 feet}}{\text{Length of rope: 100 feet}} = \text{Fall Factor: 1}$$

If the person were climbing up from below and fell before reaching the point of protection, *(see Figure 3.3 [b])* then it would be less than a factor 1 fall. For example, if he fell only 50 feet:

$$\frac{\text{Distance fallen: 50 feet}}{\text{Length of rope: 100 feet}} = \text{Fall Factor: .5}$$

Now, note the second part of the illustration. The figure has now climbed up above the point of protection by a full rope length *(see Figure 3.3 [c])*. If he were to fall now:

$$\frac{\text{Distance fallen: 200 feet}}{\text{Length of rope: 100 feet}} = \text{Fall Factor: 2}$$

Any fall from above the point of protection would be more than a Factor 1 fall *(see Figure 3.3 [d])*. If, for example he fell 50 feet above the protection:

$$\frac{\text{Distance fallen: 150 feet}}{\text{Length of rope: 100 feet}} = \text{Fall Factor: 1.5}$$

As the fall factor approaches and moves above 1, the severity becomes much worse. Note that it is not just the length of fall that matters, but the length of fall in relationship to both the amount of rope and the point of protection.

WARNING NOTE

While calculating the fall factor is useful in obtaining general estimates of the forces on a rope in a fall, these situations shown in Figure 3.3 are ideal ones. Other factors will often enter into the situation to alter the results. They include the following:

- **The rope may be running through intermediate points of protection or it may be rubbing against rock or other surfaces. This creates drag that may slow the rope's ability to stretch. This in turn effectively shortens the length of rope available to absorb the energy of the fall. The net result would be a much higher fall factor and greater impact forces on the person and parts of the system such as the anchors.**
- **In most climbing situations, a fall is not completely in free air. The Fall Factor is of little importance if the person taking the fall slams into the ground due to rope stretch or is bashed against the wall on the way down.**

FIBERS USED TO MAKE ROPE

Natural Fibers

For many years, ropes made of natural fibers, such as sisal, hemp, and manila, were standard. Then at about the time of World War II, the mass production of rope made of synthetic fibers such as nylon and polyester began.

Today, synthetic fiber ropes are considered standard for situations where the safety of a person is "on the line." In vertical sport activities such as climbing or vertical caving, synthetic fiber ropes have long since displaced those made of natural fibers. Synthetic fiber ropes are also considered the standard for rescue ropes. National organizations such as the International Association of Fire Fighters (IAFF), the International Society of Fire Service Instructors (ISFI), and the National Fire Protection Association (NFPA) have all condemned the use of natural fiber rope in life safety applications for the following reasons:

- Low resistance to abrasion.
- Limited ability to absorb shock loading.
- Will degrade in strength even with the best care.
- Can rot without outward visible signs.
- Have lower breaking strengths than ropes of the same diameter made of synthetic fibers such as nylon or polyester.
- Does not have strands that are continuous along its entire length because natural fibers are never more than a few feet long.

Synthetic Fiber Ropes

Among the advantages that synthetic fiber ropes have over natural fiber ropes are:

- Do not rot.
- Do not age as quickly as natural fiber rope.
- Each fiber runs entire length of rope resulting in higher strength.
- Can be made into more advanced rope designs than natural fibers.

There are several different synthetic fibers used in making ropes. Each fiber has distinct characteristics that make it suitable for certain uses and unsuitable for others.

Polyolefins (Polypropylene and Polyethylene)

Advantages:

- Do not absorb water.
- Float, so are useful in activities on the water.

Disadvantages:

- Relatively low tensile strength.
- Poor abrasion resistance.
- Low melting point.
- Poor shock absorbing capability.
- High stretch.
- Poor resistance to damage from sunlight.

Polypropylene or polyethylene ropes are often found in water activities. But because of the low tensile strength, low abrasion resistance, and low melting point they should not be direct loading in life support operations. For example: Polyolefin ropes are unsuitable for rappeling and rescue lowering and hauling systems.

Kevlar™ (Dupont trade name for a type of Aramid fiber)

Advantages:

- Resistance to high temperatures.
- High tensile strength.

Disadvantages:

- Easily damaged by abrasion.
- Easily damaged by continued small radius flexing (as in knotting).
- Poor shock-loading capability.

As with any new material, there has been considerable controversy over the best uses of Kevlar™ rope. However, there seems to be a consensus that it might have uses where it is not being subject to continued small radius bending. For example, rock climbers seem to have used Kevlar™ cord successfully as protection sling material, where it remains continually knotted. But the consensus seems to be that current designs of Kevlar™ rope should not be used where it will be subjected to abrasion and continued small radius flexing (as in repeated knotting and unknotting). It is generally considered unsuitable for such activities as rappelling, ascending, belaying, and rescue lowering and hauling systems.

Polyester (Sometimes known as **Dacron**™, the Dupont trade name for a type of polyester.)

Advantages:

- High tensile strength even when wet.
- Good abrasion resistance.
- Melting point of about 480°F.
- Resistant to damage from acids.

Disadvantages:

- Cannot handle shock loading as well as nylon.
- Susceptible to damage from alkalis.

Polyester fibers are found in a number of life safety applications. But because polyester does not handle shock loading as well as nylon, it is generally not found in climbing ropes.

Nylon (There are actually several different types of nylon. The two most commonly used in life safety ropes are nylon 6—also known as "Perlon"—and nylon 6,6. More on these differences in the sidebar on the right.)

Advantages:

- About 10 percent stronger than polyester in ropes of comparable diameter.
- Good shock loading capability.
- Melting point of around 480°F, 250°C (Type 6,6).

Disadvantages:

- May lose 10–15 percent of its strength when wet (will regain the deficiency when dry).
- Susceptible to certain strong acids, such as those used in storage batteries.

Ropes made of nylon yarn are commonly used in life support applications including climbing, vertical caving, rescue, and tactical operations.

Nylon Type 6, also known by its European trade name, **Perlon,** is found in most European-made ropes and in some U.S.-manufactured ropes.

Nylon Type 6,6 is found in some ropes manufactured in North America. The Type 6,6 yarn is manufactured by the Dupont and Monsanto corporations.

The major differences between Nylon Type 6 and Type 6,6 are that Type 6,6 has a slightly higher melting point and a slightly higher breaking strength than Type 6. Whether these differences in the nylon actually appear in the rope depend in large part on the individual rope design.

Rope Construction

The choice of a rope for a specific job depends in part on the fiber it is made from, but also on the manner in which it is constructed.

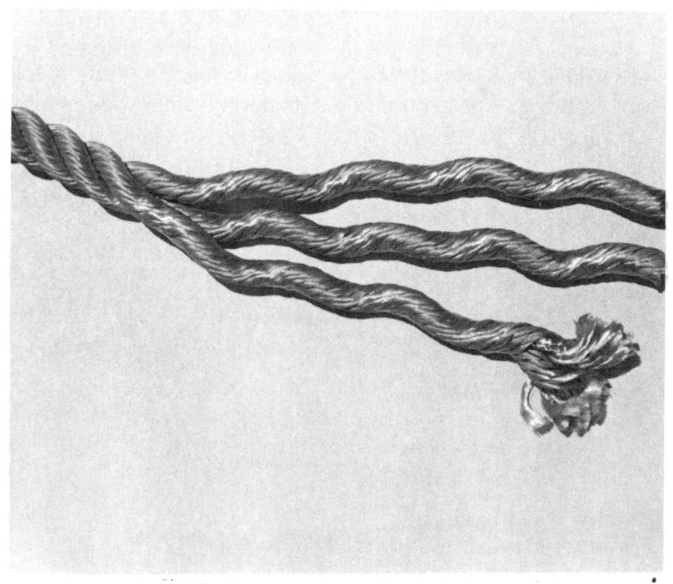

Fig. 3.4 Typical Laid Rope Construction

Laid *(see Figure 3.4)*

Laid construction (also known as "twisted" or "hawser lay") consists of twisting small fiber bundles of material and then combining them in larger bundles, which are twisted around one another, usually in groups of three. This type of rope construction resembles the designs of older types of rope constructed of natural fibers.

Characteristics:

- When loaded, the fibers tend to untwist slightly thus causing spin and kinking.
- Because each fiber may appear at the surface of the rope somewhere along its length, the load-bearing fibers are more susceptible to damage by abrasion.
- Tend to be very stretchy.
- Tend to kink unless carefully handled.

For life support activities, the tighter lay, or "mountain lay," design should be used, as opposed to the "marine lay," which is looser and more susceptible to damage by abrasion.

Ropes of laid construction are quickly being displaced in high angle work by other designs.

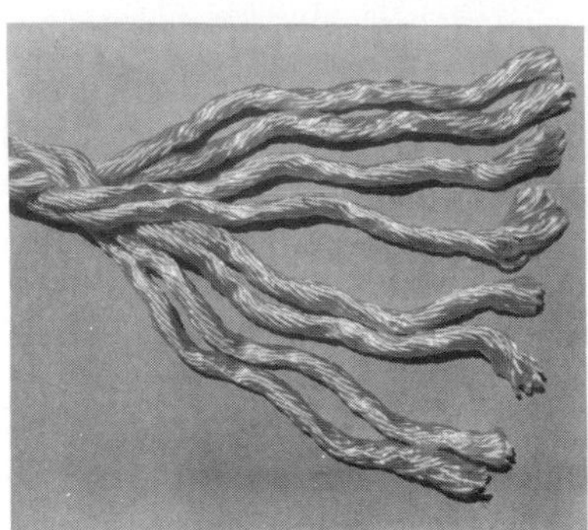
Fig. 3.5 Plaited Construction

Fig. 3.6 Single Braid Construction

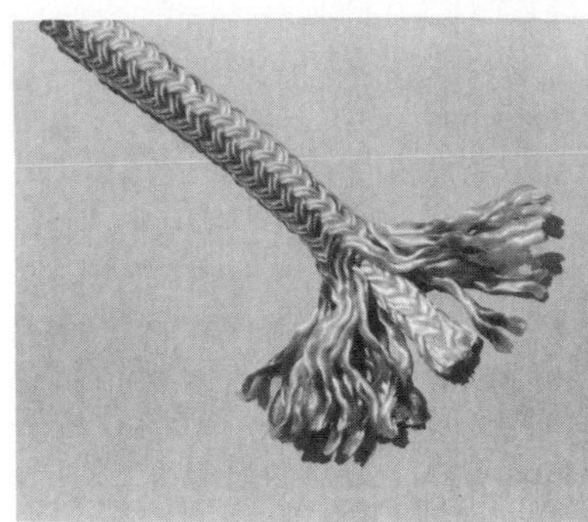
Fig. 3.7 Double Braid Construction

Plaited *(see Figure 3.5)*

These ropes usually consist of eight bundles of fibers plaited together.

Advantages:

■ Tend to be soft and pliable.

Disadvantages:

■ Prone to "picking" (the snagging and pulling out of fiber bundles).

Braided

Braided ropes are found in two different types:

Single Braid *(see Figure 3.6)*

In this type of design, the rope is constructed entirely of a single weave of three of more fiber bundles. The design is sometimes referred to as a "clothesline braid." Because the load-supporting fiber bundles in single braid construction are vulnerable to destruction when the rope is being used, single braid construction ropes have limited use in high angle operations.

Hollow Braid

This is essentially a very thick sheath. It sometimes is found with a "filler" such as scrap yarn or filament plastic. It typically is found in inexpensive hardware store type rope, and not in life safety line.

Double Braid *(see Figure 3.7)*

Double braid ropes are essentially a solid braid covered with a hollow braid combined into one construction. One braid acts as the rope core, while a second braid is constructed around it to act as a sheath and help protect the inner braid.

Advantages:

■ Soft and flexible.

Disadvantages:

■ Susceptible to contamination of core by grit and dirt.

■ Susceptible to "picking."

■ Susceptible to abrasion.

■ When cut around its diameter, sheath tends to slip down on core.

Kernmantle (also sometimes seen spelled "Kernmantel")

This term comes from a compound German word:

Kern = Core

Mantle = Sheath (or cover)

The kernmantle rope design consists of a central core (kern) of fibers which supports the major portion of the load on the rope. This core is covered by a woven sheath (mantle) that supports a lesser portion of the load. But the tight weave of the mantle protects the core from abrasion, dirt, and environmental effects, such as sunlight. The kernmantle construction results in a rope that is strong and resists damage, but is easy to handle. It also does not have the drawback of severe twisting that affects other rope designs, such as laid. There are two basic types of kernmantle ropes.

Dynamic Kernmantle

The term **dynamic** indicates a rope with high stretch. This is meant to act as sort of a shock absorber when a falling climber is caught by the rope. Some dynamic kernmantle ropes stretch by as much as 60 percent before breaking.

This stretch is created with a rope core that mechanically lets out under load, much as a spring does. The design of the core varies slightly from one manufacturer to another. Figure 3.8 illustrates one design of dynamic kernmantle construction. Depending on the specific manufacturer, these cores may be braided, or they may consist of a group of twisted bundles.

The sheath, or mantle, of dynamic kernmantle ropes tend to be relatively thin. This is so the core has room to stretch out to give the rope greater flexibility, and, in some cases, so that more yarn can be packed into the core and still maintain a specific diameter and strength.

Advantages of dynamic kernmantle ropes:

■ The elasticity is an advantage for climbing situations where long falls are possible.

■ Very easy to handle and tie knots in.

Disadvantages of dynamic kernmantle ropes:

■ Thin sheaths make them susceptible to damage from abrasion and to contamination from dirt and grit.

■ Their elasticity makes them less suitable for activities such as rappeling and ascending, and for rescue operations.

Fig. 3.8 Dynamic Kernmantle Construction

Fig. 3.9 Static Kernmantle Construction

Static Kernmantle

The term, "static" indicates a type of rope with very low stretch (no more than 20 percent of its length at break). This low stretch is created by manufacturing a rope core of fiber bundles that are nearly parallel to one another (see Figure 3.9). What stretch there is in a static rope is largely due to the inherent stretch of the nylon. Because there is so little stretch, static ropes provide a more sudden stop when catching a fall. This sudden stop subjects the climber's body, the equipment in the system, and the anchors to greater impact loading than would be seen if a dynamic rope were used.

Static kernmantle ropes also have a thicker sheath than dynamic kernmantle ropes. This thicker sheath helps protect the core from damage by abrasion and helps prevent dirt and grit from entering the core and causing damage to the inner fibers. One result of a tighter sheath is a rope that is stiffer and not as easy to handle as the dynamic kernmantle with its thinner sheath.

Characteristics of Static Kernmantle Ropes

Advantages:

- Low stretch, an advantage for certain activities.
- Good resistance to damaging dirt and grit.
- Good resistance to abrasion.
- High tensile strength.

Disadvantages:

- Not as easy to handle and tie knots in as some other rope types.
- Not designed for severe shock loading.

Choosing a Rope

A rope is the one essential element in the high angle system. But it is in essence a tool. And, as with all tools, the correct rope should be chosen to fit the job.

WARNING NOTE

All references in this manual to dynamic rope or static rope are to very specific constructions of rope. A dynamic rope is one specifically built for climbing or mountaineering and is not just a stretchy rope. A static rope is a rescue static rope made especially for rope rescue, or a sport static rope made for personal ascending or rappeling.

These special ropes will normally not be found at the average hardware store. They should be purchased only from reputable dealers and manufacturers who sell their products for these specific uses.

Dynamic vs. Static Ropes

If you are going to be doing exclusively recreational rock climbing or mountaineering, the choice will be a dynamic rope. There are numerous types of dynamic ropes that can be used depending on the climbing environment and the style of climbing. For this, you should consult an instructional manual on climbing or any of several catalogues from companies that specialize in climbing equipment.

Also, if the rope is going to be used as a belay for falls that result in severe shock loading, then a dynamic rope will be more appropriate. There is considerable debate on at what point the fall becomes severe enough to require a dynamic rope. Much of this will depend on local conditions such as the climbing environment, the equipment used, and the experience of the people involved. But certainly if a fall approaching the severity of a Factor .75 is expected, then a dynamic rope should be used. This is not so much for the sake of the rope, but to reduce the impact loads on the person attached to the rope and on the anchors.

If the activities are going to be restricted to rappeling and ascending, particularly under harsh conditions, then a static rope would probably be preferable. Most rescue personnel now use static rope in their operations.

Size and Strength

If the choice is a static rope, and the decision has been made on the type of rope fiber and construction, then there must be a decision on how strong a rope is needed, or on the **tensile strength.** The tensile strength of a rope depends very much on its cross section of material, or the amount of yarn used per foot of length.

There may be a temptation to choose the largest diameter rope available (usually 5/8 inch for a synthetic fiber, life support rope), but this could be a serious mistake and could actually hinder activity with the rope.

Problems with Large Diameter Ropes

There are several problems in using large diameter ropes, including the following:

- Higher costs because of more material in the rope.
- The larger the diameter of a rope, the more it weighs. This means difficulty in carrying the rope to where it is needed (usually up a hill or up flights of stairs).
- Handling problems. When the rope is hanging vertically, the higher weight will mean more difficulty in rappeling and in handling the rope.
- Incompatibility with other equipment. Many types of high angle equipment are not made for use with ropes above 1/2 inch diameter.

There is one obvious advantage to very large diameter ropes: their overall strength. However, in normal field use, larger diameter ropes have only slight abrasion and cut resistance over 7/16 and 1/2 inch ropes.

Determining the Safety Factor

The most realistic way to determine the needed tensile strength of the rope, and therefore the diameter, is by calculating the safety factor. The **safety factor** is a measure of conditions you are expected to encounter in the local high angle environment, plus a realistic margin of safety.

The first step in determining the safety factor is to estimate what will be the maximum load you expect to be on the rope. If this will only be the weight of one person, as in a recreational situation, plus high angle equipment, then the total weight may only be 200 pounds. If the rope is to be used in rescue, then the estimated weight will probably be more. If, for example, there will be a rescue subject plus a rescuer and a litter with assorted gear, the total weight might be 600 pounds.

If 600 pounds is the expected load, then obviously the breaking strength of the rope cannot be exactly 600 pounds. The rope will be subjected to shock loading, to reduced strength through knots and sharp bends and because of wear. So it is necessary to have a strength several times 600 pounds as a margin of safety or a **safety factor.**

In industry and construction, where the load on the rope is to be equipment or materials but not people, an accepted safety factor is 5:1. But a safety factor of 5:1 or less is unacceptable when a person is on the rope. Some individuals and groups accept a safety factor of 10:1. A safety factor that is widely published in written standards and commonly used in the fire service is 15:1.

So, if 600 pounds is the expected maximum load, with a safety factor of 10:1, the rope should have a minimum breaking strength of 6,000 pounds. If a safety factor of 15:1 is employed, then the minimum breaking strength should be 9,000 pounds.

Breaking Strength of Rope

The breaking strength of a rope is measured as **tensile strength at break.** On the face of it, this may seem like a very simple idea. But there are many different ways of conducting tests that will result in different test results on the same piece of rope. Among the different factors that will affect the outcome of a tensile test are:

- Speed of the pull. A rope that is pulled apart slowly will register a higher breaking strength than one pulled apart at a faster speed;
- Diameter of the object rope is attached to when it is pulled. A rope pulled to failure by a small diameter object will break at a lower tensile strength than a rope pulled with a larger diameter object. Similarly, a rope pulled with a knot in it will break at a lower strength than one without a knot. (This is because whenever a rope is placed under load in a sharp bend, some strength is lost. See Chapter 4, "Care and Use of Rope and Related Equipment," page 28.)

Rope Tensile Test Standards

There is one test standard developed by the U.S. Federal Government that can be used for comparisons among the rope manufacturers who use it. This is the **Federal Test Standard 191A, Method 6016.** This guide specifies guidelines such as what the rope is to be tied to during the test and the rate of pull, along with other criteria.

In addition, certain NFPA standards apply to rope that is used on the fire ground. Contact the NFPA for specific information on these standards. See Appendix III for addresses of standards-setting organizations.

Rope Colors

While many people choose a rope color for aesthetic reasons, the color of a rope can also serve a functional purpose. If, for example, there are several ropes being used together, the differing colors can help distinguish one line from another, so that the user can immediately know which rope to haul or lower.

Another way to use color is to buy a different colored rope for each year in order to quickly tell the age. Or one might use varying colors for different sizes of rope.

Process for Coloring Ropes

"Natural" nylon is off-white, so any color for a rope has to be added at some point during its manufacture. Any color added to the rope material will affect, to some degree, the rope strength and, in some cases, its resistance to damage by sunlight. How much these properties are affected depends on the method for adding the dye to the rope material.

There are two basic methods of adding color to a rope. One technique for dying a rope is known as **solution dying.** In this technique, the color is added to the raw material as the yarn is being manufactured. This usually causes slightly more strength loss in the rope than the second method of coloring, **surface-applied dying,** in which the color is added to the synthetic yarn after it has been manufactured, or to the rope after it has been braided.

Webbing

Because of its special characteristics, webbing is sometimes preferable to rope for certain situations. For example, webbing is more comfortable than rope against the body for seat harnesses. Webbing is commonly used for anchoring since it is less expensive than rope. And due to its wide, flat surface it can be more abrasion resistant in many rigging applications.

Materials

Most webbing is made of nylon or polyester, materials that have the same characteristics as those used to manufacture rope.

Construction

- **Flat Webbing—**It is constructed of a single layer of material, the same as seat belt webbing. It is less expensive, but stiffer and more difficult to work with than tubular webbing.
- **Tubular Webbing—**Because it is more supple and easier to work with, it is more often used in the high angle environment. The tubular shape is obvious if you look at it from one end and squeeze the two edges together. There are two types of tubular webbing:

  a) ***Edge-Stitched*** *(see Figure 3.10).* This is formed by folding over flat webbing lengthwise and stitching the two edges together. You should be careful when purchasing edge-stitched webbing and buy it only from a reputable dealer. Because webbing is susceptible to abrasion along its edges, some types of edge-stitched webbing may become unstitched when the thread is broken. The better design of edge-stitched webbing locks the stitches under one another so they are not as prone to this occurring.

  b) ***Spiral Weave*** (also known as "shuttle loom construction-*see Figure 3.11).* This is the design of the tubular webbing that is traditionally more common in high angle operations. It is constructed by weaving a tube as a unit.

There are hundreds of sizes and types of webbing in the marketplace. Many types are unsuitable for high angle operations. You should buy webbing for life support applications only from quality suppliers who publish tensile strengths and specifications. Be particularly cautious about the use of surplus webbing which usually has no information about tensile strength, material, specifications, date of manufacture, or history of use. Use the same procedure for determining the safety factor for webbing as for rope and other equipment.

Webbing Size

There are various widths for webbing ranging from 1/2 inch up to 2 inches. The most common size used in high angle work is the 1-inch width. For seat harnesses, the larger widths are more comfortable.

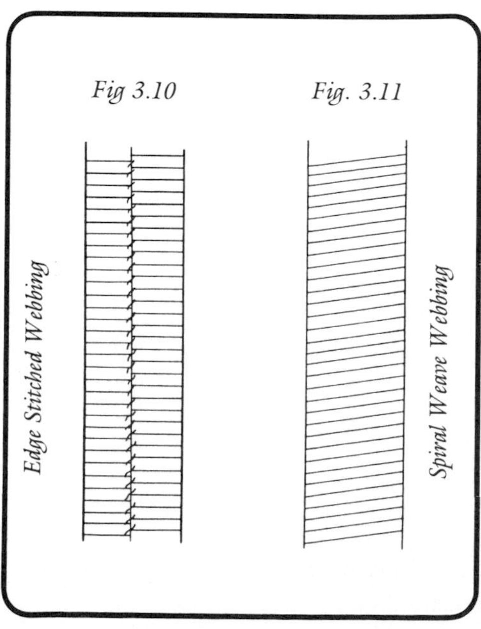

Fig 3.10 *Fig. 3.11*

Webbing Strength

While there are hundreds of sizes and styles of webbing products, the most common sold by suppliers of high angle equipment are:

- 1" tubular nylon
  Most common is MIL-W-5625, which has a tensile strength of about 4,000 lbs.
- 1" solid nylon
  Most common is Type 18, which has a tensile strength of about 6,000 lbs.
- 1 15/16" or 2" solid seatbelt type webbing–nylon or polyester.
  It has a tensile strength of 4,500–6,000 lbs. depending on material and type.
- 1 23/32" flat nylon webbing, Mil-W-4088.
  It has a tensile strength of about 9,500 lbs.

QUESTIONS for REVIEW

1. Describe why dynamic ropes are designed to stretch.
2. Describe why static ropes are designed not to stretch.
3. Calculate the fall factor for the following situations:
   a. A climber is connected at the top of a 1,200 foot cliff with a 100-foot length of rope to a secure point of protection. While at this point of protection, he falls the full length of the rope.
   b. A rescuer is connected 200 feet up a 700-foot television tower with a 100-foot length of rope to a secure point of protection. He climbs up past the point of protection by 50 feet and falls past the point of protection the full length of the rope before being caught by the rope.
4. Name four disadvantages of synthetic fiber rope for high angle activities.
5. Name at least one advantage and one disadvantage of the following fibers for high angle rope:
   a. Polyofins,
   b. Kevlar™,
   c. Polyester, and
   d. Nylon.
6. Name the two nylon types commonly used in rope for high angle activities.
7. What is the difference between mountain lay and marine lay constructions in laid ropes?
8. List one advantage and one disadvantage of plaited rope.
9. Describe why single braid ropes are generally not adequate for high angle operations.
10. Describe one advantage and one disadvantage of double braid rope.
11. Define the term "kernmantle."
12. Describe the difference between the terms "dynamic" and "static" as they apply to high angle rope.
13. Describe the situations in which the following would be used:
    a. Dynamic kernmantle ropes.
    b. Static Kernmantle ropes.
14. Name two problems in using large diameter ropes in the high angle environment.
15. Using a safety factor of 15:1, determine the minimum tensile strength rope to be used where the expected load will be 600 pounds.
16. Name two factors that affect tensile test results for rope.
17. Name the two processes used in dying rope yarn.
18. In webbing construction, describe the difference between flat weave and tubular design construction.

Chapter 4

Care and Use of Rope and Related Equipment

OBJECTIVES–

At the completion of this chapter, you should be able to:

1. Describe the need to care for rope and related equipment for high angle life support use.
2. List the steps in a rope inspection process.
3. Define the conditions under which a rope should be retired.

TERMS–*pertaining to the care and use of rope and related equipment that a High Angle Rope Technician should know:*

See **Chapter 3, *Rope and Related Equipment***

Care of Ropes

The modern high angle rope is a marvel of design and engineering. But a rope's performance, how long it lasts, and its safety still depend on how well it is cared for. The condition of a rope is in effect dependent on its history: the age of the rope, the conditions to which it has been subjected, and the care it has received. If a rope is owned and used by only one person, then that person knows the history of the rope. However, if more than one person is using the rope, then there has to be a system of tracking that rope's history. Each rope must have a rope history log.

Keeping a Rope History Log

To track the history of a rope, each line must have its own log card with pertinent information on the manufacturer, diameter, design, tensile strength, date of purchase, etc. There should be space on the log card for indicating each time the rope was used and the activity it was used in. There must be specific entries made whenever the rope was subjected to abuse that could affect its performance or safety. *(Figure 4.1 shows an example of a rope history log.)*

IT IS ESSENTIAL THAT ENTRIES FOR EACH ROPE BE MADE **EVERY TIME** IT IS RETURNED TO STORAGE. THIS DISCIPLINE MUST BE FOLLOWED IN EVERY GROUP. OTHERWISE, THE ROPE HISTORY IS INCOMPLETE.

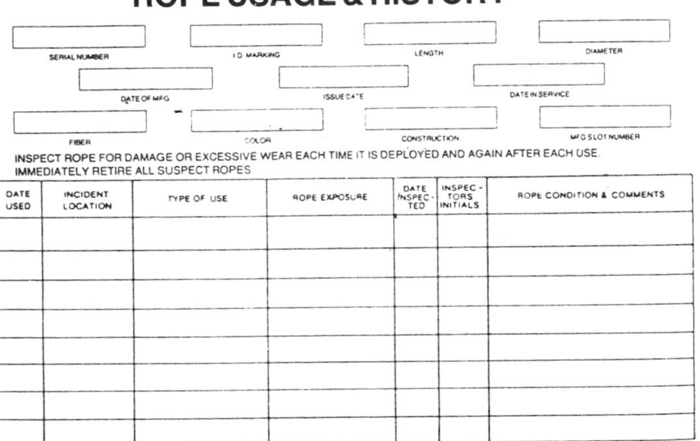

ROPE USAGE & HISTORY

SERIAL NUMBER | I.D. MARKING | LENGTH | DIAMETER

DATE OF MFG | ISSUE DATE | DATE IN SERVICE

FIBER | COLOR | CONSTRUCTION | MFG SLOT NUMBER

INSPECT ROPE FOR DAMAGE OR EXCESSIVE WEAR EACH TIME IT IS DEPLOYED AND AGAIN AFTER EACH USE.
IMMEDIATELY RETIRE ALL SUSPECT ROPES

DATE USED	INCIDENT LOCATION	TYPE OF USE	ROPE EXPOSURE	DATE INSPECTED	INSPECTORS INITIALS	ROPE CONDITION & COMMENTS

Fig. 4.1 Rope History Log

Tagging Rope

Because most groups with high angle gear have ropes of similar color, length, and diameter, there must be some way of distinguishing each individual rope so that its history can be kept. Each rope should have some distinguishing identification such as a number or letter that corresponds with its card. This identifying mark must be placed on the rope so that it is unmistakable and so that it cannot be eradicated or lost. Some examples of tagging are: *(see Figure 4.2)*

- Hot stamping the end of the rope.
- Marking the circumference at the end of the rope and protecting the mark with clear plastic tape, clear heat shrink tubing, or a protective coating such as Whip End Dip™.

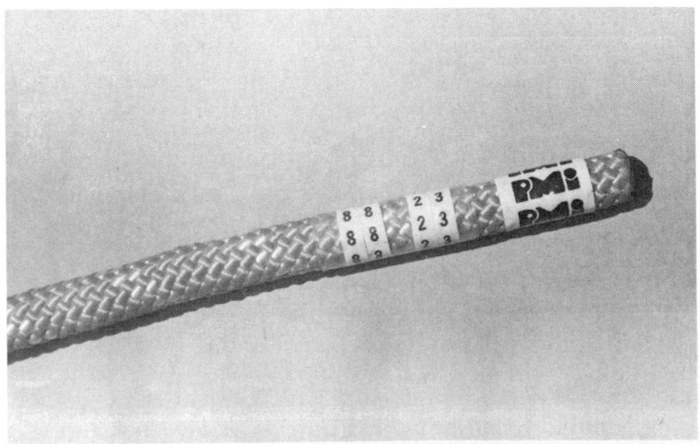

Fig. 4.2 Rope Tagging Example

Storage of Rope

In short, a life support rope must be stored in a place of its own where it is protected from harm. Rope can be damaged if it is left:

- With knots in it.

  (This will eventually weaken portions of the rope yarn.)
- In sunlight.

  (All fibers used in life safety ropes—including nylon and polyester—will degrade under prolonged exposure to sunlight.)
- Exposed to vehicle exhaust systems or to fumes or residues from storage batteries.

  (Both of these vehicle components produce substances damaging to rope.)
- On the floor.

  (Concrete floors contain damaging acids. Stepping on rope grinds in dirt and grit. Also, damaging substances can be dropped on rope.)
- Wet or in damp areas.

  (This will promote the growth of mold or mildew on the rope.)

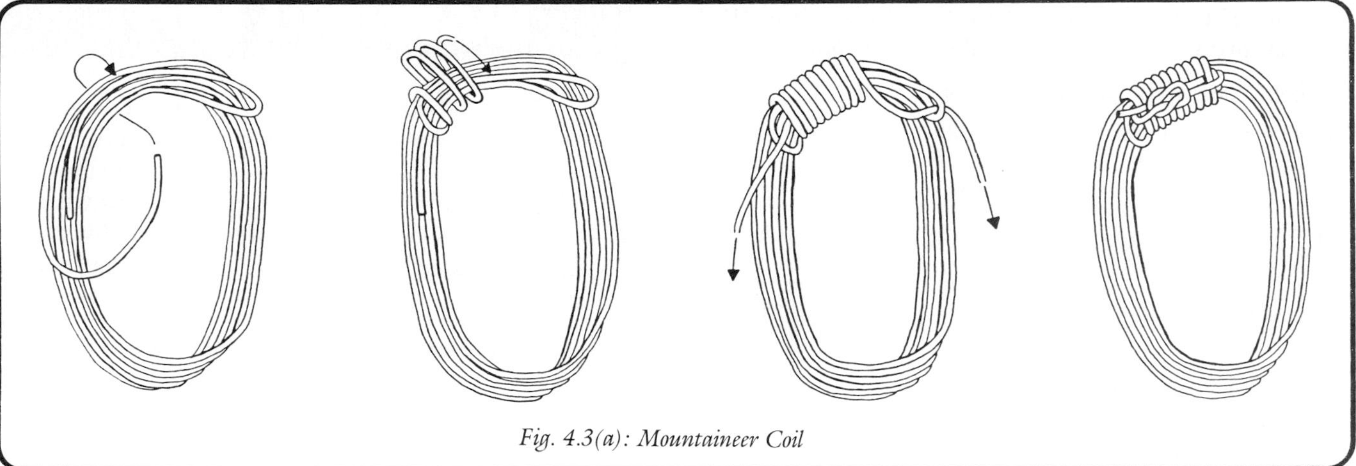

Fig. 4.3(a): Mountaineer Coil

3 Examples of Rope Coils

- In areas of high temperature.
  (Prolonged exposure to temperatures higher than humans can work in will promote degradation of rope.)
- Contaminated with dirt and grit.
  (Dirt and grit work into the core and damage the yarn. Avoid needlessly dragging a rope on the ground, and NEVER step on a rope.)

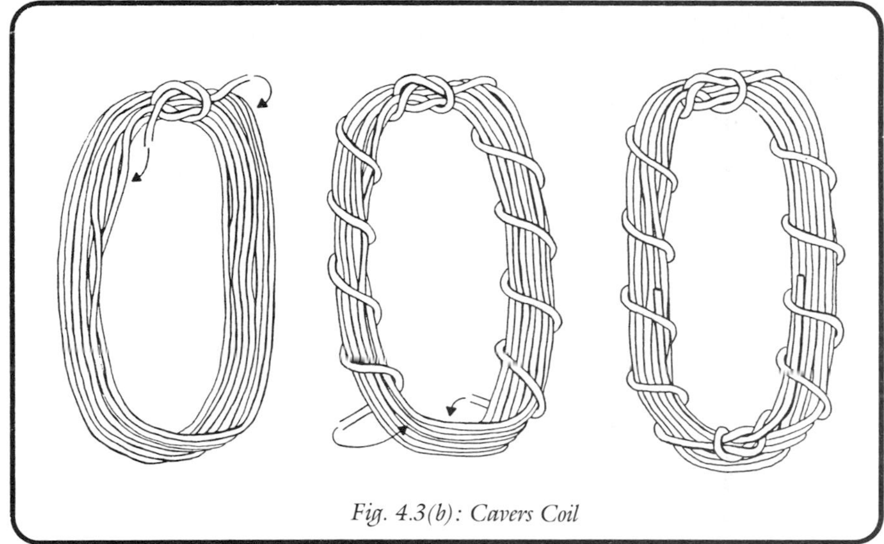

Fig. 4.3(b): Cavers Coil

Coiling/Bagging Ropes

There are a number of coils that can be used for storing and transporting ropes. The specific type of coil will depend on the circumstances or environment in which the rope is to be used. Figure 4.3 shows these types of coils along with comments on how they are used:

Figure 4.3(a): Mountaineer Coil
Figure 4.3(b): Cavers Coil
Figure 4.3(c): Skein Coil

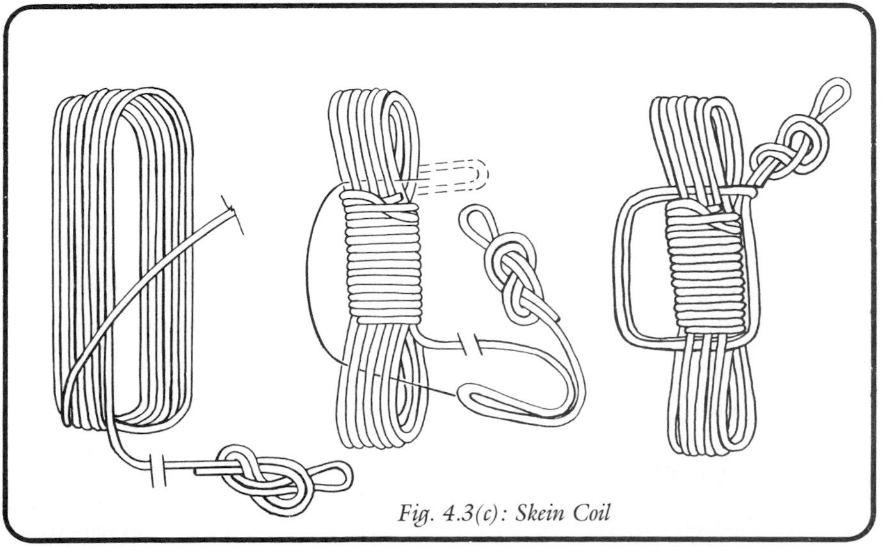

Fig. 4.3(c): Skein Coil

One of the most convenient ways of storing, transporting and at the same time protecting the rope, is **bagging.**

Among the advantages of bagging are:

- The rope can usually be flaked into the bag quicker than it can be coiled. Figure 4.4 shows a quick technique for bagging a rope.
- The bag helps protect the rope from damage while keeping it clean.
- If the bag has a shoulder strap or pack straps, then it is a convenient way in which to carry the rope.
- A bagged rope is easy to deploy. Simply secure the upper end of the rope and drop the bag over the edge. In most cases, the rope will flake out of the bag without tangles. The bottom end of the rope should be secured to the bottom end of the bag so that the bag is not lost.

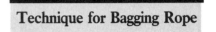

WARNING NOTE

To avoid damage to the rope, use common sense in dropping bagged rope. Do not, for example, drop a bag with 600 feet of rope down 100 feet. The 500 feet remaining in the bag when the bag hits bottom may be damaged by the impact. Match the amount of rope to the distance of the drop.

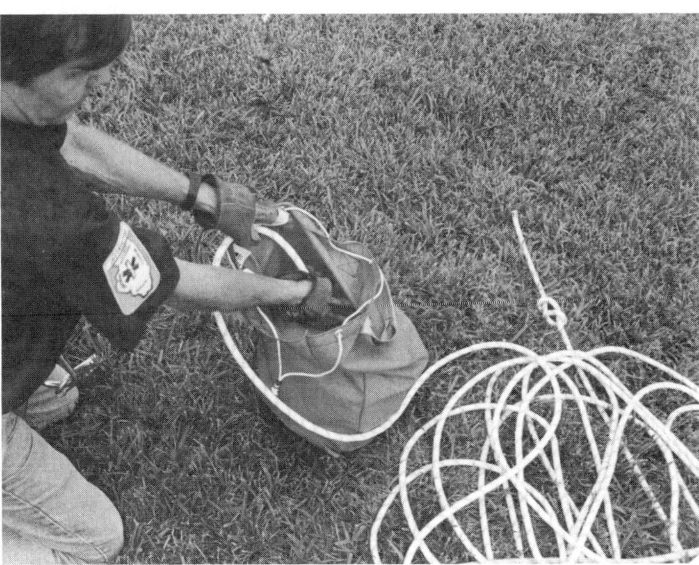

Fig. 4.4 Bagging a Rope

Technique for Bagging Rope

1. Grasp the top edge of the bag and hold the bag open and upright with your non-dominant hand *(left for right handed people).*
2. Lighly trap the rope between your thumb and index finger as it enters the bag.
3. With your dominant hand *(right hand for right handed people),* grasp the rope and pull it into the bottom of the bag.
4. Slide the dominant hand back up to the other hand, take another length of rope, and pull it down into the bag.
5. Continue with these short strokes until the rope is bagged.

How Ropes Are Damaged

Harmful Substances

Among the common substances that can destroy or deteriorate rope are:

(damaging to nylon)

- Acids, particularly those found in storage batteries.
- Bleaches.

(damaging to polyester)

- Alkali (such as found in soot).
- Many other strong chemicals. Avoid any contact with a chemical unless you know for sure it is harmless to rope fiber *(damaging to nylon and polyester).*

Overloading A Rope

Overloading a rope creates internal damage that could endanger those people who use the rope in the future. Damage from overloading usually occurs when a rope is used in activities for which it was not intended, and the load greatly exceeds the rope's safe working load. Some examples of overloading a rope include:

- Using it for towing vehicles.
- Lifting heavy objects.

A separate set of ropes, for utility use only, must be used for activities such as these two examples. These utility lines must be stored separately from life support ropes and distinctly marked, for example, **"Utility Line, NOT FOR LIFE SUPPORT OPERATIONS."**

Damage from Falling Objects

Objects, such as rocks or tools, that fall on the rope, particularly when it is under load, can cause serious damage to a rope. Any time that heavy/sharp objects have been seen falling on a rope, or when the rope has been used in a rock fall zone, then it should be inspected for damage.

Abrasion

One of the most common ways of destroying a rope, or shortening its life, is through abrasion. This kind of damage is usually avoidable. Damage from abrasion commonly occurs when the rope is under tension and is lowered and raised across a rock or over the edge of a building. Abrasion also often happens when a person is doing "bouncy" rappels or ascending, causing the rope to "saw" back and forth across a rock or hard object.

A less common form of abrasion, but one that can damage a rope from end to end very quickly, is the use of a rappel device or other hardware with a sharp metal edge. Check for burrs on all hardware from time to time and de-burr or retire the equipment as necessary.

Techniques for Avoiding Abrasion

There are numerous ways for avoiding abrasion on a rope, many of them using simple and inexpensive equipment. Every person and every team that owns a rope should carry equipment for preventing edge abrasion.

Rope Pads

Rope pads are among the simplest and least expensive techniques for protecting rope from abrasion. Among the commonly used type of pads are:

- Canvas pads

  Heavy duty canvas can easily be made into protective pads. For greater protection and durability, a square of heavy duty canvas can be folded twice, stitched around the edge to prevent fraying, and then cross-stitched *(see Figure 4.5)*. For added convenience, large grommets can be set in two corners. These grommets can be used to attach the pad so it will not slide away as the rope is moving. A complete edge protection kit should include a variety of pads ranging from approximately 2 x 3 feet to 2 x 6 feet or larger.

- Fire hose

  Sections of discarded fire hose can also be converted to effective rope pads. In order to avoid having to feed the rope through the hose, modify the hose in the following manner:

  1. Split the hose down the center.
  2. To prevent the rope from slipping out, secure the edges of the hose with a closure such as snaps or Velcro™ *(see Figure 4.6)*.
  3. Set a hole or grommet in each end so the hose can be anchored to prevent it from slipping down the rope.

- Commercial rope protectors

  They are similar to the fire hose pads, but are constructed of PVC coated nylon *(Figure 4.7)* or nylon lined with canvas.

- Improvised techniques

  If no premade edge protection is available, then other materials may be pressed into service to protect the rope from abrasion. Among improvised rope pads are:

  - Packs
  - Turnout coats
  - Clothing, and
  - Carpet squares

  (NOTE: Wool carpet squares are preferable, since synthetic material carpet may melt under heat fusion with the rope. If only synthetic material carpeting is available, use the jute backing as a surface against the rope.)

- Edge rollers

  Edge rollers are one of the most effective techniques for protecting ropes from abrasion. They are usually more expensive than the techniques mentioned above, but they have the added advantage of greatly reducing friction of rope over an edge. This is particularly important in hauling systems, where edge friction makes the tasks of raising much more difficult and puts great stress on equipment.

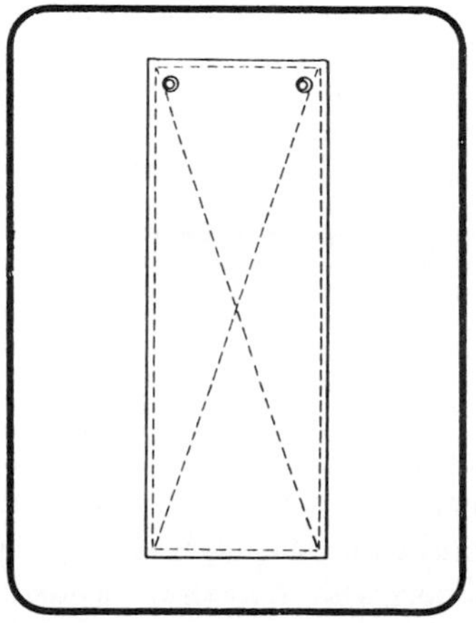

Fig. 4.5 Canvas Rope Pad

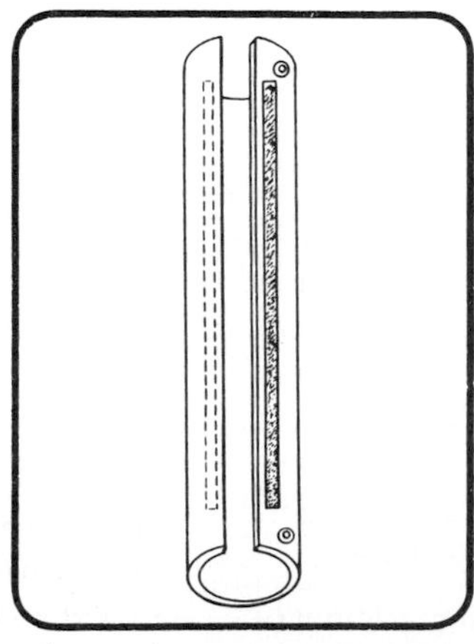

Fig. 4.6 Fire Hose Rope Pad

Fig. 4.7 Commercial Rope Protector

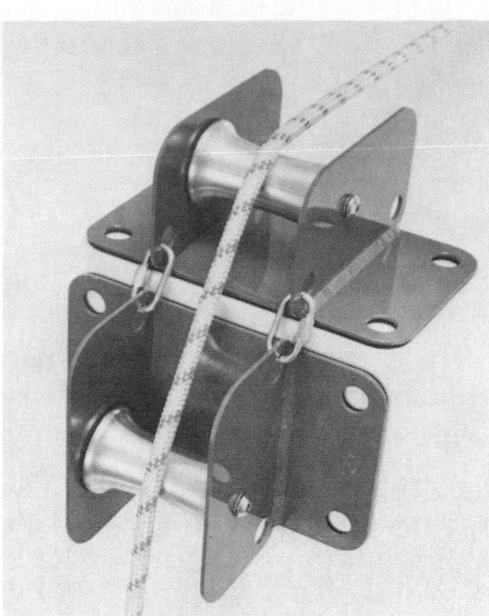

Fig. 4.8 Edge Roller

Edge rollers are available in two main types:

A. Single unit rollers. A single roller is set into a frame with a flat base *(see Figure 4.8)*. It usually takes two or more of these single units to provide adequate edge protection. They generally must be stabilized by anchoring them with their attachment points. When they are stabilized, these units perform well on irregular surfaces, such as cliffs and other natural situations.

B. Roof Rollers. The roof roller consists of a unit of two rollers set in a 90° frame *(see Figure 4.9)*. They are designed for edge protection on buildings and other structures where 90° angles are present.

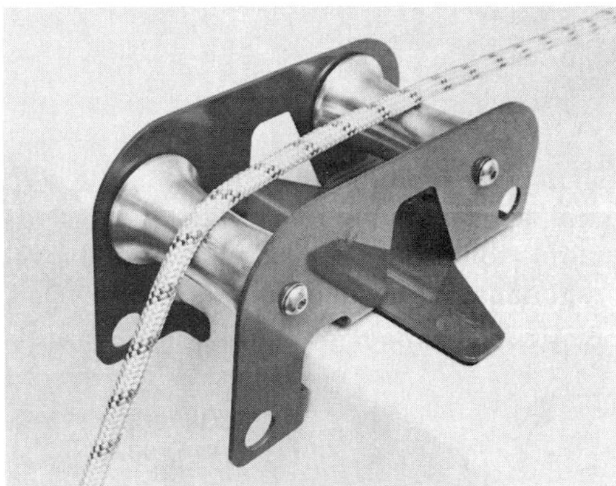

Fig. 4.9 Roof Roller

Heat Fusion

Heat fusion is the result of two pieces of synthetic material rubbing together. It is very destructive of rope and can cut a line as surely as if it were cut with a knife. Heat fusion usually occurs when one rope runs across another, or across webbing, and one piece of synthetic material remains stationary, while the other moves across it rapidly at one spot. Common situations where heat fusion occurs include:

- Two ropes under tension where one remains stationary while the other, being lowered, runs across the first.
- A loaded rope running across an anchor rope or webbing also under load.
- A rappler holding a rope against his own seat harness webbing while performing a rapid rappel.

Damage to ropes because of heat fusion can happen quickly, without warning, and can be catastrophic. Everyone working in the high angle environment must constantly be on the alert for heat fusion and take steps to avoid it.

Ways to prevent heat fusion due to rope cross include:

- Rigging ropes so that they do not make a contact to create heat fusion.
- Holding ropes away from one another with pulleys or edge rollers.
- Padding the stationary rope where one rope runs across it.

NOTE: Heat fusion occurs when one rope is stationary and the other moves across it **in one spot** so that the heat builds up. If both ropes are moving constantly so that one spot is not subjected to the heat buildup, then destructive heat fusion is not likely to occur. In a Munter Hitch (see page 83) for example, the rope is running across itself, but all surfaces are moving. So when correctly used, the Munter Hitch is not likely to cause heat fusion.

Rope Damage through "Flash" Rappels

All rappel devices operate through friction of the rope across the device. This results in heat buildup which increases with the speed of the rappel. Fast rappels must be avoided because they can damage rope through heat buildup. They also indicate poor technique and/or lack of control on the part of the rappeler.

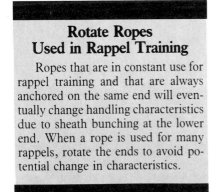

Rotate Ropes Used in Rappel Training

Ropes that are in constant use for rappel training and that are always anchored on the same end will eventually change handling characteristics due to sheath bunching at the lower end. When a rope is used for many rappels, rotate the ends to avoid potential change in characteristics.

Strength Loss through Knots

All knots reduce the overall strength of rope, but some knots cause a greater loss than others. The general rule is this: knots with tight bends, such as bowlines, cause greater strength loss than knots with more open bends, such as the Figure 8 family of knots.

Effects of Bending a Rope

Whenever a rope is placed under load in a sharp bend, some strength is lost *(see Figure 4.10)*. The rope fibers on the outside of the bend receive a greater share of the load, while those on the inside of the bend receive very little of the load or none at all.

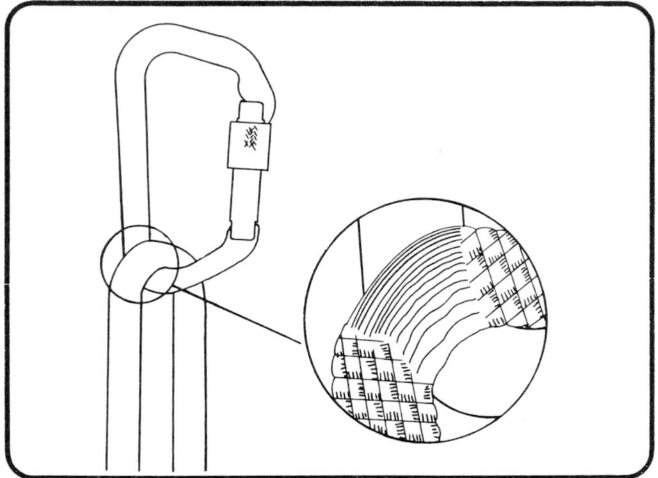

Fig. 4.10 Effect of Bending Rope

Common situations where ropes receive this kind of stress include when ropes have knots or kinks or when they run over a sharp bend, such as in a carabiner or small pulley.

The 4:1 Rule

When choosing mechanical devices such as pulleys, their effect on the strength of the rope can be estimated by using what is known as the **4:1 Rule.**

It has been determined that strength loss in a nylon rope does not become significant until the rope has a bend less than four times the diameter of the rope. This means, for example, that a half inch rope should not have a bend that is less than two inches in diameter. Otherwise, some strength loss will occur.

To choose a pulley using the 4:1 rule, compare the diameter of the rope to the diameter of the pulley sheave. If the rope diameter is 1/2 inch, then the pulley should have a diameter of at least 2 inches.

Inspecting a Rope

Rope inspection is an ongoing process to be done before, during, and after rope use. It is conducted two ways: **looking** and **feeling.** After each use, the rope should be thoroughly inspected by looking and feeling along every inch of its length.

Visually inspect the rope, looking for:

- Discoloration.
  An obvious change from the rope's original color. Discoloration, particularly brown, grey, black, or green, could indicate chemical damage.
- Glossy marks.
  Could indicate heat fusion damage.
- Exposed core fibers (white in most static rope).
  Indicates damage to the sheath.
- Lack of uniformity in diameter/size.
  May indicate damage to the core.
- Excessive fraying.
  May indicate broken sheath bundles.
- Inconsistency in texture and stiffness.
  (Hold the rope in a loop and see if it is a uniform radius around the entire bend.)
  An inconsistency in the bend may be the result of a soft spot that indicates core damage.

Run the rope slowly through bare hands, feeling for:

- Stiffened fibers.
- Obvious changes in diameter.
  If enough core strands are broken, there will be a localized change in the diameter of the rope usually indicated by a depression or hourglass shape that can be felt. Some types of damage will result in "puffs," core fibers protruding from the sheath.
- Contamination with dirt and grit.
  The rope should be washed.

Retiring a Rope

Unfortunately, the only tests that currently exist to reliably measure rope strength will also destroy the rope. Thus, the ability to determine if a rope should be retired is essential. That ability is the result of education in rope use and construction combined with experience and good judgment.

The following are general guidelines that can assist in deciding when to retire a rope:

- **Sheath Wear**—More than half of the outer sheath yarns are broken.
- **Shock Loading**—The rope has been subjected to severe shock loading.
- **Overloading**—The rope has been subjected to the kind of overload for which it was not designed. Examples of overloading for life support rope would include towing a vehicle or hauling heavy equipment or materials.

- **Chemical Contamination**—Unless the chemical is specifically **known to be harmless,** it should be considered a contaminant.
- **Lack of Uniformity in Texture**—Soft, mushy places or hard spots.
- **Age**—The rope is simply "worn out" from use.
- **Lack of Uniform Diameter**—The rope necks down to a smaller diameter similar to an hourglass shape.
- **Loss of Faith**—The rope was used by persons you suspect may not have taken proper care of it.

But the bottom line is this:

WHEN IN DOUBT, THROW IT OUT.

Compared to many other types of equipment, rope is an inexpensive tool. The cost of replacing a rope is certainly less expensive than a severe injury or loss of life.

Establishing Responsibility

As with other life safety devices, such as breathing apparatus in the fire service, ropes used by a team must be assigned a **chain of responsibility.** Someone must be responsible for knowing where they are, how they have been used, who has used them, and what condition they are in. Someone must be responsible for inspecting them after each use, for keeping a log for each rope, and, where appropriate, for removing them from service.

Washing Rope

All rope will eventually become dirty after use, so one element of a rope inspection program should be to determine when the rope needs a bath. While it most obviously affects the appearance of rope, the most serious effect of soiling is more serious and hidden. Particles of grit and dirt eventually work their way into the core of rope to damage the load-supporting yarn as it stretches and flexes. (Stepping on a rope forces more of this damaging material into the rope core.) Furthermore, the dirt on the surface of a rope will accelerate wear on hardware such as rappel devices, much as would sandpaper.

Rope Washing Devices

Commercial devices are available that are specifically designed for washing ropes. They operate very much like the hose washing devices that are used by fire departments. As water jets spay into the center of the device, the rope is pulled slowly through it *(see Figure 4.11).* Rope washing devices are most effective against larger particles of dirt, but may not make the surface of the rope appear completely clean. This can be done only with further steps.

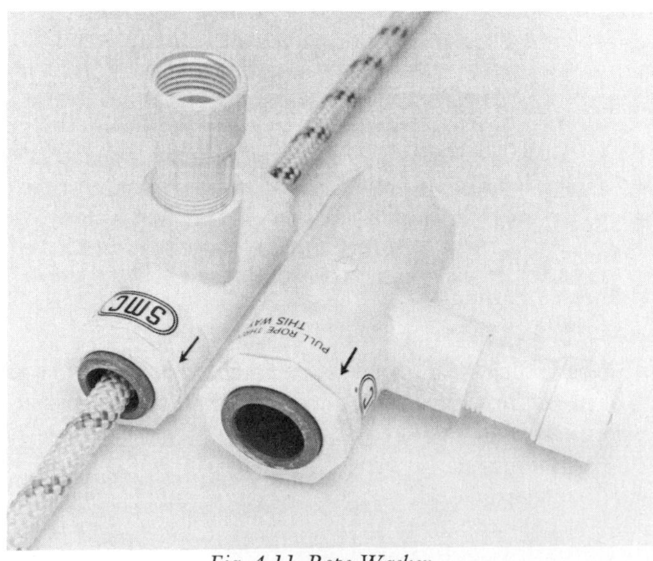

Fig. 4.11 Rope Washer

Cleaning Ropes with a Washing Machine

Washing machines can thoroughly and effectively clean rope, but they must be used with care and with certain specific precautions to prevent damage to the rope.

- Front-loading, tumbling type machines are preferable. Their tumbling-type action usually causes less tangling of the rope than the agitator action of top loading machines. The machine should not have a plastic window, since in a spin cycle the rope could be damaged by heat fusion from striking the plastic. Top loading machines can also potentially damage the rope through abrasion against the agitator.
- Coil the rope to prevent tangling. One commonly used coil for washing is the "chain coil" *(see Figure 4.12).*
- Use cool water. While hot water may not seriously damage rope, it will likely shrink the yarn and change the rope's handling characteristics.
- Use gentle soaps and follow package directions for their use. Soaps that indicate they are "safe for all synthetics" will most likely be safe for rope cleaning. Still, to make certain, some rope owners use only the most gentle cleaners such as Woolite™ or Ivory™.

WARNING NOTE

Some persons use fabric softeners to give a soft feel to the rope sheath. But a rope treated with fabric softener will also be slippery, so there will be less friction on the descender, and the possibility of more difficulty in controlling a rappel.

One study indicates that the soaking of rope in a *heavy concentrated solution* of fabric softener may be destructive to the rope yarn and cause significant strength loss to the rope.

- Do not use bleaches.
- Carefully dry the rope without heat. Hang the rope loosely out of direct sunlight and allow it to air-dry.

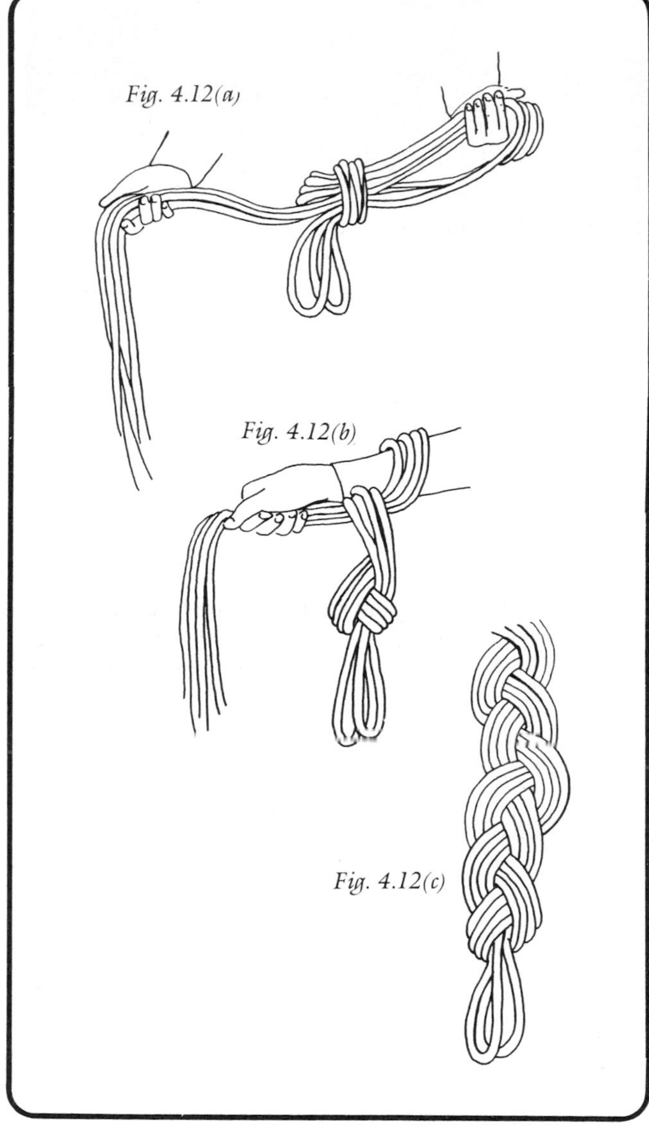

Fig. 4.12 Chain Coil

Special Cleaning Problems

Despite careful handling, ropes may become spotted with oil, grease, or mildew. There is no indication that any of these substances destroy rope fiber, but they are unattractive and may stain clothing or high angle gear. Petroleum substances may cause other contaminants to stick to the rope.

These substances can often be removed by soaking the rope in cool soapy water and scrubbing the affected areas with a fingernail brush.

Avoid strong solvent-based cleaners. Many solvents that loosen grease and grime will also dissolve nylon. Contact the rope's manufacturer for specific types of cleaning problems.

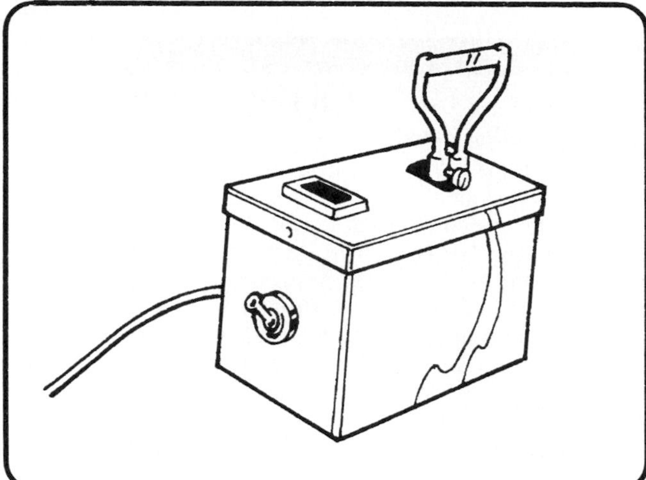

Fig. 4.13 Hot Cutter

Dressing Rope Ends

Cut rope ends should always be carefully dressed. Frayed ends have a sloppy, unprofessional appearance, become snagged, and eventually grow in size. This is particularly true with laid rope.

The most effective method of cutting synthetic fiber rope is using an electric hot cutter *(see Figure 4.13)*. Before the rope is cut, the spot where it is to be severed should be firmly taped.

If a hot cutter is not available, then take the following steps:

1. Firmly tape the spot to be cut to prevent fraying.
2. Cut down the center of the tape.
3. Immediately fuse the cut ends with heat, such as a lighter or other small flame. Taper the shape of the melt slightly *(see Figure 4.14)*. It should not be in the shape of a mushroom, which would get snagged when pulled through hardware or rock. **CAUTION: Molten nylon will burn your fingers.** (An alternative to melting the rope ends is to seal them with a liquid vinyl material such as Whip End Dip™.)

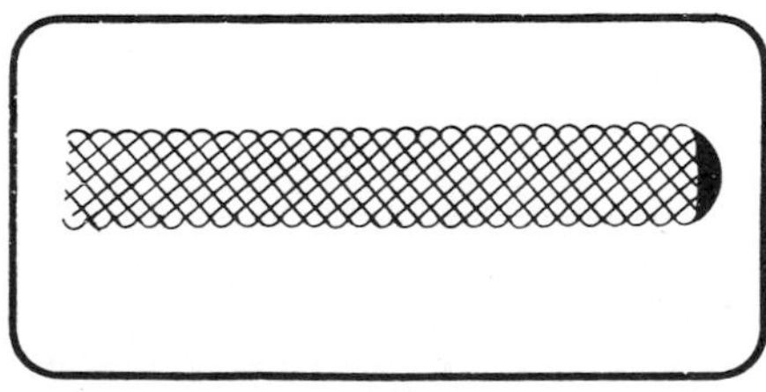

Fig. 4.14 Shape of Melted Rope End

Care of Webbing

In general, webbing should have the same care as rope: protect it from abrasion, and damaging substances, and inspect after each use.

Note that webbing is more susceptible to damage from shock loading. As with rope, webbing should be protected from friction heat damage.

HINTS FOR ROPE HANDLERS

Avoiding Tangles

Rope will not usually come out of a coil without tangling. To ensure that the line runs smoothly in the operation, it should first be "stacked." This simply consists of taking the rope off the coil and laying it on the ground on top of itself with the end to be used first (bottom end usually) on the top *(see Figure 4.15)*. The stack should be made out of way of the operation, but located so that the rope will run off to where it is needed without entangling debris, other rope, or persons. Rope stacked in this manner will usually slide off the pile smoothly. But it is a good idea to assign a rope handler to feed the rope to the person using it to ensure there are no tangles or kinks.

Bagged rope is less likely to tangle. But friction devices such as Figure 8 Descenders will twist the rope, creating tangles between the '8' and the bag that will increase as more rope is fed through the descender. So when using devices such as the Figure 8 that twist the rope, pull the rope out of the bag and stack it on the ground. Friction devices such as the Brake Bar Rack twist the rope very little, so the rope can be left in the bag when using 'Racks.'

Throwing/Dropping Rope

Most coils will not automatically feed out when dropped over the edge. To get a rope over the edge without tangles:

1. Stack the rope as described above. Be sure that the top end (at the bottom of the stack) is secured so it will not slip over the edge.
2. At the bottom end of the rope (top of the stack) take several loose coils in your throwing hand *(see Figure 4.16)*.
3. Add three or four loose coils in the opposite hand. The remainder of the rope should be coming off the top of the stack.
4. Loudly shout, "Rope" so that any persons below will be warned of falling rope.
5. With a side arm motion pitch the loose coils in your throwing hand out horizontally.
6. Allow the momentum of the falling rope to take the coils from your second hand. The rope should now pay out as it is pulled down by its own weight. Be sure to control the rope's speed through your **gloved** hand or, on very long drops, through a descender or belay device.

Fig. 4.15 Stacking Rope

Fig. 4.16 Throwing Rope

QUESTIONS *for* REVIEW

1. Name two conditions in which rope should not be stored.
2. Name two advantages in bagging rope over coiling it.
3. Name two examples in which rope damage may result from overloading.
4. Describe two general techniques for reducing rope damage from abrasion.
5. Name two situations in which heat fusion may damage a rope and, with each case, ways to prevent it.
6. Using the 4:1 rule, determine the smallest bend that can occur in a 1/2 inch rope before it begins to reduce strength of the rope.
7. Describe two precautions to take when cleaning a rope in a washing machine.

Chapter 5

Basic High Angle Hardware

OBJECTIVES–

At the completion of this chapter, you should be able to:

1. Describe the function of the basic hardware used in high angle operations.
2. Describe how carabiner designs relate to the specific job they are to perform.
3. Define what is meant by correct **manner of function** for carabiners.
4. Select carabiners that are appropriate for the jobs they are to perform.
5. Describe the function of descenders and how they are designed for specific purposes in the high angle environment.
6. Select descenders appropriate to the jobs they are to perform.
7. Describe the function of hardware in anchoring.
8. Describe the function of belaying and what hardware might be preferable alternatives to body belaying.
9. Select hardware to be used for belaying.
10. Describe the function of pulleys in the high angle environment and how the 4:1 rule applies to their function with the rope.
11. Select pulleys appropriate to the job they are to perform.
12. Describe the function of edge rollers.

TERMS—*pertaining to basic high angle hardware that the High Angle Rope Technician should know:*

Ascenders—Rope grab devices used by individuals to ascend a fixed rope or, with specific types of ascenders, used in the creation of hauling systems. There are two categories of ascenders: (1) **handled ascenders** which are normally used for no more than one person's body weight and (2) **cams** which are used both as personal ascenders and for hauling systems.

Belay—The securing of a person with a rope to keep him from falling a long enough distance to cause harm.

Belay Plate—A metal plate containing one or more slots for a rope and used with a carabiner to create friction on the rope for a belay.

Bolts—Metal devices used to create permanent anchors in rock. Because they permanently deface the rock and take time to place, they have limited application in high angle activities.

Brake Bar Rack—A descending device consisting of a "U"-shaped metal bar to which are attached several metal bars that create friction on the rope. Some "racks" are restricted to use for personal rappeling, while others may also be used for rescuer lowering. Also commonly known as a **Rappel Rack.**

Cams—(1) A generic term for ascenders that grip the rope through pressure. Some cams are spring-loaded to assist in this function. (2) Devices used in climbing for protection or in anchoring and which lodge in a rock crack through offset cam action.

Carabiners—Metal snap links used to connect elements of a high angle system. Sometimes (outside of the United States) spelled **karabiner.** Also known as **'biners** or **crabs.**

Descenders—Metal devices that, through friction with the rope, create braking action for a controlled rappel or lowering.

Edge Rollers—In-line, free-turning rollers that are anchored at an edge of a cliff or building to reduce rope friction.

Figure 8 Descender—A device used for rappeling and, in some cases, for lowering. It is in the general shape of an "8," with a large ring to create friction on the rope and a smaller ring for attaching to a seat harness.

Locking Carabiner—A carabiner with a locking sleeve on its gate side that secures the gate shut.

Manner of Function—The method in which a particular piece of equipment was designed to be used.

Non-Locking Carabiner—A carabiner without a means of securing its gate shut.

Piton—A slender metal wedge, with an eye for attachment, that is driven into a rock crack for climbing protection or for anchoring.

Pulley—A device with a free-turning, grooved metal wheel (sheave) used to reduce rope friction, and with side plates to which a carabiner may be attached.

Rappel Rack—See **Brake Bar Rack.**

Rope, webbing, and other software are critical to the high angle system. But another vital link in the system is a category of equipment known as **hardware.** Hardware includes a variety of gear, usually constructed of metal, that performs specific functions in the high angle environment.

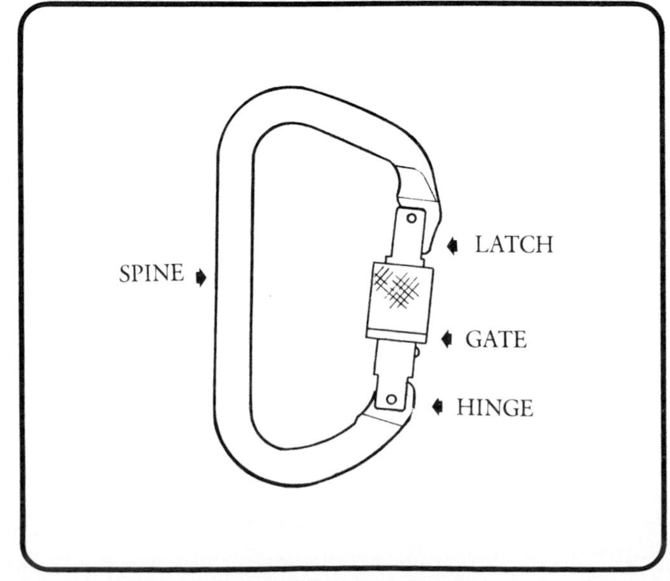

Fig. 5.1 Basic Parts of a Carabiner

CARABINERS

Carabiners are metal connectors that link the elements of the high angle system. They are also sometimes called "biners," "snap links," and "crabs." (In Europe, the United Kingdom, and some parts of Canada, the word is spelled with a **"K"— karabiner.**)

The basic parts of a carabiner include the spine, hinge, gate, and latch *(see Figure 5.1)*. However, many rock climbers prefer carabiners that can be opened at any time. So some carabiners specifically designed for climbing can be opened even when loaded. Though carabiner latches are designed in a variety of ways, depending on the manufacturer, there are two basic types *(see Figure 5.2)*. One type consists of a pin in the top of the gate which slips into a slot in the spine. The other basic type of latch consists of a sort of claw that slips into a slot.

Fig. 5.2(a)

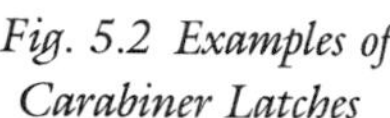

Fig. 5.2 Examples of Carabiner Latches

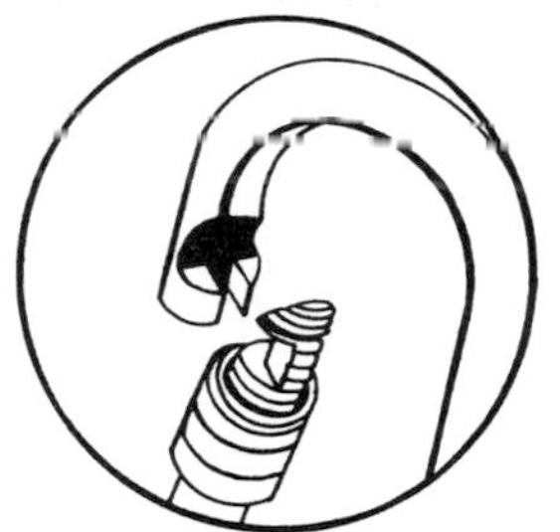

Fig. 5.2(b)

Basic Carabiner Shapes

Carabiners are manufactured in a wide variety of shapes and are usually designed for specific uses. They were originally designed as a simple oval *(see Figure 5.3 [a])*. When the oval carabiner is placed under load, the stress is the same on both sides, equally on the spine and the gate. The problem with this is that in any carabiner the weakest part is the side with the gate.

The design that takes advantage of the strength of the spine is the "D"-shaped carabiner *(see Figure 5.3 [b])*. Note that the spine side of the carabiner is longer than the gate side, with the top and bottom of the carabiner flaring toward the spine. This means that when under load, material such as a rope clipped into the carabiner will tend to slip into position on the spine side. The result is greater stress on the stronger side of the carabiner.

WARNING NOTE

Strength ratings for "D" carabiners usually represent the ideal situation, when material pulls only on a small area of the carabiner next to the spine. If material pulls across a wider area of the carabiner top or bottom (such as wide webbing does) then there will be greater stress on the gate side of the carabiner. Consequently, the carabiner may fail at a load lower than its rated strength.

Since the development of the "D" carabiner, there have been other designs such as the "Modified D" *(see Figure 5.3[c])*.

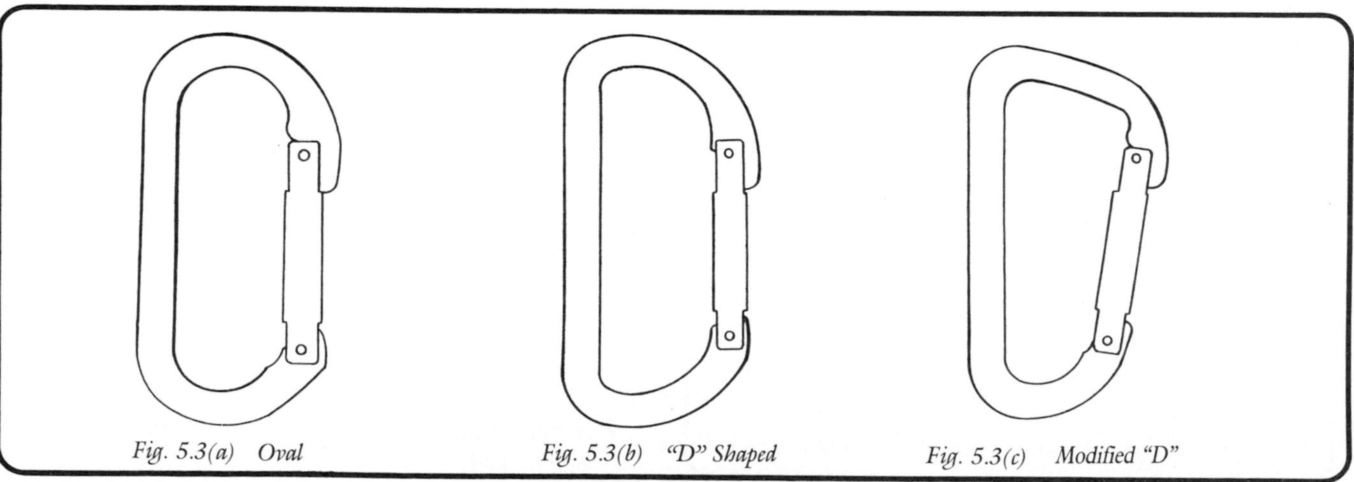

Fig. 5.3(a) Oval *Fig. 5.3(b) "D" Shaped* *Fig. 5.3(c) Modified "D"*

Fig. 5.3 Carabiner Designs

Carabiner Sizes and Strengths

Because they must carry their equipment with them as they climb, rock climbers have a great concern for weight. Consequently, carabiners made primarily for rock climbers tend to be lightweight. Some of these are constructed of hollow aluminum and have strength ratings as low as 3,000 lbs.

Carabiners for more demanding use and where more than one person's body weight may be involved, such as rescue, need higher strength ratings. These carabiners are made of solid aluminum or of steel.

Another difference among carabiners relates to size. Again, because climbers place a premium on weight and bulk, their carabiners tend to be relatively smaller. But in situations where large amounts of material must be connected inside the carabiner, such as in rescue activities, the carabiner will need to be larger.

Carabiner Gate Opening

Along with differences in carabiner sizes and weights, there are also differences in the widths of gate openings. For some activities, the carabiner gate opening will have to be larger than normal. In rescue activities, for example, a carabiner may have to be clipped over a litter rail, which can be as large as 1 inch in diameter. Only a few carabiner designs have gates that will open this wide.

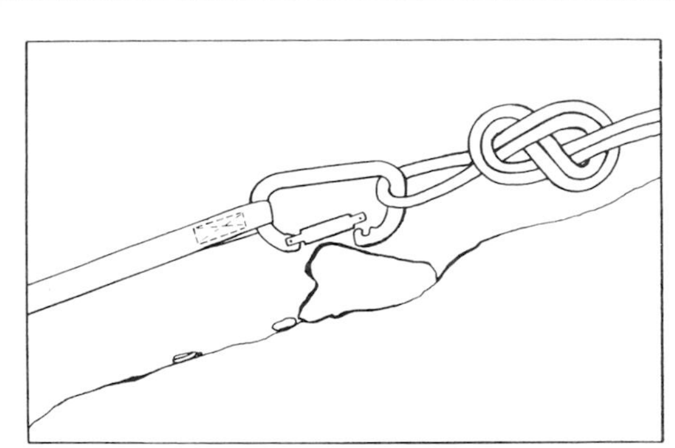

Fig. 5.4 Gate Opened by Rock/Building Edge

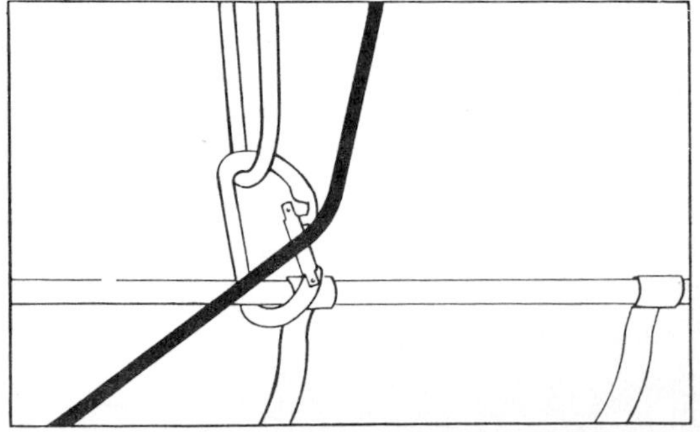

Fig. 5.5 Gate Opened by Rope/Webbing

Accidental Gate Opening

The main job of a carabiner is to maintain its link with the other elements of the high angle system. To do this, the carabiner gate must remain securely closed. If it does not remain closed, then the connecting elements will come unlinked and the system will fail.

Unfortunately, there are several ways in which carabiner gates may come open accidentally. Among the most common are situations where:

- The carabiner is pressed against a rock or edge of a building, forcing the gate open *(see Figure 5.4)*.
- A rope or piece of webbing is pulled across the carabiner gate, forcing it open *(see Figure 5.5)*.

If there is any chance that a carabiner gate may come open, and only non-locking carabiners are available, then two carabiners should be set together **reversed and opposed.** That is, the pair of carabiners should be set with their gates reversed to one another and their tops and bottoms opposed to one another *(see Figure 5.6)*.

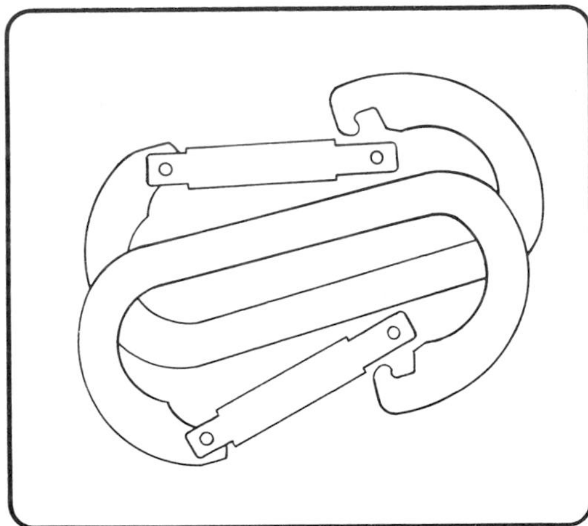

Fig. 5.6 Carabiner Reversed and Opposed

Locking Carabiners

While two reversed and opposed carabiners are usually very secure, it is often quicker and more convenient to use a single locking carabiner, *(see Figure 5.7)*. Though specific designs will vary with the manufacturer, locking carabiners usually fall into the following categories:

- A locking-sleeve moves on screw threads over the nose of the carabiner to ensure closure.
- A sleeve turns around a pin on the gate to move up and close over the nose.
- A spring-loaded sleeve makes a quarter turn to unlock. (While this is a very convenient carabiner to use, it should be employed with caution, since it comes open very easily.
- A sleeve moves downward over the hinge to hold the gate locked.

Some Carabiners Are Not As Strong When They Are Unlocked

Not all carabiners are as strong with their locking knobs in the unlocked position as when they are locked. Some carabiner designs need the locking knob to hold the claw-type latch together in the locked position for the carabiner to have its full-rated strength. Tests have shown that some of these types of carabiners can fail at a much lower rating than their advertised strength if the gate is unlocked.

In addition to the locked strength of a carabiner, you should also know its unlocked strength.

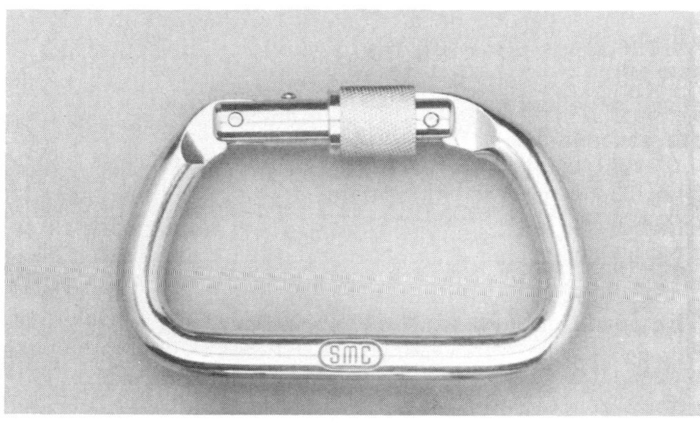

Fig. 5.7 Locking Carabiner

WARNING NOTE

Locking carabiners can and do come open after being locked.

The following are common ways they can come open, along with possible countermeasures:

- Carabiner gate unlocks by rubbing against face of cliff or building.
  Countermeasure: turn carabiner away from face.
- With some carabiner designs, vibration can cause the locking sleeve to unscrew.
  Countermeasure: place carabiner so the gate is at the bottom and gravity keeps the locking screw closed.
- Gate is opened by rope or webbing running across it.
  Countermeasure: move carabiner out of contact with the line or place padding between it and the rope or webbing.

Whatever type of carabiner is used, it is the responsibility of those using it to maintain vigilance to make certain that the carabiner stays closed.

Aluminum Vs. Steel Carabiners

Carabiners are constructed from one of two metals:

Aluminum Alloy

Advantages:

- Significantly lighter weight.
- Does not rust.
- Less expensive than steel.

Disadvantages:

- Locking mechanism on some aluminum designs may eventually wear out.
- May suffer permanent damage as the result of severe shock loading.
- Some aluminum carabiners are not as strong as comparable steel designs.

Steel

Advantages:

- Locking mechanism on some steel-locking carabiners may hold up better than the locking mechanism on aluminum carabiners.
- May hold up better under severe shock loading.

Disadvantages:

- Heavier. This becomes a significant factor when one has to carry more than a few carabiners for any distance.
- More expensive than aluminum.
- Will rust, consequently steel carabiners require more maintenance.

Additional Concerns for Locking Carabiners

If a carabiner chronically becomes dangerously unlocked without cause, then it should be retired from service. Carabiners are designed to be locked only to light finger tightness. In their concern for safety and in their anxiety at being in the high angle environment, some people will over-tighten a locking carabiner and then be unable to unlock it. This situation commonly occurs when a person tightens down hard on his seat harness carabiner while it is loaded (he is hanging in the harness). Not only can you find yourself in an embarrassing situation, but you can damage the carabiner locking mechanism.

If a carabiner locking mechanism becomes "frozen" through overtightening, then the following procedure may release it:

1. If the carabiner is not already on a seat harness, attach it to one. Have the wearer move to a secure position, away from the edge of any drop.
2. Attach the carabiner via a sling to a convenient anchor mode.
3. Reload the carabiner by sitting down with it attached to the anchor point.
4. In many cases, the locking nut can then be easily loosened.
5. If it still cannot be loosened, try tightly wrapping a short piece of webbing around the lock nut to gain leverage.
6. If this does not work, then the careful use of a pair of pliers may be the only remaining option.

General Points on Buying Carabiners

Look first at the quality of the product, then at the price. As with all hardware, a few cents more, wisely spent, may extend the useful life of equipment for several years.

WARNING NOTE

All equipment used in the high angle environment is designed to be used in a specific manner of function. This is particularly true of carabiners. Any equipment, such as carabiners, not used in the manner of function may result in failure of the equipment and in severe injury or death. Details on carabiner manner of function are explained below.

Using Carabiners in their Manner of Function

The carabiner manner of function is designed for being loaded along its long axis, or lengthwise *(see Figure 5.8)*. As mentioned earlier, the weakest point of a carabiner is the gate. Consequently, side loading stresses the gate, places an unnatural force on the carabiner, severely reduces its strength, and may cause it to fail.

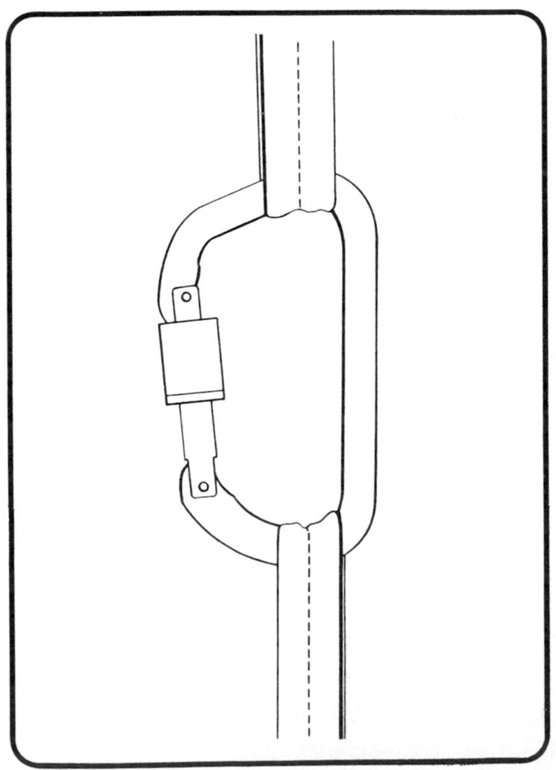

Fig. 5.8(a) Acceptable Loading

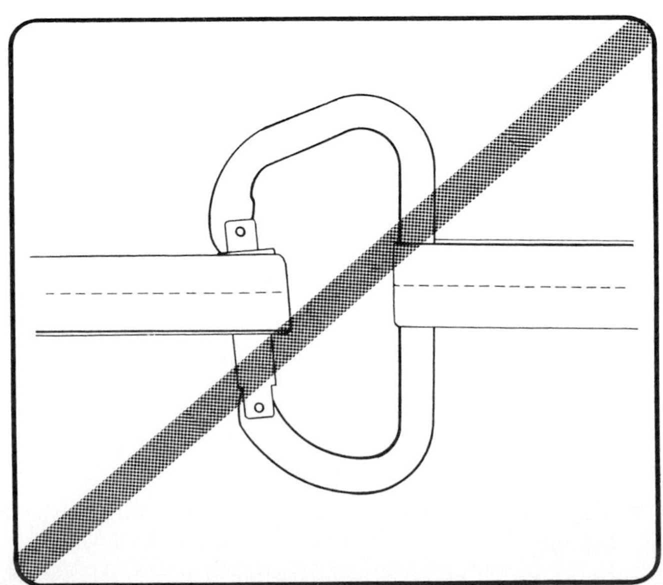

Fig. 5.8(b) Unacceptable Loading

Carabiner Brake Bars

To avoid cross loading, you should avoid using carabiner brake bar systems *(see Figure 5.9)*. Brake bars are solid bars of metal, with a hole drilled in one end and a slot cut in the other. They are sometimes fitted to oval carabiners to create rappel systems and, on occasion, lowering systems. Although carabiner brake bar systems are not as common as they once were, they are still found among some persons in the high angle environment. Because they subject carabiners to forces for which they were not designed, carabiner brake bar systems should be avoided. (Brake bars are used on Brake Bar Racks [see page 45]. When used on Brake Bar Racks—for which they were designed—brake bars do not create an unnatural stress on the "rack.")

Three-Way Loading

As mentioned before, carabiners are designed for loading along their long axis. However, some designs of seat harnesses have tie-in points consisting of right and left attachments. When a carabiner is clipped into these attachments and then loaded through a rope to a descender or other point, the carabiner becomes loaded from three directions. This can place potentially dangerous side-loading forces on the gate.

Fig. 5.9 Carabiner Brake Bars

Possible Solutions to Three-Way Loading:

a) Use a large pear-shaped carabiner for the seat harness. Place the two harness attachment points in the large end of the carabiner and the third point, such as the descender, in the small end. You must still be careful that the carabiner does not rotate out of this configuration and become sideways so it loads the gate.

b) Use a triangular or semicircular screw link in place of the seat harness carabiner. These devices are designed for loading from any direction. They are described below.

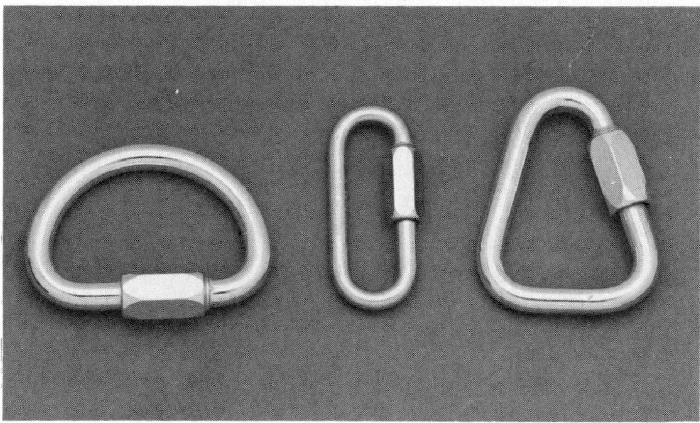

Fig. 5.10 Screw Links

Screw Links

Triangular or semicircular screw links are increasingly being used in the place of carabiners where three-way loading is necessary. In the place of the swing-open gate found on carabiners, screw links have a screw-locking sleeve that closes the opening (See Figure 5.10). When screwed closed completely, the screw link has the same strength regardless of direction(s) of loading.

Advantages:

- Strong regardless of direction of load.
- Inexpensive.

Disadvantages:

- Must be screwed all the way closed for strength.
- Because of the number of turns required to close the locking sleeve, they are slow to open and close.

Care for Carabiners

Problem: Sticking gates or gates that close sluggishly.

Solution: This may be due to contamination by dirt or to corrosion. The mechanism may be cleaned by blowing out with an air hose. The gate mechanism should be lubricated with graphite or other dry lubricant. **Do not use oil or grease-based lubricants** because they will attract more dirt and grit.

One other possibility is that the gate mechanism has been damaged. If this is the case, the carabiner should be discarded.

Problem: Carabiner gate with broken spring.

Solution: Carabiner should be discarded.

Problem: Broken latch mechanism.

Solution: Carabiner should be discarded.

Problem: Carabiner is bent (usually the result of being loaded with gate not completely closed, or subjected to overloading).

Solution: Should be discarded.

DESCENDERS

NOTE: A more complete coverage of the **use** of decenders is included in Chapter 9, "Rappeling." The following is a brief review of the equipment used in rappeling and lowering.

Descenders, also generally called rappel devices, are braking devices that are placed on the rope and attached to a person to allow him to descend the rope at a controlled speed. Some of the same equipment can also be used as braking devices to lower people or equipment under control. Such lowering operations may be required during rescue activities. Although there are many different descenders on the market, they all work on the same principle: they create friction by the rope's running through them to result in a controlled descent. In most cases, the user controls the speed of the descent by pulling on the side of the rope that is below the descender.

Guidelines for purchasing a descender:

- The descender should create enough friction so that you have absolute control over the descent without using brute strength. You should be able to go as slowly as you like and be able to stop at any time.
- It must have a lockoff function so that you can secure yourself and remain stopped on the rope with hands off the device.
- It must be strong enough to have an adequate safety factor (see page 24 for information on how to calculate a safety factor).
- It must be adequate for the length of descent needed.
- If desired, it must double as a lowering device.

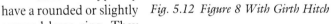

Types of Descenders

Fig. 5.11 Conventional Figure 8 Descender

Figure 8s

Figure 8 descenders are one person rappel devices roughly in the shape of an "8," but with rings of unequal size. The smaller ring, or lower one when in use, is clipped into a seat harness with a carabiner or screw link. The larger ring, or upper one, is the one through which the rope passes to create friction *(see Figure 5.11)*. All Figure 8s are fabricated of metal. Most are either of machined or forged-aluminum alloy, which is then anodized for greater resistance against wear from the rope.

Some Figure 8s are constructed of steel. Their primary advantage is their resistance to wear. They will outlast several aluminum Figure 8s, so they are particularly appropriate for heavy duty uses such as in training departments. Their disadvantages are that they weigh much more than the aluminum 8s and they cost more.

Figure 8s are made by a number of different manufacturers and come in a variety of shapes and sizes.

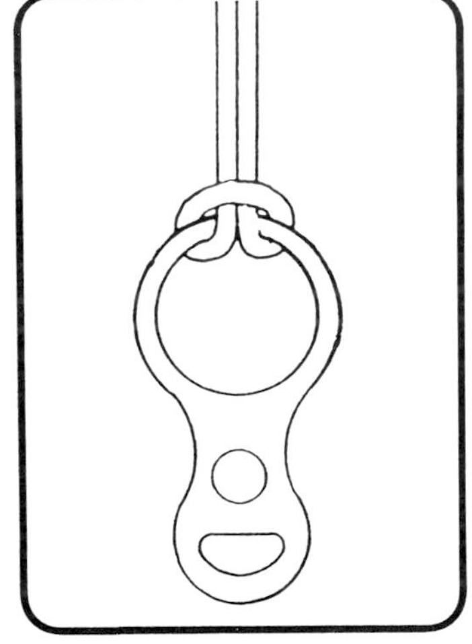

Fig. 5.12 Figure 8 With Girth Hitch

Conventional Figure 8s

These are mostly found in small sizes and typically have a rounded or slightly squared large ring. They are generally used in recreational activites such as climbing.

ADVANTAGES:

- Fairly compact and lightweight. Generally favored by climbers and others for whom size and weight are primary considerations.

DISADVANTAGES:

- Smaller models will not dissipate heat easily.
- Smaller models will not take larger diameter ropes.
- All models will twist the rope.
- In all models rope can slip around the large ring to form a girth hitch *(see Figure 5.12)*. If this occurs during a rappel, it may trap the user on the rope in a situation from which it will be difficult for him to extricate himself alone. (See Chapter 10, "Ascending," for possible ways to extricate oneself from this kind of problem.)

- In all models, once on the rope, the rappeler cannot create a wide range of friction on the device.
- In all models long rappels (over approximately 150 feet) are more difficult to control.

Figure 8 with "Ears" (Sometimes called "Rescue 8s")

The "ears" are projections fabricated into the large ring *(see Figure 5.13)*. They are specifically designed so that the rope contours better around the large ring and does not slip over it to form a girth hitch. This style of Figure 8 is found in larger sizes.

ADVANTAGES:

- Rope will not slip around large ring to form girth hitch.
- Because these models are in larger sizes, they dissipate heat better.
- Will accept large-size ropes.
- Easier to lock off.

DISADVANTAGES:

- Bulkier and slightly heavier than the smaller models of the conventional Figure 8.
- As with all Figure 8s, will twist rope.
- As with all Figure 8s, once rappeler is on rope, he cannot create a wide range of friction.
- As with all Figure 8s, more difficult to control on longer rappels.

Brake Bar Racks

Brake Bar Racks (also called Cole Racks or Rappel Racks) are descending devices that offer a great amount of control and the ability to vary greatly the amount of friction *(see Figure 5.14)*. They can also be used for very long rappels.

The Brake Bar Rack consists of two primary elements:

1) An inverted "U" shaped frame, one leg of which is longer than the other, made of cold-rolled steel. The end of the longer leg has an eye through which a carabiner can be clipped.
2) A series of bars with a hole drilled in one end so that they slide freely on the long side of the frame. On their other end, the bars are notched so that the end of the bar will clip into the short side of the frame. The rope is woven through the bars. Under tension, the rope keeps the bars in place on the frame.

Friction for rappeling may be controlled by the control hand on the rope below the rack, by varying the bar spacing with the other hand, and by varying the number of bars engaged on the rope. (For more detailed information concerning the use of the rack for rappeling, see Chapter 9, "Rappeling").

Fig. 5.13 Figure 8 With Ears

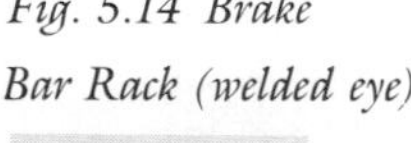

Fig. 5.14 Brake Bar Rack (welded eye)

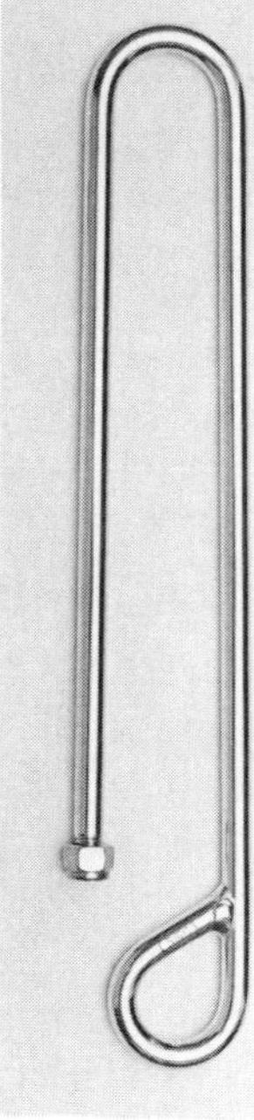

The bars are the only element of the device that wear out. They must be replaced from time to time.

Depending on the requirements for friction, the bars can be purchased in a variety of sizes and materials:

Larger bars:

- Create greater friction.

Hollow steel bars:

- Dissipate heat better.
- Last longer.
- Are more expensive.
- Give less friction.
- Are prone to rusting (unless they are stainless steel).

Aluminum bars (most commonly used):

- Give more friction than steel.
- Are less expensive than steel.
- Wear out sooner than steel.
- Leave aluminum marks on rope.

In the most common configuration, the rack is arranged with a 1-inch diameter grooved "top bar" at the top to keep the rope in the middle of the bars as it runs through the device. Usually five additional aluminum 3/4 inch diameter bars fill out the remainder of the rack.

A short version of the rack is available with five bars. But it is not recommended for rescue work since it provides less friction and has less versatility than the six bar rack.

For rappeling, the Brake Bar Rack works most efficiently with the open side of the rack towards the ground. Some seat harnesses have a horizontal "D" ring as a clip-in point which will cause the open side of the rack to be facing the rappeler's side. To adapt to this there are versions of the rack with a quarter turn in the eye so that the rack will remain with the open side towards the ground when you are wearing this design of harness.

Personal vs. Rescue Versions of the Rack

The rack frames are designed in differing configurations for the specific needs of the user. In the personal version of the rack, the eye at the end of the long side is formed by twisting the steel bar around itself *(see Figure 5.15)*. Because large loads could cause this type of eye to untwist, this style of rack should not be used for loads of more than one person.

For larger loads, such as in rescue operations, there is the rack with the welded eye *(see Figure 5.14)*.

Still another version of the rack is the RSI Rescue Rack™. The RSI Rescue Rack™ differs from other versions of the rack in that the legs of the "U" are of equal length. In use, the Superrack™ has the ends of the "U" pointing up and secured with nuts, while the attachment to the seat harness is in the curve of the "U." The RSI Rescue Rack™ has three fixed and three slotted bars.

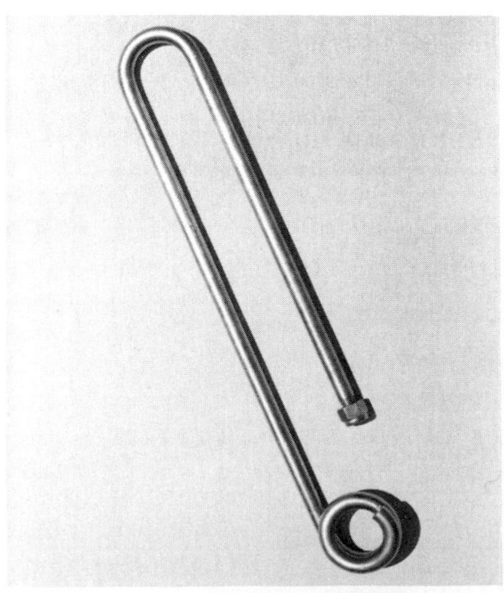

Fig. 5.15 Brake Bar Rack (coiled eye)

Characteristics of the Standard Rack

Advantages:

- Friction can be varied greatly even after the rappel has begun.
- Welded eye version is very strong.
- Can be used as a lowering device.
- Does not twist rope.
- Can be easily attached to rope without detaching the rack from the seat harness.
- Will take large-size ropes.
- Can use two ropes at the same time.

Disadvantages:

- Bulkier and heavier than some other rappel devices.
- May take slightly longer to lace onto rope.

Other Descenders

There are a number of other descenders available. Among them are:

- Stop (Petzl).
- Sidewinder (Lirakis).
- Ruapehu (Ruapehu Mountain Equipment Co.).
- Wonder Bar (Forrest).
- Sky Genie (Descent Control).

As with all gear you use in the high angle environment, it is your responsibility to obtain specific information on the characteristics of each device and training in its use.

Emergency Descent

Emergencies may occur in the high angle environment where it is necessary to descend immediately, even though a person has no hardware specifically designed for rappeling. One improvisation that has commonly been used in the past is the "carabiner wrap." This consists of wrapping the rope several turns around the carabiner to create friction and attaching the carabiner to the seat harness.

The carabiner wrap **is not recommended** as a rappeling technique. There have been numerous injuries and some deaths in rappeling attempts using the carabiner wrap. These failures have often been the result of the rope wraps slipping out of the carabiner gate.

There are other alternatives for emergency descent that are preferable to the carabiner wrap. For a review of emergency rappel techniques, see Chapter 9, "Rappeling."

Ascenders

Ascenders are devices used to travel up ("ascend") a fixed rope. They all work on the same principle: when correctly attached to the rope, their cam action allows you to slide the ascender freely in one direction (up). But they lock in place when you apply a downward force, such as body weight, to them.

At least two ascenders are used and are attached to the user with rope or webbing sling. You ascend the rope by raising one ascender with a hand while being supported by the other ascender as it locks on the rope. By alternating this action, you are able to proceed up the rope. This is the basic principle of ascending. There are dozens of variations in this technique, many of them using three ascenders for added security. (See Chapter 10, "Ascending," for specific details on ascending techniques.)

There are also several different brands of ascenders available, but they can be classified into two basic types: "cams and handled ascenders."

Cams

Cams operate primarily by the force of a cam action wedging the rope in them *(see Figure 5.16)*. The load is attached directly to the cam, and when the cam is loaded, it activates and grips the rope. Initially you may find cams to be a little more difficult to operate than other types of descenders, since they must be taken apart to be placed on the rope. But once you have assembled them on the rope, they tend to stay there.

Because of their higher strength, cams are commonly used in rescue and hauling systems. They tend to hold better on wet, muddy, or icy ropes than do other types of ascenders. The teeth on the cams are designed to hold well on the rope while causing a minimum of damage to the rope sheath. Under high loads or severe shock loading, any cam may cause damage to rope. The Rescucender™ cam is designed to cause less damage to rope than the Gibbs under the stress of rescue and hauling systems.

Currently, there are two brands of cams available:

- Rescucender™ (Rock Exotica, Inc., Centerville, Utah).
- Gibbs (Gibbs, Inc., Salt Lake City, Utah).

The Gibbs are available in two basic types:

- "Free Running" (there is no spring involved in the mechanism).
- Spring-Loaded (a spring action assists in setting the cam). Recent improvements in the spring design have made it more durable and easier to manipulate than the older design. However, the older model spring-loaded Gibbs was designed so that its spring could be disengaged so the cam could be used free running; the newer model spring loaded Gibbs does not have this option.

They are available in two sizes:

- Regular—For rope diameters 1/2 inch (12.7mm) and smaller.
- Large—For rope diameters 5/8—3/4 inch.

And they are available in a variety of metals:

- Cast cams, aluminum shell (light weight, for recreational activities).
- Forged cams, stainless steel shells (heavy duty, longer wear).

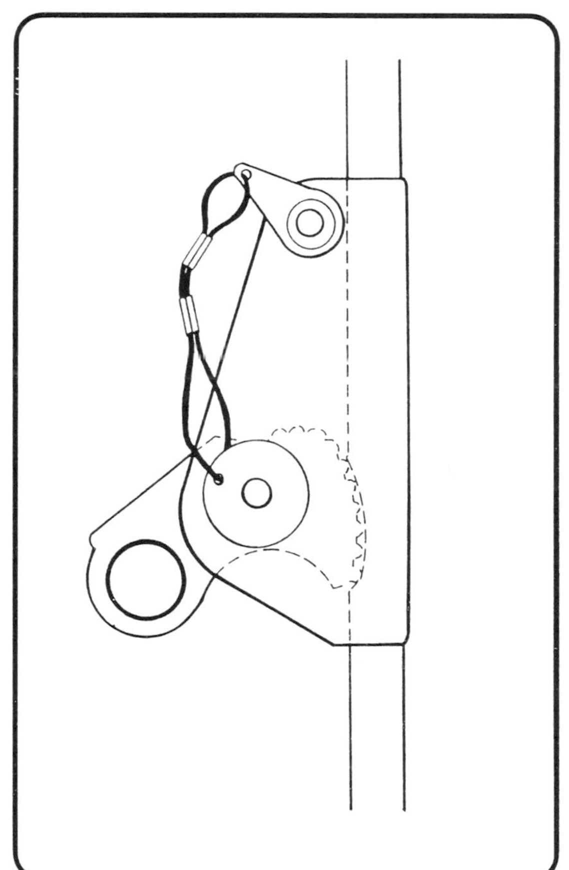

Fig. 5.16 Typical Cam

WARNING NOTE

Any cam device can cut the rope at much less than the rated strength of the rope.

CHARACTERISTICS OF CAMS

Advantages:

- Hold better than other ascenders on wet, icy, or muddy ropes.
- Strong.
- Cam action does not tear rope sheaths as easily as some other types of ascenders. Thus, cams are preferable for high load applications, such as in rescue and hauling systems.

Disadvantages:

- Must be taken apart to be put on/take off rope.
- Takes two hands to put on/take off rope.

Handled Ascenders

Handled ascenders are designed for ascending with only one person's body weight (not for hauling systems). They work in part through a cam action, which pivots inside the ascender frame, and also grip the rope with a toothed cam *(see Figure 5.17).* They have a handle which is equipped with a safety catch that releases the cam so that the ascender can be placed on a rope using one hand.

Handled ascenders are used in pairs, usually with right and left-handled models. For additional security, many people also include a third ascender in their system. Most handled ascenders cannot be used on rope above 1/2-inch diameter.

Among the handled ascenders currently being manufactured are:

- Clog (Wales).
  The frame is fabricated from rolled aluminum.
- CMI (West Virginia).
  Three models:
  - 5004, designed for ropes 1/2-inch and smaller;
  - 5003, designed for 5/8-inch ropes;
  - "Shorti," a compact design with no handle, but with a cam design similar to the 5004.

All three models are constructed of extruded aluminum and have toothed cams.

- Jumar (Switzerland).
  The frame is fabricated from cast aluminum.
- Petzl (France).
  Three models:
  - Expedition.
  - The basic Jammer, a compact, handleless design, but with a toothed cam.
  - Croll, a design with a toothed cam, designed to lie flat against the chest attached between a seat harness and a chest harness.

All Petzl models are fabricated from rolled aluminum.

- SRT (Australia).
  The frame is machined from extruded aluminum. There are several models made for different size ropes:
  - 8 to 11 mm rope
  - 8 to 16 mm rope

Questions to Ask When Purchasing Handled Ascenders:

- Can they be operated easily (placed on/taken off the rope) with one hand?
- Do they have a secure safety catch to prevent them from accidentally coming off the rope?
- Can they be easily used while wearing gloves?
- Are they comfortable in the hand while using them as a rope grip device?

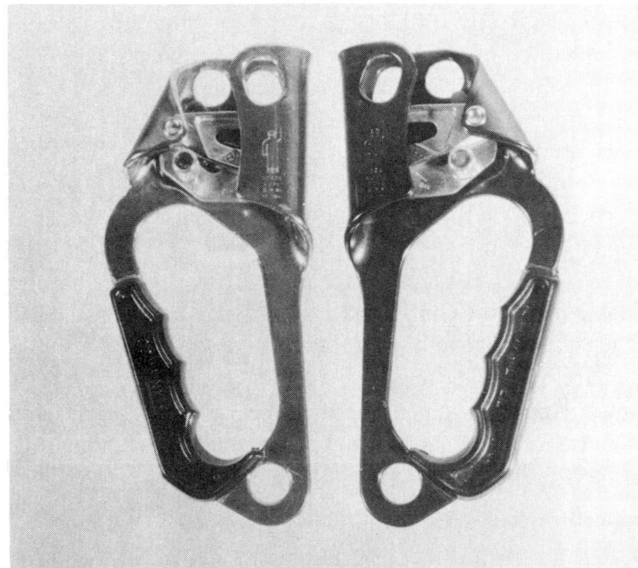

Fig. 5.17 Typical Handled Ascenders

WARNING NOTE

Handled ascenders may cut through rope sheath when loaded as low as 1200 pounds. Consequently, handled ascenders should never be loaded with more than the weight of one person and should never be shock-loaded.

HARDWARE FOR ANCHORING

Rock Anchors (Artificial Anchors)

Artificial anchors comprise a family of hardware that is used to create anchors where there are no natural anchors (such as trees or rocks) around which rope or webbing can be placed.

Artificial anchors go directly **into** the rock (and in some cases, into ice or snow). Most often, they are placed in cracks or gaps in the rock, but in some cases they are physically driven into the rock to become permanent emplacements.

The ability to place reliable artificial anchors is an art involving a number of subtleties, such as the character of the rock itself, the nature of the cracks or gaps in the rock, and the direction of loading on the anchor. This kind of knowledge is dependent very much on hands-on instruction and experience. These finer points of placing artificial anchors is beyond the scope of this manual. Those wishing to develop these skills should seek qualified instruction and literature (some of which is listed in the Appendix).

Pitons

Pitons have been used in mountaineering for many years as a means of anchoring. They are a thin metal spike with an eye to which a carabiner or webbing can be attached *(see Figure 5.18)*. Pitons are driven into a rock crack with a hammer.

Fig. 5.18 Typical Piton

WARNING NOTE

It is very difficult to drive pitons without permanently defacing the rock. Consequently, in some areas such as parkland, the use of pitons is prohibited. Also, many in the rock-climbing community consider the use of pitons to be bad etiquette.

Bolts

Bolts are a means of establishing a permanent anchor in a wall, usually when no other anchors are feasible. While there are different designs of bolts, most work on the same principle *(see Figure 5.19)*.

1. First, a hole is driven into the rock using either a separate drill or, in some designs (called "self-driving"), a drill that is a part of the bolt itself.
2. The bolt is set in the hole by causing an element in it to expand, either through a screw-like action, or by hammering on it.
3. A "hanger" is attached to the bolt through which a carabiner can be attached.

WARNING NOTE

The setting of a bolt creates a permanent change to the rock. Either the bolt remains fixed in place or is broken out (chopped), which means defacing the rock. In many managed areas, such as parkland, the setting of bolts is forbidden. Also, many in the rock-climbing community consider the use of bolts to be bad etiquette.

Bolts set previously by unknown persons and at an unknown time are, as a result, an unknown factor. Because of the varying techniques for setting bolts and because the rock may weather around them, bolts of unknown quality should not be trusted for anchors.

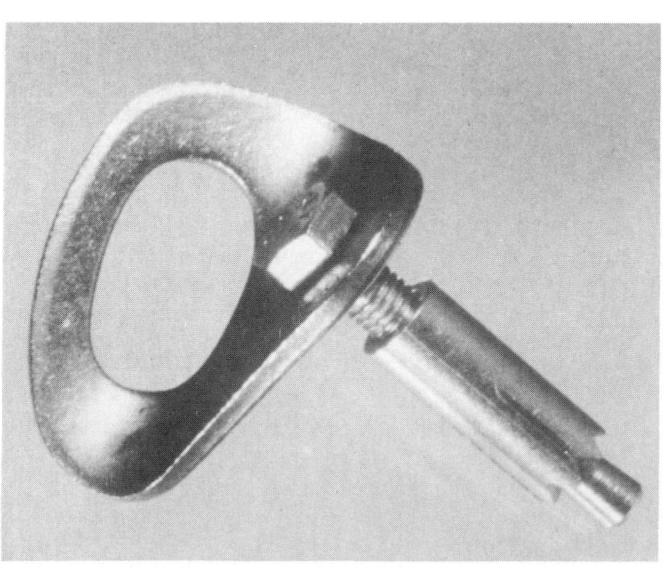

Fig. 5.19 Typical Bolt

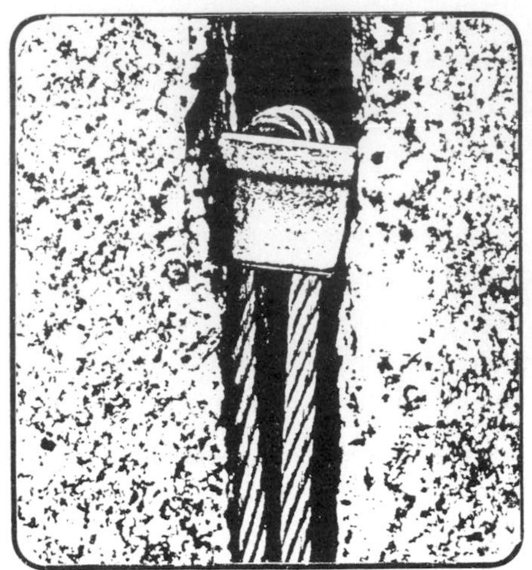

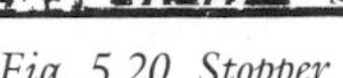
Fig. 5.20 Stopper

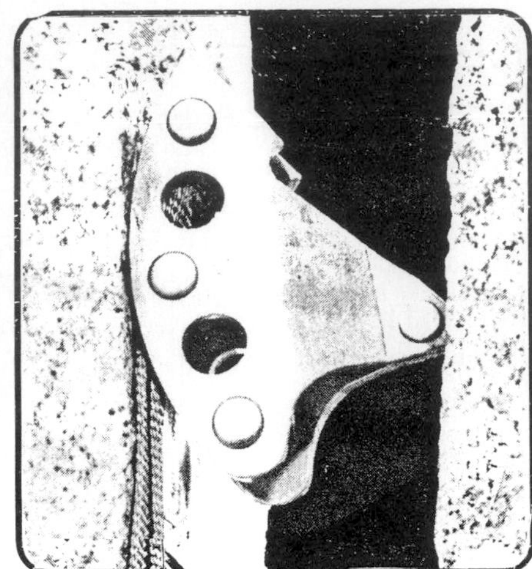
Fig. 5.21 Typical Cam (protection)

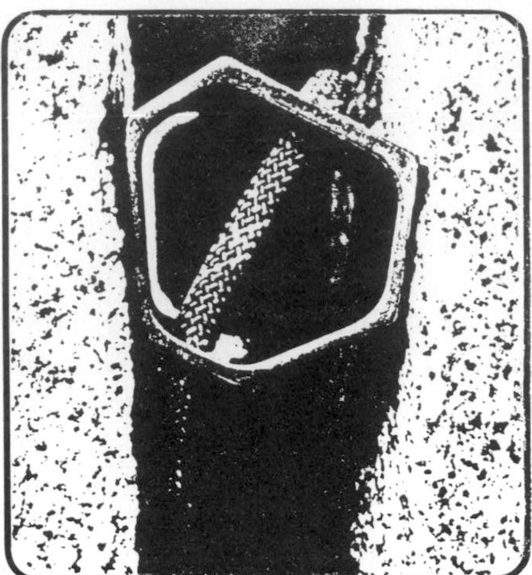
Fig. 5.22 Hexcentric

"Clean" Hardware

Because pitons and bolts deface the rock, in recent years, the climbing community has for the most part stopped using them. Climbers now tend to use types of anchoring hardware that does not deface the rock and which is collectively known as being "clean." Though clean hardware generally does not deface rock, it is often not as secure as pitons and bolts.

These devices are manufactured by a number of different companies and some may have very specialized uses. Some with very similar designs may go by different names. Some of the basic types of "clean" hardware for anchoring include:

- Nuts, chocks, or stoppers *(see Figure 5.20)* work by being wedged in the rock crack that bottlenecks.
- Cams *(see Figure 5.21)* anchor in the rock crack by an offset cam action.
- Hexcentrics *(see Figure 5.22)* can work both by a wedging and a cam action.
- Spring-Loaded camming devices (Friends™, Camalots™) work by a spring-loaded, opposed cam action *(see Figure 5.23).*

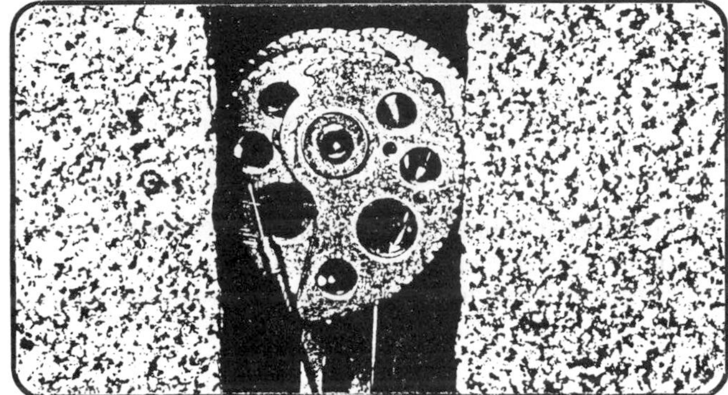
Fig. 5.23 Spring-Loaded Caming Device

Belaying

Belaying is, in essence, the securing of a person with a rope to prevent him from falling a long enough distance to injure himself. The difference in belaying techniques relates mainly to the manner in which the rope is held, or belayed.

The oldest of the belay techniques is the **body belay.** This involves the running of rope around the body of the person doing the safety (the belayer), usually around his waist. In this way, the rope can be brought tight when the person on the end of the rope (the climber) falls. The rope is held by friction around the belayer's body. There are significant disadvantages to body belaying:

- The force of the fall may cause the belayer to lose control of the rope and drop the climber.
- The force of the fall can easily injure the belayer.
- The belayer can become entangled in the rope.
- If the belayer does catch the climber, he must hold the climber until the climber can become secure.

Because of these problems, alternatives to body belaying have been developed.

Belay Plates

Belay plates are very simple devices consisting of a small metal plate with one or two holes *(see Figure 5.24).* A bight of rope is fed through the plate and secured with a carabiner on the opposite side. The carabiner is then clipped into an anchor. When the two strands of rope are pulled apart, a high degree of friction is created on the rope. This stops the fall of a climber.

Among the leading designs of brake plates are:

- Cosmic (Clog, Wales). Has a groove in front of the plate that is designed to increase rope friction smoothly.
- Sticht (Salewa, Austria). The oldest of the belay plate designs. "Sticht Plate" has almost become a generic term. It comes in several different configurations depending on the specific needs of the user.

WARNING NOTE

A belay plate may <u>not</u> be the most appropriate device for catching loads of more than one person's body weight.

Before using any device or system to belay a rescue load, you should test it under conditions similar to those you will encounter in an actual rescue situation to make certain that you will be able to catch the load when it falls (see example belay practice system on page 79).

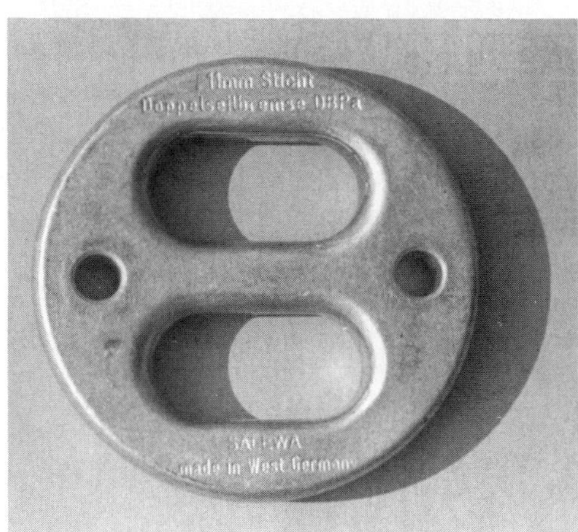

Fig. 5.24 Typical Belay Plate

Other Belay Alternatives

Another system for belaying requires only a large locking carabiner, something to anchor it to, and a rope. This is the Munter Hitch (also known as the **Half Ring Bend** or **Italian Hitch**).

The Munter Hitch is used by a number of rescue persons who believe they can successfully catch rescue loads using the device. Recent data suggest that a rescue load may be difficult to catch when it is **hanging in free air,** that is when the rescue load and the rope do not touch an edge, face or directionals. These elements in a high angle system add friction that help absorb the force of a falling load.

No belay device or technique is perfect for all rescue loads and in every rescue environment. You should use great caution in choosing any belay device for rescue loads and test it under realistic conditions that you will encounter in your own rescues before putting it to use. (See Chapter 8, "Belaying," for more information concerning the Munter Hitch.)

Pulleys

Pulleys are designed primarily to reduce rope friction. But this quality makes them useful in a number of functions in the high angle environment:

- To change direction of a running rope. Among the ways this may be useful are:
  - a) To position a rope more conveniently, such as to an area where people using the rope will be less exposed to falling, where there is less rockfall, or where they might have more room.
  - b) To reduce abrasion on a rope. A pulley could be used to hold a rope up from a rock, or to bring it away from other rope or webbing.
- In hauling systems to develop mechanical advantage. *(See Chapter 15, "Hauling Systems.")*

In certain situations, such as "big wall" climbing, lightweight pulleys are often used for such activities as hauling gear bags. Because weight is a primary consideration and the pulley is used for low-stress activities, these lightweight pulleys are often made of plastic or nylon.

But in certain high angle systems, and particularly in rescue hauling, the rope is under such stress that heat friction can cause a pulley to melt and fail. So where such life support is involved, only all-metal pulleys should be used.

Other Characteristics of Pulleys for High Angle Activities *(see Figure 5.25)*

- The **sheave** (wheel) should have a diameter that is at least four times the diameter of the rope. *(See Chapter 3, "Rope," for a discussion of the 4:1 rule.)*
- **Sideplates** should be moveable so the pulley can be placed on the rope anywhere along its length without having to feed the end of the rope through the pulley.

The sideplates should extend beyond the edge of the sheave far enough to protect the rope from abrasion.

Sideplates are generally the weakest part of a pulley. Consequently, pulleys for higher tensile strength rope (above 1/2 inch) should have steel sideplates.

- **Axles** should have rounded bolt heads that will not snag rope, other gear, or rock.
- **Bearings.** Most pulleys for high angle work have one of two types:
  - a) Bronze Bushing
    - *Advantages*
      - Less expensive than ball bearing.
      - Can be taken apart to be cleaned.
      - Very strong.
    - *Disadvantages*
      - Can be contaminated by dirt and grit.
  - b) Ball Bearing
    - *Advantages*
      - Turns slightly freer than does the bronze bushing.
      - Sealed, so is not contaminated by dirt and grit.
    - *Disadvantages*
      - Slightly more expensive.
      - Does not take stress, such as sudden blows, as well as the bronze bushing. They become damaged when the hardened steel balls dent the bearing races (technically called **Brinelling**).

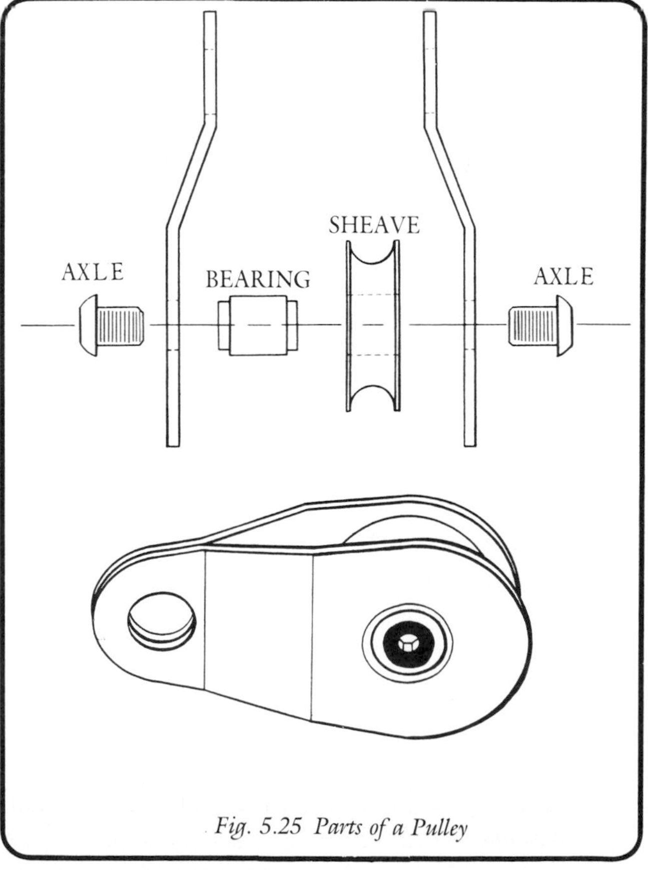

Fig. 5.25 Parts of a Pulley

Specialized Pulleys

There are a number of pulleys for the high angle environment that are designed for specific tasks. Among them is the knot-passing pulley *(see Figure 5.26)*. Their large sheaves are designed so that knots that connect lengths of rope will easily pass over them.

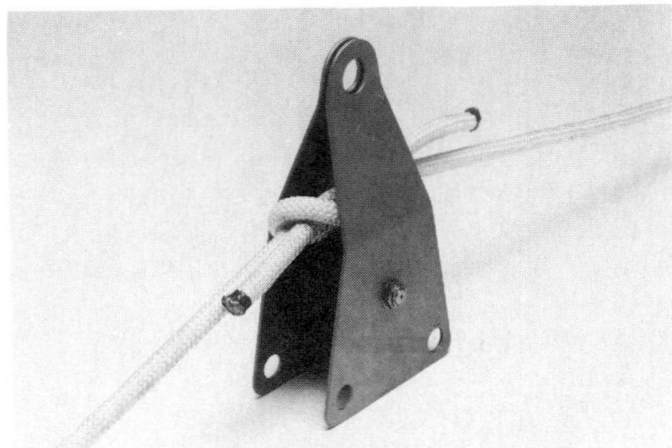

Fig. 5.26 Knot Passing Pulley

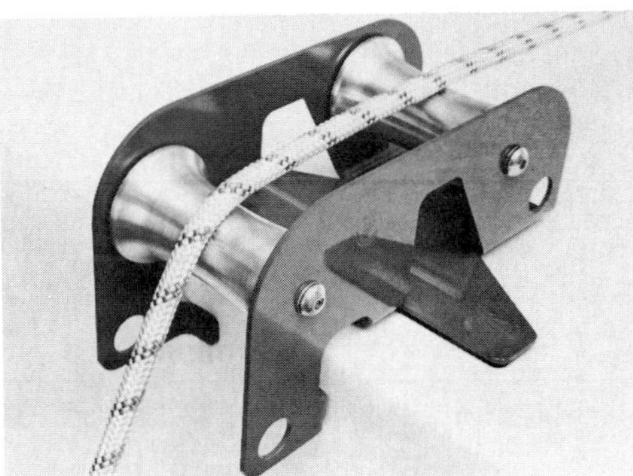

Fig. 5.28 Roof Roller

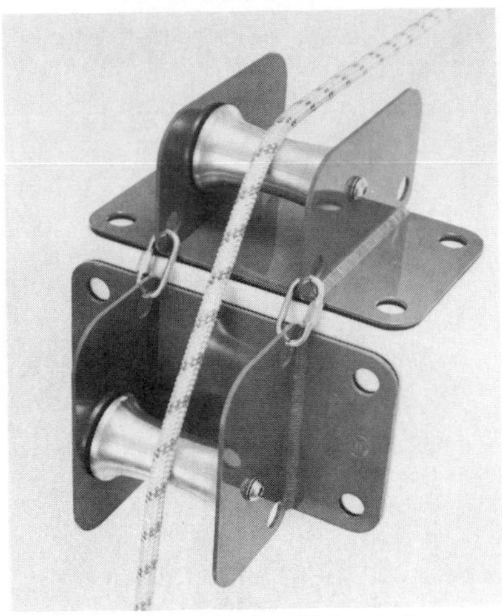

Fig. 5.27 A Pair of Edge Rollers

Edge Rollers

Edge rollers are important in reducing friction of rope over an edge. This not only helps with the work being done, as in hauling systems, but helps prevent abrasion damage to the rope.

Two types are available:

- Single unit sets designed for cliffs and other uneven surfaces usually found in natural terrain. They are normally used in sets of two or more that are linked together *(see Figure 5.27)*.
- Roof Rollers, used as one unit for 90-degree edges such as those found on buildings *(see Figure 5.28)*.

QUESTIONS *for* REVIEW

1. What are the drawbacks of an oval carabiner?
2. Name two ways in which non-locking carabiners can accidentally come open.
3. What does the term **reversed** and **opposed** mean as applied to carabiners? How is this technique used?
4. Name three ways that a locking carabiner can accidentally come open and, in each case, a way of preventing it.
5. What is meant by "manner of function" as applied to a carabiner?
6. What would be two possible solutions to three-way loading of a carabiner on a seat harness?
7. Other than a rope or other gear slipping out of it, what is the danger if a carabiner is loaded with the gate open?
8. Name three criteria to be used when choosing a descender.
9. Give at least one advantage and one disadvantage in using a Figure 8 descender.
10. What is the functional danger of using a Figure 8 **without** ears?
11. Name one advantage and one disadvantage in using the Brake Bar Rack.
12. Name at least one difference between a cam and a handled ascender.
13. What is a major difficulty in using a cam ascender?
14. What is the potential danger in high loading of a handled ascender?
15. What purposes do artificial anchors serve?
16. What is the major drawback in using pitons?
17. What is the danger in using a bolt set by unknown persons at an unknown time?
18. Why are some anchoring devices known as "clean?"
19. Name three types of clean hardware.
20. Name three disadvantages of body belaying.
21. Name two alternatives to body belaying.
22. Name two situations in the high angle environment where pulleys would be of assistance.

Chapter 6

Knots

OBJECTIVES–

At the completion of this chapter, you should be able to:

1. Describe why it is necessary to have proficient knot skills before entering the high angle environment.
2. Discuss why it is necessary to continue practicing tying.
3. Describe the qualities of a good knot.
4. List the ways in which knots affect rope.
5. Describe the functions of the following knots:
   a) Simple Overhand.
   b) Simple Figure 8.
   c) Figure 8 on a Bight.
   d) Figure 8 Follow Through.
   e) Figure 8 Bend.
   f) Water Knot.
   g) Barrel Knot.
   h) Grapevine ("Double Fisherman's") Knot.
6. Tie correctly, confidently, and without hesitation the following knots:
   a) Simple Overhand.
   b) Simple Figure 8.
   c) Figure 8 on a Bight.
   d) Figure 8 Follow Through.
   e) Figure 8 Bend
   f) Water Knot.
   g) Barrel Knot.
   h) Grapevine ("Double Fisherman's") Knot.

TERMS—*pertaining to knots that the High Angle Rope Technician should know:*

Back-Up Knot—A knot used to secure the tail of another knot. Also known as a **Safety** or Keeper Knot.

Bend—A knot that joins two ropes.

Bight—The open loop in a rope formed when it is doubled back on itself.

Foundation Knot—A simple knot that is tied as the first step in tying a more complicated knot. Examples of foundation knots include the overhand and the simple Figure 8.

Safety Knot—See Back-Up Knot

Stopper Knot—A knot that helps provide security in ropework. Examples would be a simple figure 8 tied in the bottom end of a rope to prevent a person from rappeling off the end or tied in the top end of the rope to prevent it from accidentally slipping through equipment.

The ability to tie knots correctly, confidently, and without hesitation, and to know how they are used are necessary skills for the high angle technician. If you go into the high angle environment without these knot skills you may be a danger to yourself and others around you.

WARNING NOTE

An improperly tied knot, or the incorrect application of a knot, could result in serious injury or death.

Knots are the link for many of the elements in the high angle system. The following are some of the situations in which knots are used in the high angle environment:

a) In anchoring.
b) For tying ropes together.
c) For tying webbing together.
d) For tying loops in rope and webbing.
e) For tying people directly into ropes.
f) For certain belay systems.
g) For emergency situations, such as emergency seat harnesses.
h) For backing up other knots.
i) To keep rope ends from pulling out of equipment.
j) For personal safety, such as to keep from rappeling off the end of a rope.
k) To create emergency ascenders.
l) For tying safety lines.
m) For improvisation, where other elements of the system fail.
n) To extricate yourself from unexpected difficulties.

Knots must be continually practiced so that you remain proficient at knot tying and can tie a specific knot instantly when it is necessary. To maintain good high angle skills, you should possess at least two lengths of rope, each several feet long, so that you can continually practice knot tying.

Group training sessions typically should begin with a review of knots. Those persons who do not maintain their knot tying skills should be denied group certification and not be allowed in the high angle environment. The failure to learn simple but essential skills such as knot tying may indicate lack of motivation.

Because many activities take place under severe environmental conditions, every high angle rope technician should be able to tie knots under stress, in the dark, when cold, using only one hand, and with diminished physical ability.

There are thousands of knots. But the number of knots explored in this chapter have been reduced to those necessary for the major situations encountered in the high angle environment.

NOTE: Certain specialized knots are not reviewed in this chapter, but in those sections associated with the special skills that use the knot. For example, the Munter Hitch ("Half Ring Bend") is reviewed in Chapter 8, "Belaying," and the Prusik Knot is included in Chapter 10, "Ascending."

The Qualities of a Good Knot

While knots vary in their specific use, all good knots have certain characteristics in common:

- They are relatively easy to tie.
- It can easily be determined whether they are tied correctly.
- Once tied correctly, they remain tied.
- They have a minimal effect on rope strength.
- They are relatively easy to untie after loading.

How Knots Affect Ropes

Every knot diminishes the strength of rope to some degree. The reason for this is that in any sharp bend of a rope (less than four times the diameter of the rope), the rope fibers on the outside of the bend carry the majority of the load on the rope. The fibers on the inside of the bend will carry very little of the load or none at all *(see Figure 6.1)*. Some knots, such as bowlines that have sharper bends cause more of a strength loss in a rope than knots such as Figure 8s that have more open bends. Ultimately, the kinds of knots, along with other elements of a high angle system, must be taken into consideration when deciding on a safety factor for a rope. (See Chapter 3, "Rope and Related Equipment," regarding a Safety Factor.)

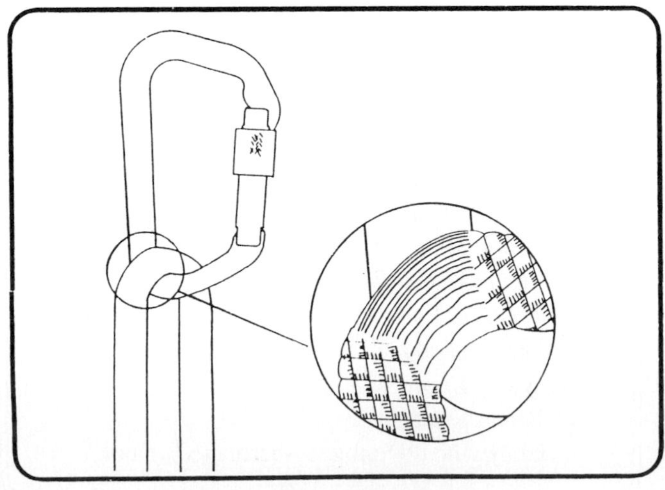

Fig. 6.1 Effect of Bending Rope

Knots should be removed from a rope before it is put away and stored, for the following reasons:

- If left in the rope over a long period, knots may cause a permanent loss of strength in the rope yarn.
- If left in the rope over a long period, knots will tend to "set" and become more difficult to untie.

Backing Up Knots

It is good practice to back up knots with a **Backup** or **Safety** knot. While an Overhand Knot is often used as a backup, a more secure backup is is the Barrel Knot. A backup knot is a particular concern when the rope is stiffer such as static kernmantle ropes or the knot is going to be flexed a great deal.

The backup should be tied up close to the main knot it is safetying.

NOTE: For clarity, many of the knots in this manual are not shown with backups. It should be assumed, however, that in each case they would have a backup.

THE KNOTS

Overhand Knot (for rope and webbing)

APPLICATIONS:

- As a "foundation knot" for beginning other knots such as the Water Knot.
- As a backup to secure other knots.

Figure 6.2 [a] through 6.2 [c] show the tying of a simple Overhand knot.

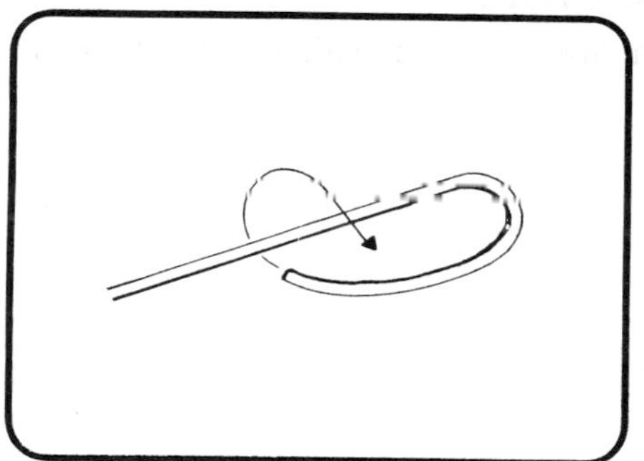

Fig. 6.2(a)

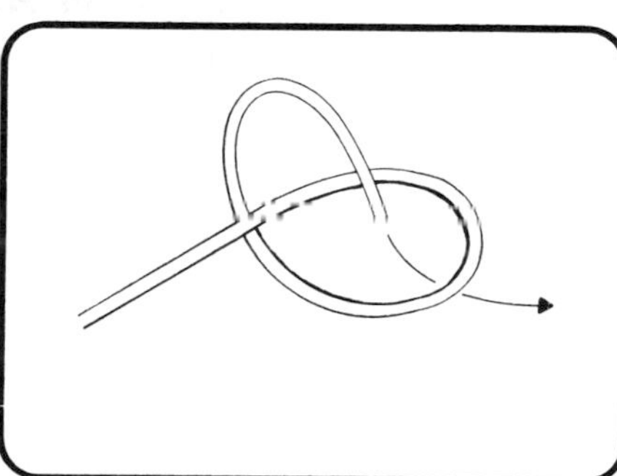

Fig. 6.2(b)

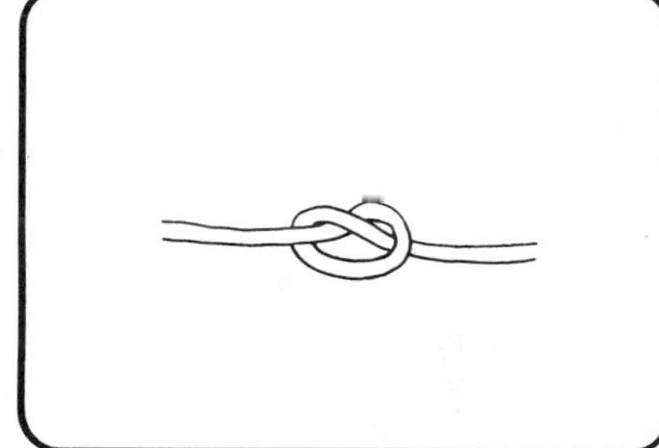

Fig. 6.2(c)

Fig. 6.2 Tying a Simple Overhand Knot

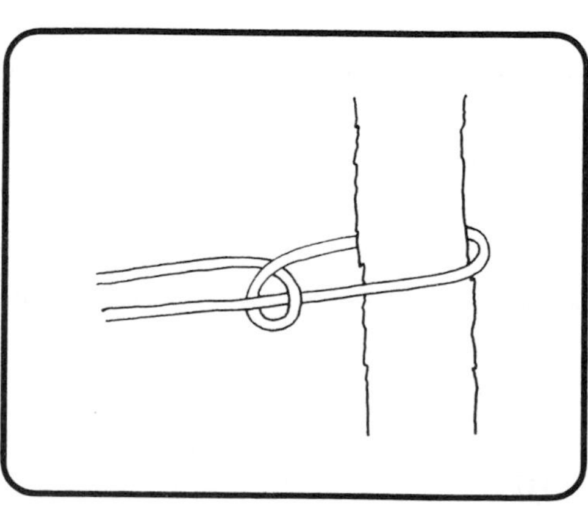

Fig. 6.3 Half Hitch

CAUTIONS:

- **Do not mistake the Overhand knot for a Half Hitch *(a half hitch is shown for comparison in Figure 6.3).***
- **When used as a backup knot, the Overhand knot must be pulled down tightly and close to the knot it is backing up *(as is shown in Figure 6.8)* backing up a bowline.**

THE FIGURE 8 FAMILY

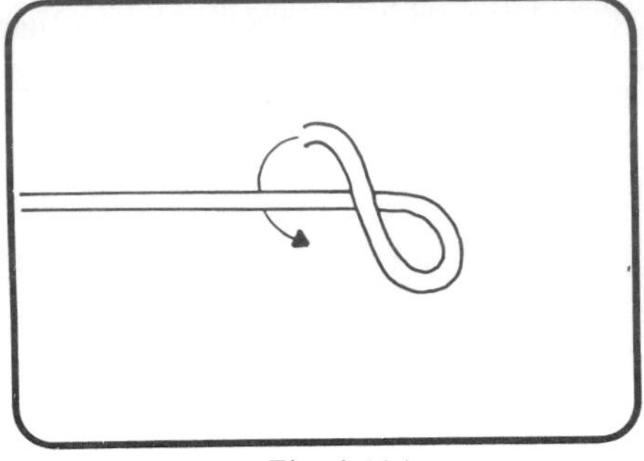

Fig. 6.4(a)

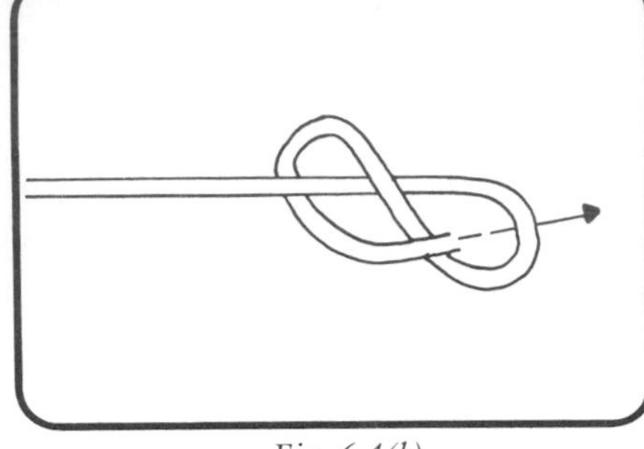

Fig. 6.4(b)

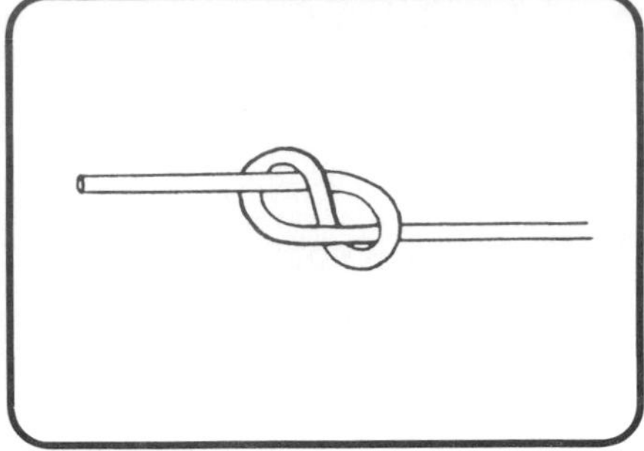

Fig. 6.4(c)

Fig. 6.4 Tying a Simple Figure 8 Knot

The Figure 8 family of knots will enable a person to deal with many of the knot tying needs in the high angle environment. The Figure 8 knots are increasingly being preferred by high angle rope technicians because:

- When tied correctly, they tend to be secure, less likely to come apart under loading and flexing.
- It is easy to tell if they are tied correctly.
- They diminish rope strength less than some other knots.
- It is easy to remember how to tie them.

WARNING NOTE

As with all knots, Figure 8 knots should be contoured (the strands aligned and uncrossed) and compacted (all ends pulled down so the knot is compact). This ensures that the knot has its greatest holding power, while reducing the rope strength as little as possible.

Simple Figure 8 (for rope)

USES:

- As a **stopper knot** for certain types of security, such as:
  a) To be tied in the bottom end of a rope to prevent a person from rappeling off the end.
  b) To be tied in the top end of rope to prevent it from accidentally slipping through equipment.
- As a foundation knot for beginning the Figure 8 Follow Through or the Figure 8 Bend.

  Figures 6.4 [a] through 6.4 [c] show the tying of a simple Figure 8 knot.

CAUTION: Do not confuse this knot with the Simple Overhand. Compare it with the Overhand in *Figures 6.2 [a] through 6.2 [c]* and note the extra step in tying the simple Figure 8. When you hold this knot up by either end of the rope, it should have the rough appearance of an "8."

Figure 8 on a Bight (for rope)

NOTE: A "bight" is simply the loop formed when the rope is doubled back on itself. This can occur in the middle of the rope, or at the end, as shown in Figure 6.5 [a], which is the first step in tying this knot.

USES:

- As a secure loop in a rope for clipping into for such things as:
  a) Safety lines.
  b) Persons being lowered.
  c) Litter and other rescue equipment.
  d) Anchor lines.

Fig. 6.5 Tying a Figure 8 on A Bight

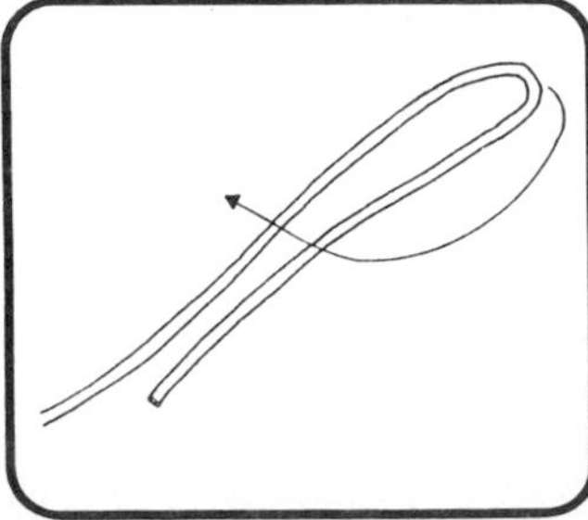

Fig. 6.5(a)

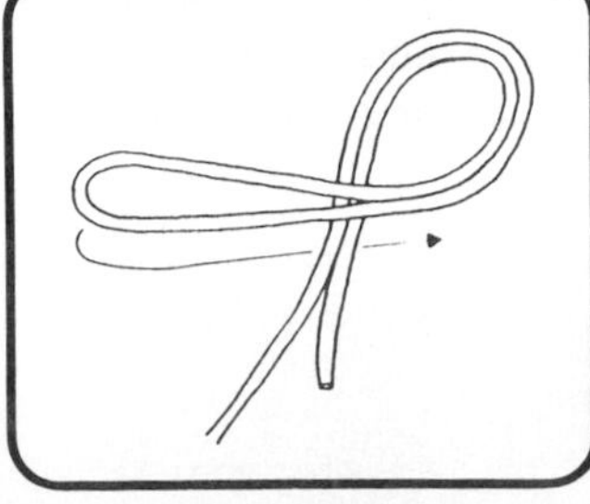

Fig. 6.5(b)

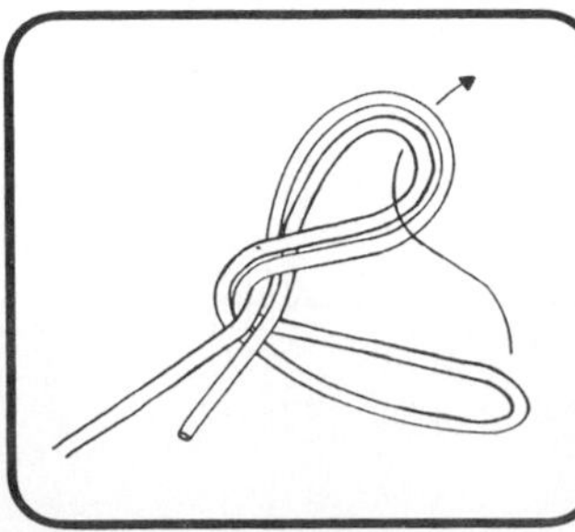

Fig. 6.5(c)

Fig. 6.5(d)

CAUTIONS:

1. Be certain that you tie a *Figure 8* on a bight and not an *Overhand* on a bight which is illustrated in Figure 6.6.

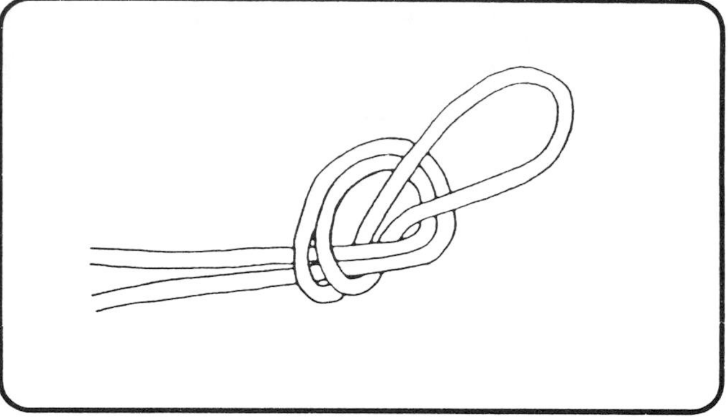

Fig. 6.6 Overhand on Bight

2. There is a preferable way to tie a Figure 8 on a Bight, which is shown in Figure 6.5. Not as preferable is the way shown in Figure 6.7. This second method loads the inside rope strand to create tighter bends and diminishes the strength of the rope more than the preferred method.

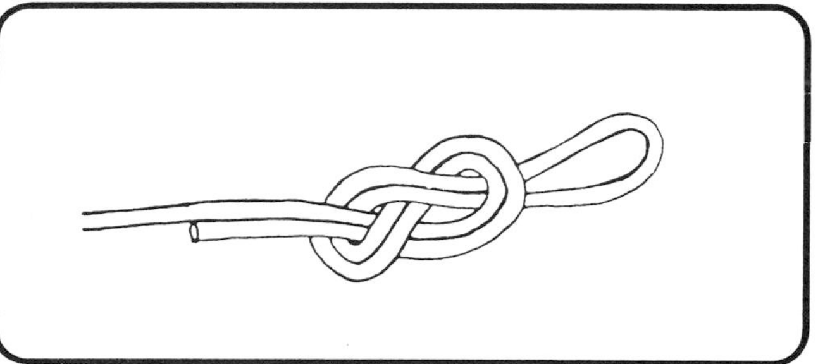

Fig. 6.7 Less Desirable Figure 8 on a Bight

WARNING NOTE

IF you decide to use a bowline, make certain that it is tied correctly and:

- **Make certain that the tail of the knot ("bitter end") is backed up with an Overhand or other "keeper" knot;**
- **Avoid the use of a Bowline on a moving rope. If the Bowline snags, the knot may capsize (changing the direction of the load on it) and fail.**

OPTIONAL APPROACH

Many authorities experienced in the high angle environment feel that the Figure 8 family of knots is preferable to the bowline family for the following reasons:

- The Figure 8 knots are more likely to be tied correctly.
- The Figure 8 knots are more likely to be remembered.
- It is easier to tell quickly if a Figure 8 knot is tied correctly.
- A Figure 8 knot remains stable if loading on it comes from a direction different from that which was intended.
- A Figure 8 knot is more likely to remain tied after repeated loading and unloading.
- A Figure 8 knot is less likely to capsize when pulled across an obstruction.
- A Figure 8 knot tends to weaken the rope less than a Bowline.

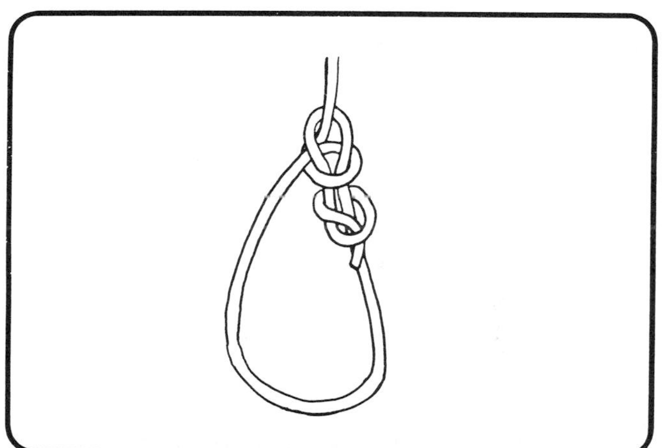

Fig. 6.8 Simple Bowline With Overhand Backup

However, due to the long tradition of using the bowline, there is often a passionate attachment to its use. *(A simple Bowline is shown correctly tied, along with a overhand knot backup, in Figure 6.8.)*

If local policy dictates the use of a bowline for tying a loop in a rope, instead of a Figure 8, then instruction in the tying of a bowline should be obtained from qualified personnel.

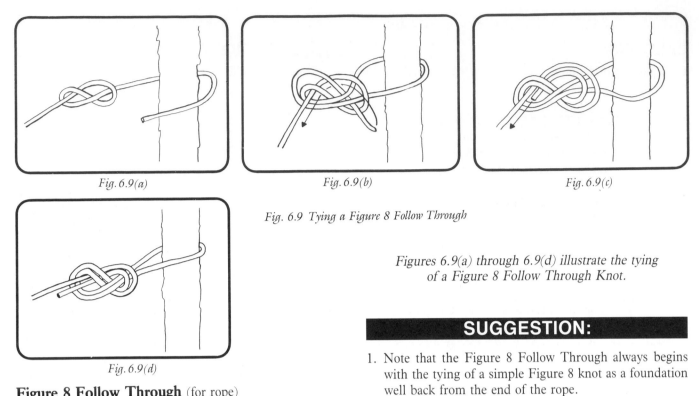

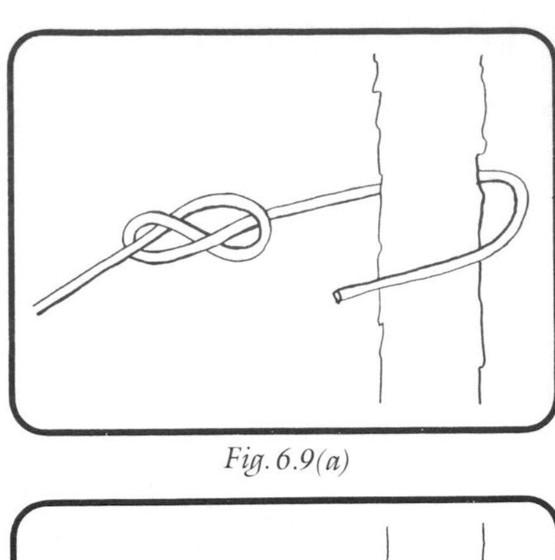
Fig. 6.9(a)

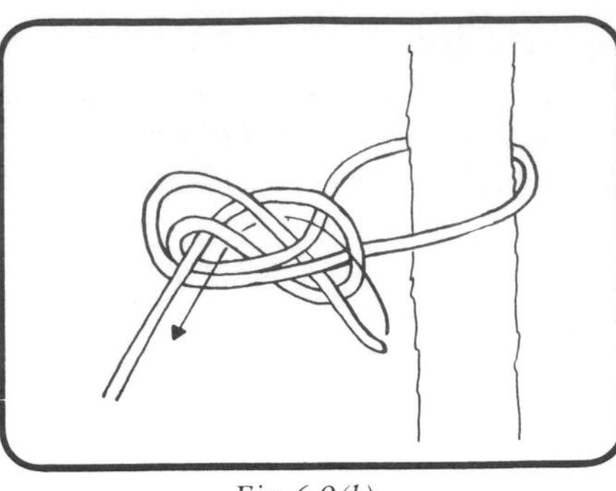
Fig. 6.9(b)

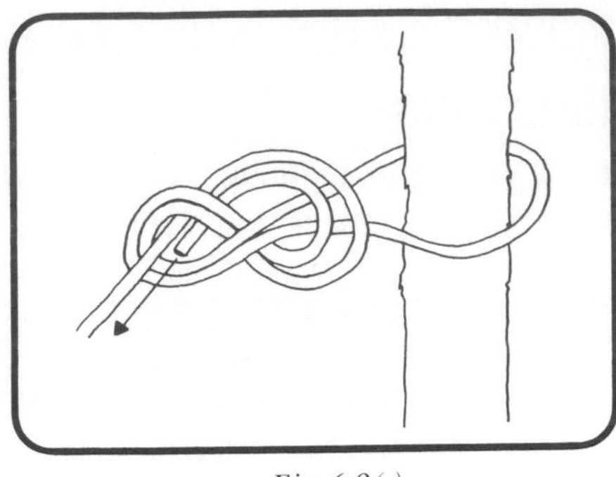
Fig. 6.9(c)

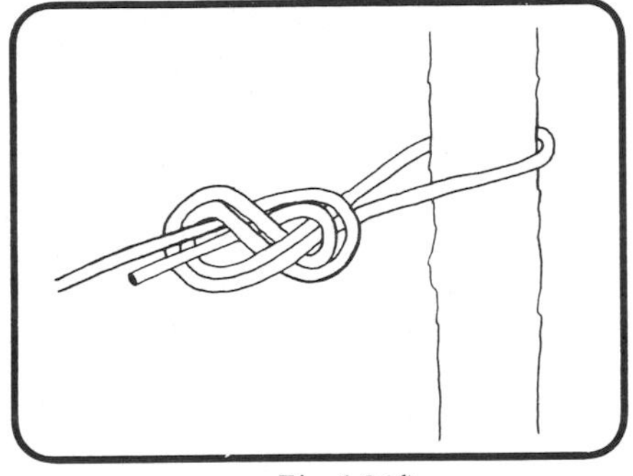
Fig. 6.9(d)

Fig. 6.9 Tying a Figure 8 Follow Through

Figures 6.9(a) through 6.9(d) illustrate the tying of a Figure 8 Follow Through Knot.

Figure 8 Follow Through (for rope)

USES:

- For a loop at the end of a rope in situations where a Figure 8 on a Bight cannot be tied. An example would be a situation where you would want to anchor to a tall object, such as a tree, but a simple loop cannot be gotten **over the top.** Therefore, a Figure 8 Follow Through is tied **around it**.

SUGGESTION:

1. Note that the Figure 8 Follow Through always begins with the tying of a simple Figure 8 knot as a foundation well back from the end of the rope.
2. After the simple Figure 8 is tied, pass the end of the rope around the anchor point, then follow back through parallel to the first knot. Follow every contour of the first knot with both rope ends going **in the same direction.**
3. Do not confuse this knot with the Figure 8 Bend *(shown below).*

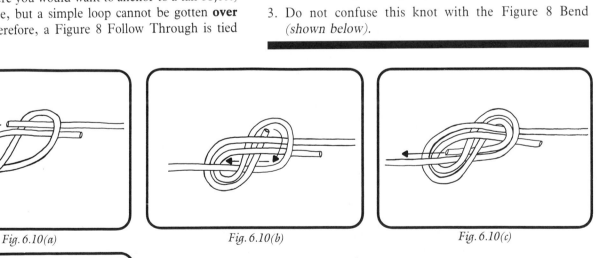

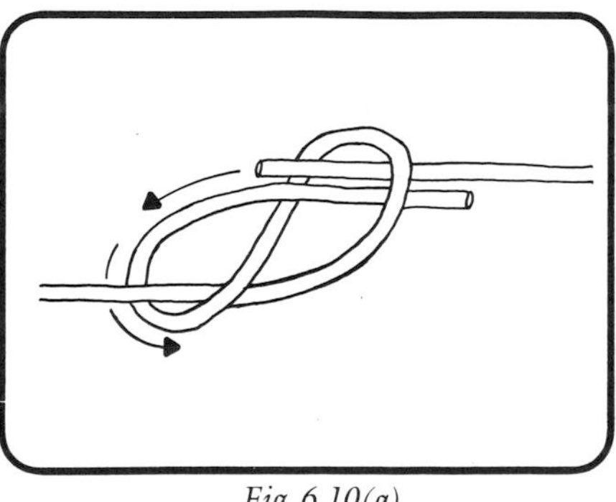
Fig. 6.10(a)

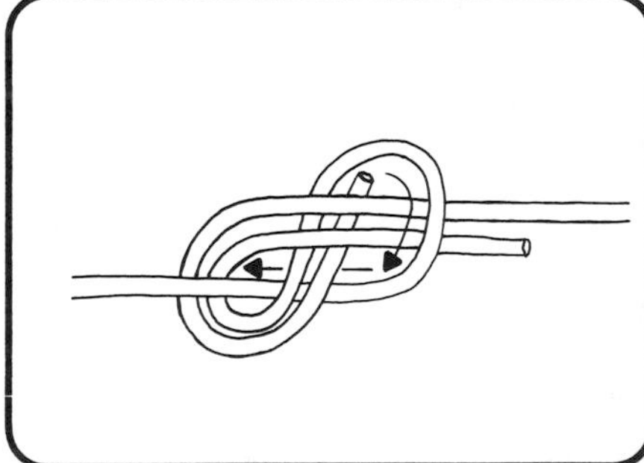
Fig. 6.10(b)

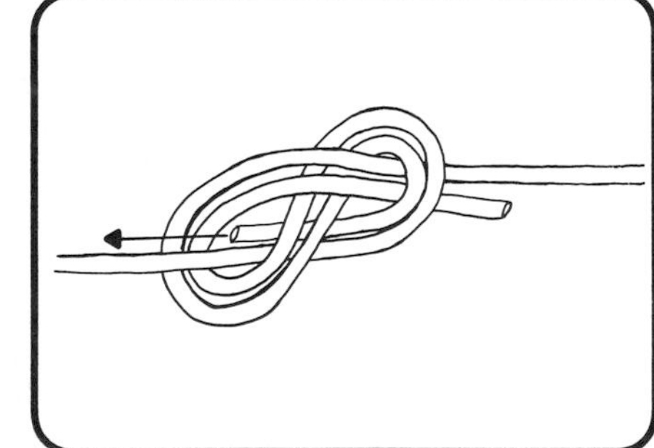
Fig. 6.10(c)

Fig. 6.10(d)

Fig. 6.10 Tying a Figure 8 Bend

Figure 8 Bend (for rope)

NOTE: The term **bend** as applied to knots refers to the joining of two ropes together.

USES:

- For joining two ropes together—
  a) For connecting two long pieces of rope.
  b) For creating a loop of rope by joining both ends of one rope together.

Figures 6.10(a) through 6.10(d) show the tying of a Figure 8 Bend.

SUGGESTION:

First try tying this knot using two ropes of different colors. This will make it easier to distinguish the different strands of rope.

CAUTIONS:

1. Note that the Figure 8 Bend always begins with the step of tying a simple Figure 8 knot as a foundation.

2. The next step is to exactly follow the contour of the first knot with the rope ends approaching at *opposite* directions.

3. Do not confuse this knot with the Figure 8 Follow Through *(shown in Figures 6.9 [a] through 6.9 [d]).*

OPTIONAL APPROACH

Another knot that can be used to securely join two rope ends to form a longer rope or to form a loop is the Grapevine Knot (also known as the Double Fisherman's). While this knot is very secure, it may be more difficult to learn and to tell if it is tied correctly. The Grapevine Knot can be difficult to untie after it is loaded, particularly in softer lay ropes. This knot can only be used in rope.

Figures 6.11(a) through 6.11(d) illustrate the tying of a Grapevine or Double Fisherman's Knot.

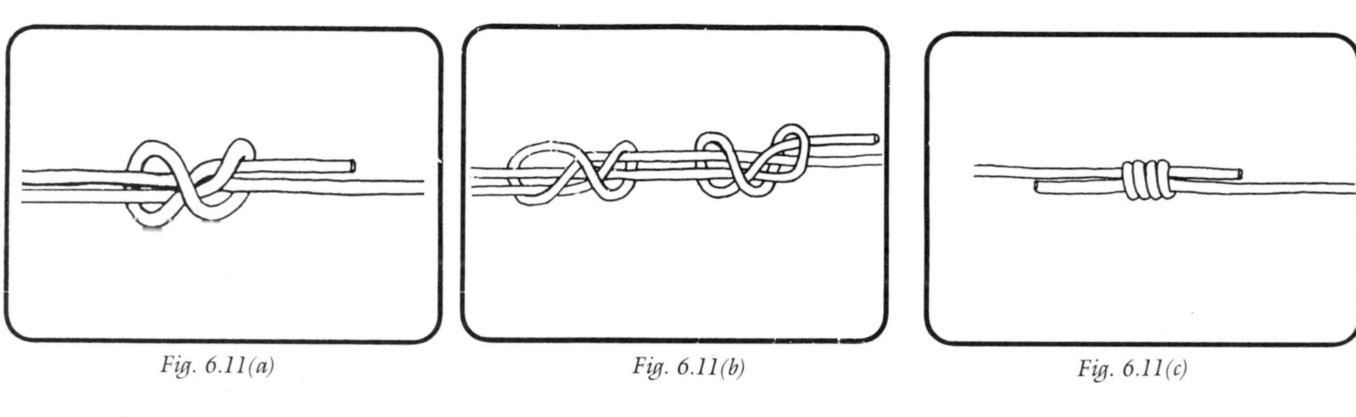

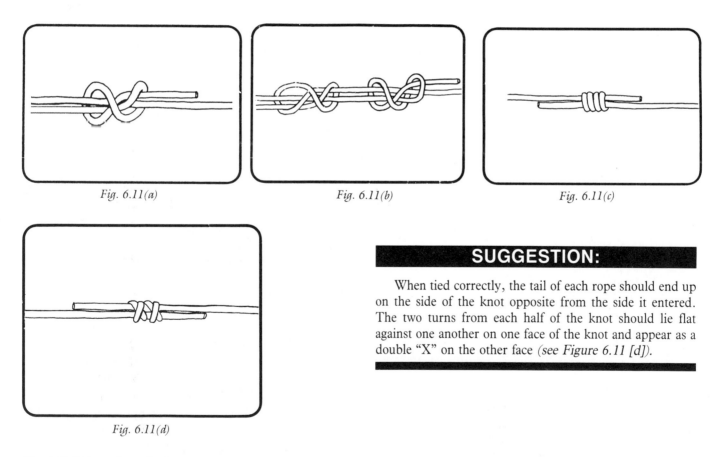

Fig. 6.11(a) *Fig. 6.11(b)* *Fig. 6.11(c)*

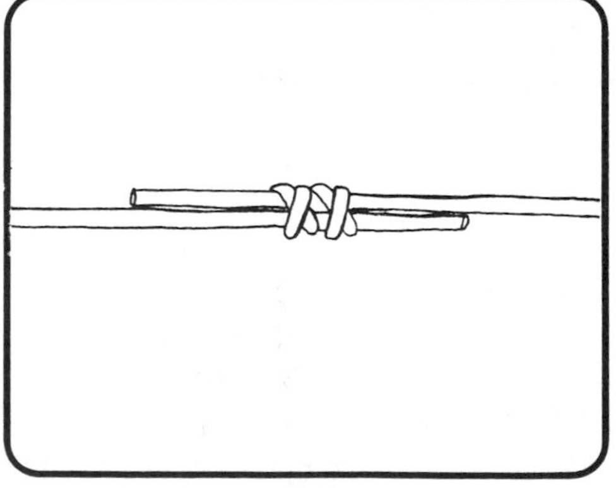

Fig. 6.11(d)

Fig. 6.11 Tying a Grapevine Knot (also known as Double Fisherman's Knot)

SUGGESTION:

When tied correctly, the tail of each rope should end up on the side of the knot opposite from the side it entered. The two turns from each half of the knot should lie flat against one another on one face of the knot and appear as a double "X" on the other face *(see Figure 6.11 [d])*.

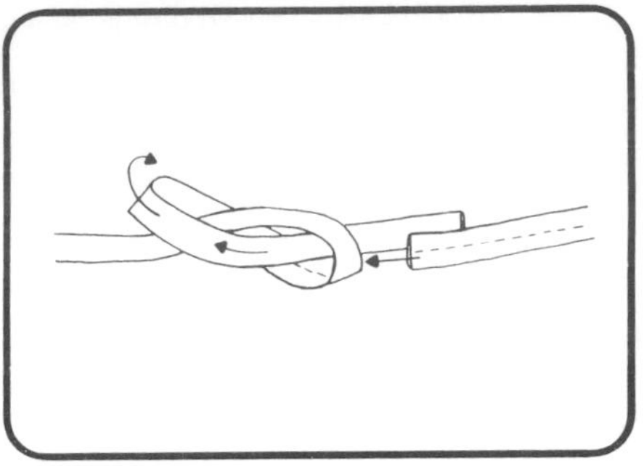

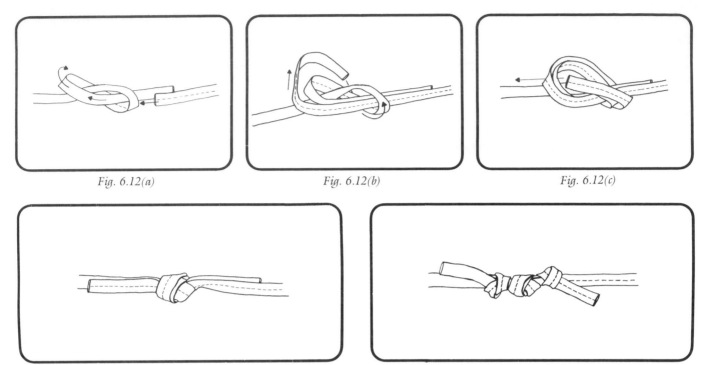

Fig. 6.12(a)

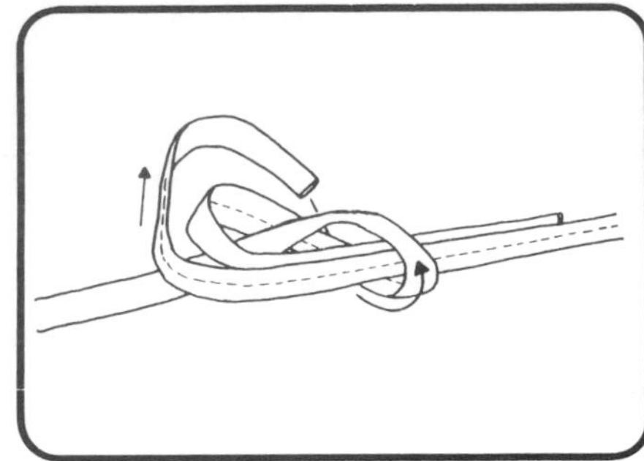

Fig. 6.12(b)

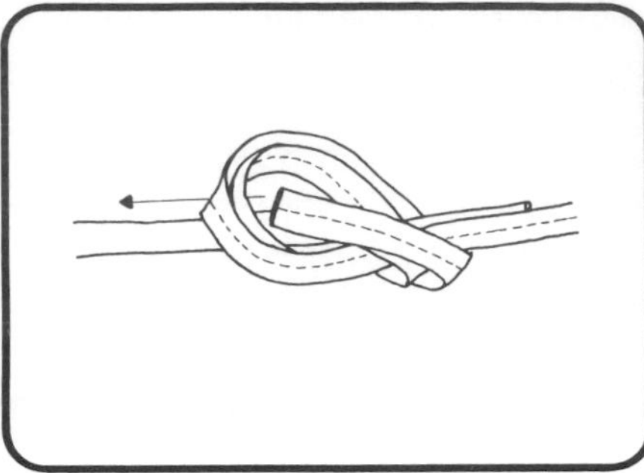

Fig. 6.12(c)

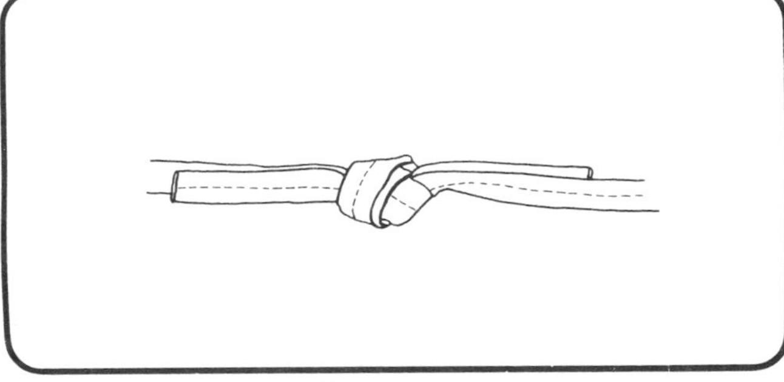

Fig. 6.12(d)

Fig. 6.12(e)

Fig. 6.12(a) - (d) Water Knot

Fig. 6.12(e) Water Knot Backed-Up with Overhand Knots

Water Knot (for webbing only)
(also known as the "Tape Knot," the "Overhand Bend," or the "Ring Bend.")

USES:

- For tying webbing together
  a) For joining two different pieces of webbing to form a longer piece.
  b) For tying the two ends of one piece of webbing together to form a loop.

WARNING NOTE

The Water Knot is to be used only for webbing. It is not to be used for rope. Because of the flat nature of webbing, it has the quality of contouring over itself. Rope does not have this quality and a Water Knot in rope may easily come out.

Figures 6.12(a) through 6.12(d) illustrate the tying of a Water Knot in webbing.

SUGGESTION:

First try tying this knot using two pieces of webbing of different colors. This will make it easier to distinguish the different pieces of webbing as you tie the knot.

CAUTIONS:

1. Always have at least two inches of webbing in the ends of water knots *after they are tied and pulled tight.* Though it contours well in a Water Knot, webbing tends to be slippery. Ends that are too short tend to slip through under stress. For additional insurance, back up both sides of the knot with an Overhand knot. *(See Figure 6.12 [e].)*

2. A Water Knot in webbing should be inspected frequently since over time it tends to work loose.

3. Be certain the webbing follows flat through the knot. A twist in the webbing inside the knot will allow the knot to slip at relatively low loads.

Barrel Knot (for rope)

USES:

- For backing up other knots.

The Barrel Knot is, essentially, one half of a Double Fisherman's Knot.

Figures 6.13(a) and 6.13(b) illustrate the tying of a Barrel Knot.

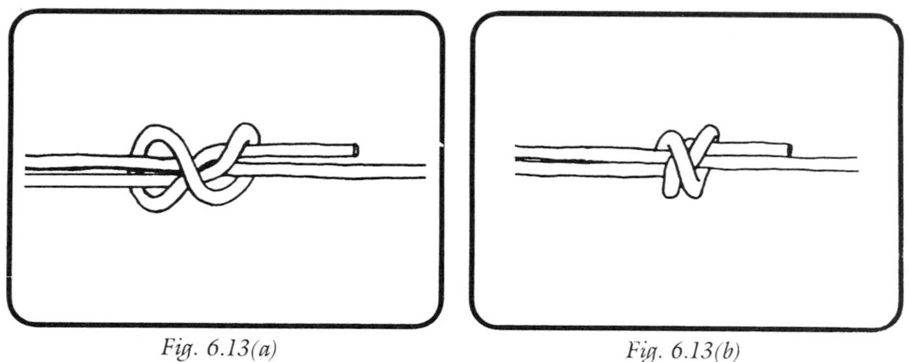

Fig. 6.13(a) *Fig. 6.13(b)*

Fig. 6.13 Tying a Barrel Knot

QUESTIONS for REVIEW

1. List at least five situations in which knots would be used in the high angle environment.
2. List at least three qualities of a good knot.
3. Describe how knots diminish the strength of a rope.
4. Give two reasons why knots should not be left in ropes when they are stored.
5. Describe a use for each of the following:
   a) Overhand knot.
   b) Simple Figure 8.
   c) Figure 8 on a Bight.
   d) Figure 8 Follow Through.
   e) Figure 8 Bend.
   f) Water Knot.
   g) Barrel Knot.
   h) Grapevine ("Double Fisherman's") Knot.

★ ACTIVITIES ★

1. Tie the following knots:
   a) Overhand.
   b) Simple Figure 8.
   c) Figure 8 on a Bight.
   d) Figure 8 Follow Through.
   e) Figure 8 Bend.
   f) Water Knot.
   g) Barrel Knot.
2. Tie the knots listed above either while blindfolded, while holding your hands under a table, or in another manner that prevents you from seeing the rope as you tie the knot.

Chapter 7

Anchoring

OBJECTIVES–

At the completion of this chapter, you should be able to:

1. Describe the purpose of anchoring.
2. Know what creates forces on anchors and how these forces can be magnified or reduced.
3. Estimate the forces on an anchor system.
4. Demonstrate how to avoid overloading of an anchor system.
5. Define where anchors should be placed.
6. Discuss the purpose of directionals.
7. Set up a directional.
8. Specify what knots are best employed in anchor systems.
9. Describe the purposes of the Tensionless Hitch.
10. Use the following knots in an anchor system:
    a) Tensionless Hitch.
    b) Figure 8 on a Bight.
    c) Figure Eight Follow Through.
    d) Water Knot.
11. Describe how rope and webbing are employed in anchoring systems.
12. Discuss what areas on a building would present suitable anchor points and which would present unsuitable anchor points.
13. Establish secure anchors on buildings.
14. Discuss what parts of a vehicle would present suitable anchor points and which would present unsuitable anchor points.
15. Establish secure anchors on vehicles.
16. Describe in what circumstances multiple anchor points are needed.
17. State what angles on a multi-point anchoring system would create unacceptable forces.
18. Describe the concept of self-equalizing anchors.
19. Correctly tie a load-sharing anchor using two anchor points.
20. Correctly tie a self-equalizing anchor using a webbing loop on two anchor points.
21. Correctly set a self-equalizing anchor system using three or more anchor points.
22. Discuss the concept of a picket system.
23. Establish a picket system.

TERMS—*relating to anchoring that a High Angle Rope Technician should know:*

Anchor—In the high angle environment the means of attaching the rope and all other portions of the system to something secure.

Anchor Point—A single secure connection for an anchor. It will range in size from a piece of hardware wedged in the crack of a rock to a large tree or rock.

Anchor System—Multiple anchor points rigged in such a way that together they provide a "bombproof" anchor.

Artificial Anchors—The use of specifically designed hardware to create anchors where good natural anchors do not exist.

Backing Up—The creation of an additional independent anchor, or anchors, to sustain the high angle system should initial anchors fail. Backing up may be to the same anchor point if it is very solid, or to additional anchor points.

Bombproof—An anchor that will not fail.

Directional—A technique for repositioning a rope at a more favorable angle than would exist using only its anchor.

Pendulum—To swing on a rope.

Self-Equalizing Anchor—An anchor established from two or more anchor points that: a) maintains near equal loading on the anchor points despite direction changes on the main line rope, and b) reestablishes equal loading on remaining anchor points if any one of them fails. Also known as **SEA**.

Anchors and the High Angle System

Anchors are the means of securing the ropes and other elements of the high angle system to something solid. The place where the anchors are connected is the anchor point. Anchor points can take a number of forms. In the outdoors, they might be trees, rocks, or other natural anchors. Or they might be what are termed "artificial anchors" specifically placed there by persons working in the high angle environment. On buildings, the most secure anchor points are structural parts such as integral beams and columns. In some situations, the only available anchor points may be on vehicles.

Anchoring is to the high angle system as a foundation is to a building. Without suitable and secure anchors, the remainder of the high angle system—ropes, hardware, and other gear—are in danger of failure, no matter how well they are established. Just as a solid foundation is the primary concern before construction of a building is begun, a suitable and secure anchor system is essential before the remainder of the system can be put in place.

Anchor Points

The anchor point is the single secure connection for an anchor. The specific kind of anchor point will depend on the specific kind of high angle environment in which you are working.

Natural Anchors

The most commonly used natural anchors are trees and rocks around which webbing or rope can be wrapped. However, both of these have potential for failure. Before using them as anchors, you should examine trees for potential rot. But even trees with sound wood may not be good anchors if their root systems are shallow or thin or if they are in wet soil. Boulders weighing tons can be pulled over by the stresses of anchor systems. Trees or rocks that are so massive that they definitely will not fail as anchors are sometimes called **BFTs** or **BFRs**.

Other possibilities for natural anchors include pickets in earth (described later in this chapter) and snow and the use of various anchoring techniques in ice.

Anchors on Structures

In the urban setting, it will often be necessary to establish anchors on buildings. This can be difficult because of the questionable nature of some of the potential anchor points found on older buildings. And on modern buildings, it is often difficult to find any anchor points at all.

Some examples of deteriorated material are:

- Corroded metals.
- Weathered stonework.
- Deteriorated mortar in brickwork.

Some examples of inherently weak structural features are:

- Vents constructed of sheet metal.
- Flashing.
- Gutters and downspouts.
- Brickwork without bulk, such as small chimneys.

When you rig anchors on buildings, choose anchor points that are inherently part of the building's structure or specifically constructed to support high loads.

Some examples are:

- Structural columns.
- Projections of structural beams.
- Supports for large machinery.
- Anchors for window-cleaning equipment.
- Brickwork with large bulk, such as corner walls.

Artificial Anchors

Artificial anchors are special types of hardware specifically designed for creating anchor points or "protection" where there are no natural anchors. Many artificial anchors are inserted into rocks or spaces in rock. They include such hardware as nuts, chocks, hexcentrics, and cams. Artificial anchors usually are not as secure as good natural anchors. For them to be safe and effective anchors, they need to be placed by a person who has a great deal of skill and practice in their use.

Among the artificial anchors, bolts are commonly used in rescue operations. But their placement is time-consuming, and they also require a great deal of practice and training for correct placement. They also create permanent damage to the rock, which is frowned on in managed areas such as park land. (See Chapter 5, "Basic Vertical Hardware" for more details on artificial anchors.)

Placement of Anchors

Whatever their nature, the placement of secure anchors is very much dependent on good judgement, which is developed through experience and practice. Though their specific nature may vary from place to place, there are certain characteristics common to all anchors.

Strength of Anchors

Anchors must be able to sustain the greatest anticipated force on the high angle system as calculated through the **safety factor.** (See Chapter 3, "Rope and Related Equipment," on how to calculate a safety factor.) Anchors that are so strong that they will withstand any force that the high angle system will deliver to them are said to be **bomb proof.**

If the potential anchor point will not sustain the anticipated forces, then it must be abandoned for another more substantial one, or teamed with one or more other anchor points (multiple anchors are described later in this chapter).

This ability of an anchor to withstand the necessary forces will depend on a number of factors, including:

- Condition of anchor.

  For example, a live tree will usually withstand greater forces than a dead one.

- Structural nature of the anchor point.

  For example, a load-bearing structural column in a building will generally withstand greater forces than a handrail.

- Location of force on anchor point.

  For example, a tree with the force pulling on it near the ground will generally withstand greater force than one with the stress higher up.

Direction of Pull on Anchor Point

Always consider the direction of pull that will be on the anchor point. Try to set anchors that will be in line with the direction of pull. Also consider what will happen if the direction of pull changes. Some anchors are rigged so that they are strong only when the pull comes from one direction. If the direction of pull changes, then the anchor could weaken or fail.

Positioning of Anchors

The position of an anchor will have an effect on the high angle activity. This is particularly true in rescue activity. Under the ideal conditions, the anchor should be close to and directly above the subject to be rescued. However, there would be circumstances where it might be preferable to have the anchor off to the side. Some examples are:

- Conditions where rocks or other dangerous objects might fall on the rescue subject or on the rescuers.
- Where there are conditions between the anchor point and the rescue subject that could endanger rescuers or damage equipment such as rope.

  Example—There is a flashover and fire from a window.

  Example—The presence of a hostile or deranged person.

- Where there are no suitable anchors directly above.

In any high angle activity, it is often desirable that segments of the high angle system, such as the belay rope, be off to the side in order to avoid rope cross or tangles. But if the primary anchor (the one bearing the greatest load) is off at an extreme angle, there could be problems for those people attempting to manage the ropework. They might, for example, have to make a wide swing on the rope (a **pendulum**) to reach their objective. However, there are some anchoring techniques that may solve such a problem.

Directionals

A directional is a technique for bringing a rope into a more favorable angle. There are numerous ways of creating directionals. You have to judge each method on its advantages and disadvantages. As shown in the illustration in Figure 7.1, there are two trees wide apart that could serve as strong anchors at the top of a cliff. But let's say that you want to use their rope to reach spot "X" on the ground which is between the trees.

a) If you anchor the rope to either one of the two trees, as it has been to the tree on the right in Illustration 7.1 [a], it would be difficult for you to reach X without a significant pendulum.

b) Now, let's say as shown in Figure 7.1 [b], you add a secondary anchor, a **directional**, to the smaller tree on the left. In the end of this second rope, you tie an Overhand Figure 8 knot, and clip a locking carabiner into the Figure 8 knot. Finally, you clip this locking carabiner across the main anchor rope. Thus, the main rope is now at a better angle for you to reach "X." One additional advantage with this approach is that the main rope runs freely through the carabiner so that as the exact position of the activity moves back and forth, the angle can change slightly.

c) A less desirable alternative would be to tie a Figure 8 Overhand knot in the main line and clip it directly into the carabiner that is on the directional *(shown in Figure 7.1 [c])*. This would prevent the main line from sliding freely through the carabiner. This would also create a multi-point anchor, but without the advantages of example [b]. When the rope moved to one side or the other, it would create a great deal of loading on one anchor, but very little loading on the other anchor.

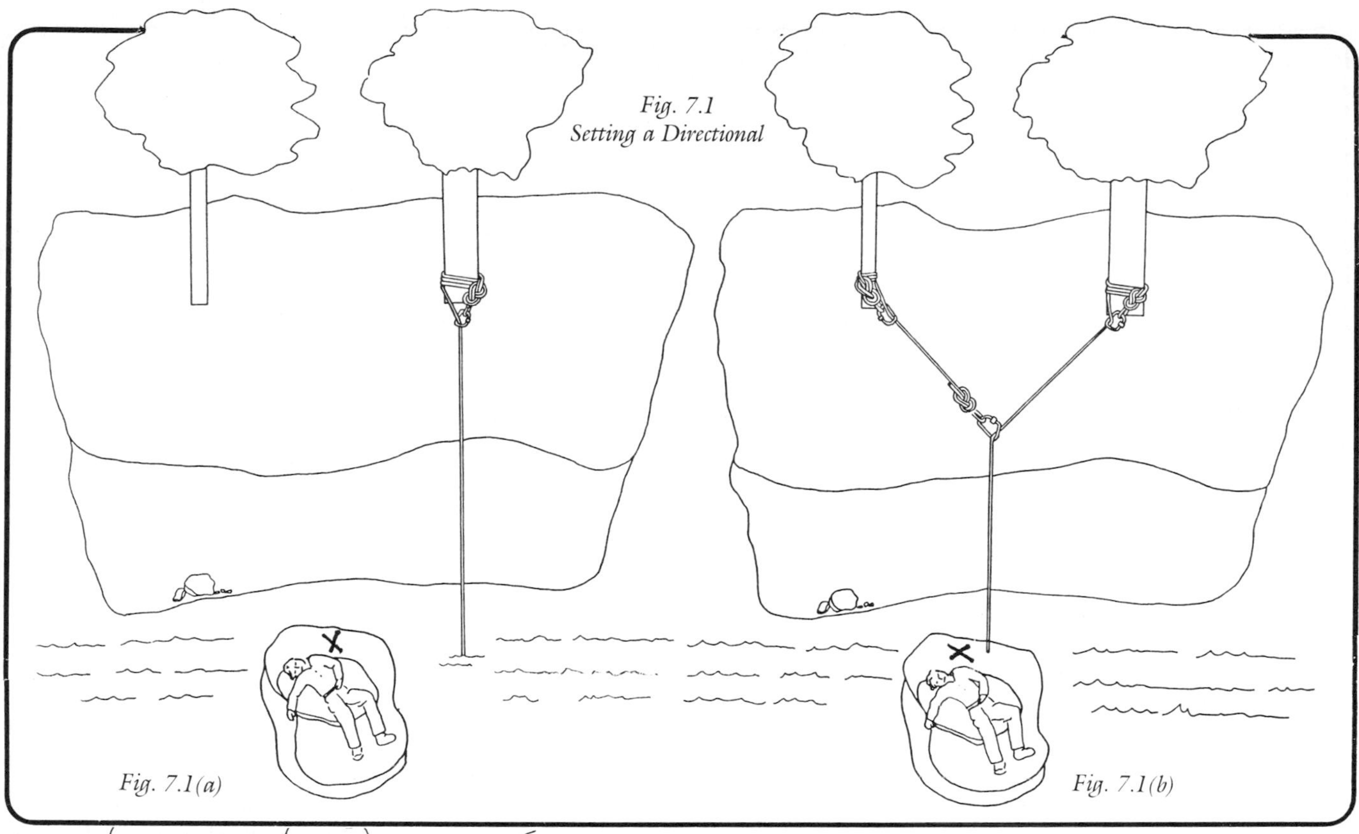
Fig. 7.1
Setting a Directional

Fig. 7.1(a)

Fig. 7.1(b)

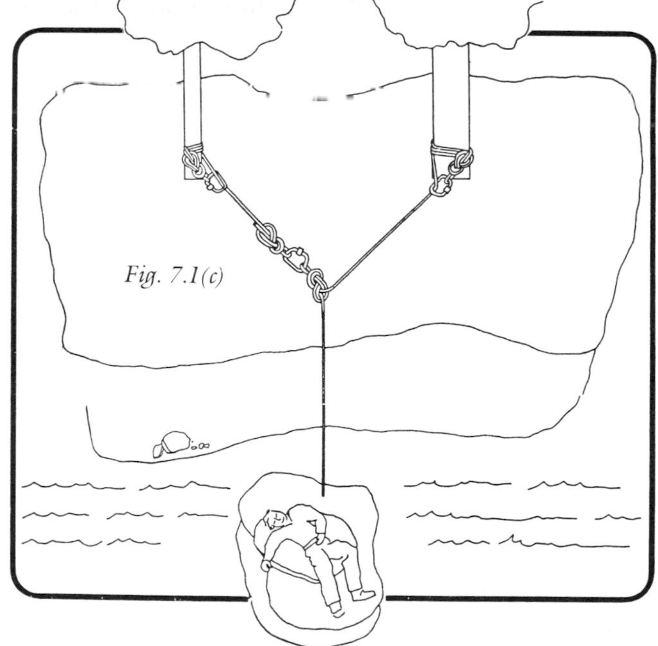
Fig. 7.1(c)

Location of Directionals

When you establish anchors and directionals you must keep in mind how safe and accessible their locations are for the high angle personnel who work with them. In Figure 7.1 above, for example, you could possibly make some improvements by changing the location of the anchors and directional. If there were anchor points, such as the trees, far enough distant back from the edge of the drop, then you could rig the anchor system and the directional on the top.

As a result:

a) Those people rigging the anchors and the directional would be in less danger of falling.

b) The rigging would be more accessible to the people and therefore would be more in their control.

c) With the rope running over the edge, not all the weight would be directly on the anchor; part of it would be taken by the edge of the drop. (The drawback would be possible abrasion on the rope.)

WARNING NOTE

1. Depending on the angle the primary anchor rope makes with the directional rope, there will be greater forces on the anchor system than if there were only a single anchor. In this illustration, for example, there will a greater force on the directional anchor than on the main anchor. A directional is in effect creating a multiple anchor system. See discussion later in this chapter on how multiple anchors create forces on the system depending on the angles the ropes create.

2. Remember, that should the directional fail, there will be a significant *pendulum*, the main line will drop, and the main line anchor will be shock-loaded.

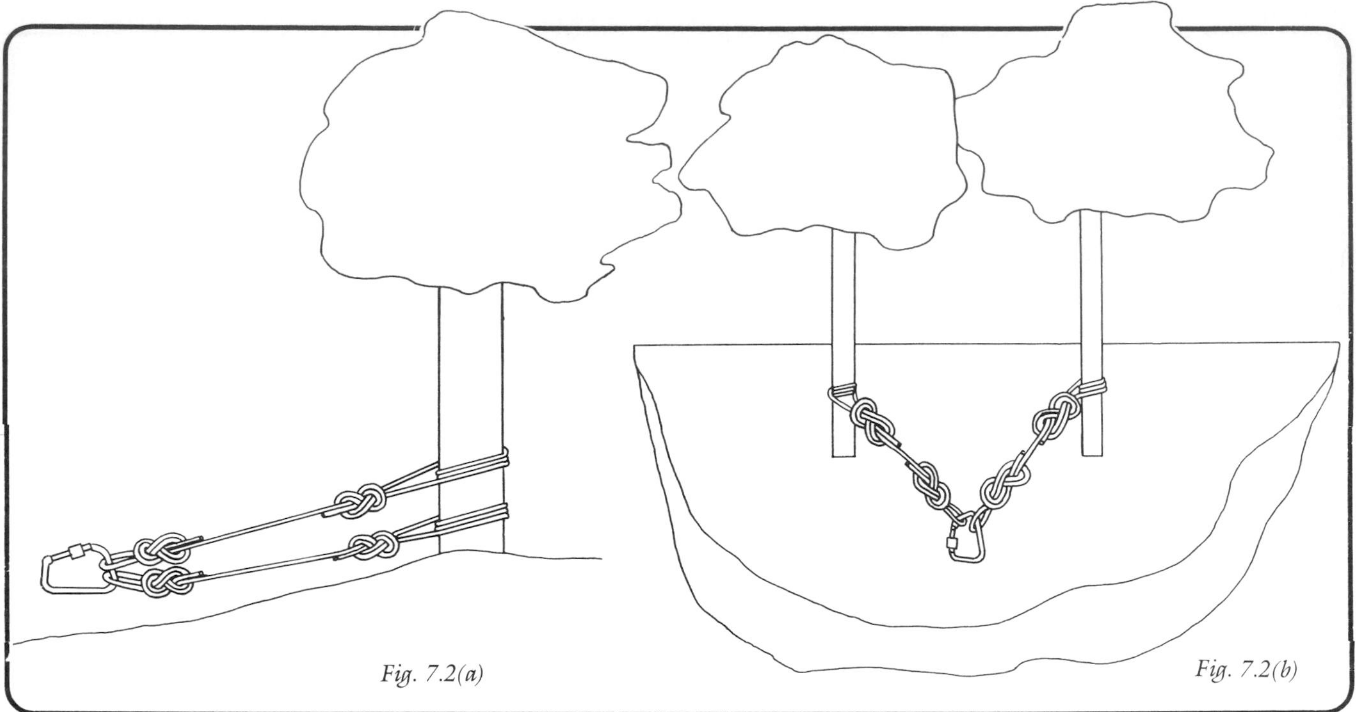

Fig. 7.2 Backing Up Anchors

Backing Up Anchors

Anchor systems present a number of opportunities for failure:

- Uncertain strength of anchor points:
  You can rarely say for certain what kind of stress an anchor point will take.
- Failures in human judgement/experience:
  Knots may be tied incorrectly, carabiner gates may be left unlocked, for example.
- Equipment failures:
  Abrasion of rope and webbing, and stressing of both hardware and software can occur.

Because of the potentials for anchor system failure, and because the rest of the high angle system depends on anchors, it is good practice to back up anchors. Backing up is the creation of redundant anchors for safety. Two primary ways of backing up an anchor are:

- Rigging to the same anchor point *(see Figure 7.2 [a])*. Only do this when you are absolutely certain that the one anchor point can sustain any forces subjected to it by the high angle system (it is "bomb proof"). This type of backup is done when there is the possibility of failure in other portions of the anchor system (carabiners, knots, slings, etc.).
- Backing up to a separate anchor point *(see Figure 7.2 [b])*. This requires a multiple anchor system. If the direction of loading will be shifting from side to side, you should make the multiple anchor system self-equalizing (explained later in this chapter).

Specifically the method of backing up the anchor system and the number of anchor points you will need depend on a number of variables.

a) The condition of the anchor points.
   If there is the potential for failure of one of them, or of the equipment that is attached to it, then you need more than one anchor point.

b) The nature of the high angle operation taking place.
   If there is a main line and a belay line, for example, then you may need two separate anchor systems. If both the main line and the belay line originate from the same anchor point, then there is the danger of causing line tangles or damage to the rope from line cross (see Chapter 3, "Rope and Related Equipment," on avoiding damage from rope cross).
   The belay system would also have to be substantial enough to catch a fall, which means you need a substantial second anchor.

c) The loads and stresses involved in the system.
   This will vary in intensity depending on the uses of the anchors:
   1. Supporting only equipment.
   2. Supporting the weight of only one person.
   3. Rescue lowering operations.
   4. Hauling systems.
   5. Highlines (a system of using a rope suspended between two points to move persons or equipment).

MATERIALS FOR ANCHORS

Using Rope for Anchors

One of the simplest procedures in establishing an anchor is to connect the main line rope directly to the anchor point. In urgent situations, where time is critical, this may be the best solution. It also means a simpler anchor system, with less chance for failure that might be in a more complicated system.

However, if you rig the main line rope directly to the anchor point, it will reduce your flexibility and limit your ability to make modifications in the anchor system. These modifications may be necessary because of the changing conditions that can occur in rescue situations.

One potential solution is to use a shorter piece of rope for the anchor system that is of the same diameter or larger than the main line, and attach the main line to it:

1. Attach the shorter piece of rope to the anchor point.
2. Tie an Overhand Figure 8 Knot in the ends of both ropes where they will meet.
3. Clip the Figure 8 knots together with carabiners.

If a group involved in high angle activities has been established for some time, many of these rope pieces will become available. Often, when a main line rope has developed a small bad section, that small section can be cut away, while the remaining pieces of the rope can be used as anchor lines.

Knots for Anchor Ropes

One of the most attractive knots for anchor ropes is the **Tensionless Hitch**, also known as the **no-knot knot**. It has advantages as an anchor knot for three reasons:

- It is simple.
- It reduces stress on rope and equipment.
- It gives the flexibility to deal with changing conditions.

PROCEDURE FOR TYING THE TENSIONLESS HITCH

(See Figure 7.3)

1. Take several wraps around the upright with the rope. The number of wraps depends on the diameter of the anchor point. A smaller diameter anchor point will require more wraps, but on any anchor point, there should be at least three. The objective is to have enough rope turns around the object so that the running end of the rope will have no tension on it. Instead, the tension is absorbed by the friction on the turns of rope around the object.

2. There should be no rope cross in the turns. If the anchor is above the load, the rope should spiral up with the running end at the top. This should be done in case the load has to be lowered slightly. In effect, that could be done by using the tree or vertical object as a lowering device.

3. Tie a Figure 8 on a Bight Knot in the running end of the rope and clip a locking carabiner into the Figure 8 Knot.

4. Clip the carabiner across the standing end of the rope at the bottom of the spiral. The spiral should be adjusted (easier if two people are doing it) so that there is no slack in the spiral, yet no sharp angle where the carabiner is clipped across the standing end of the rope.

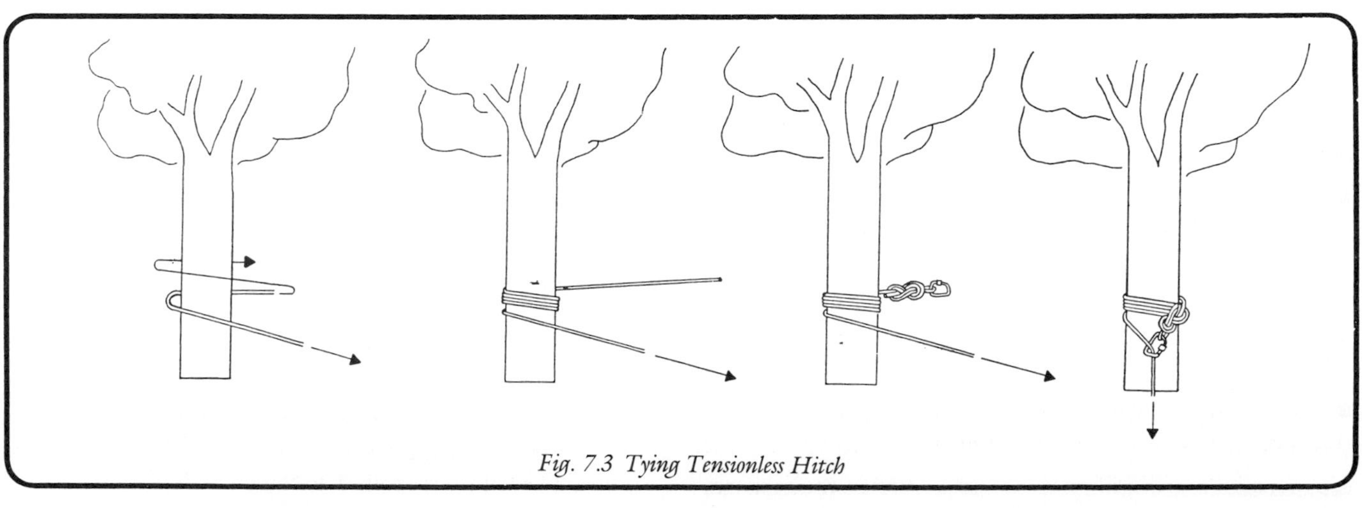

Fig. 7.3 Tying Tensionless Hitch

Where The Tensionless Hitch May Be Used:

You can use the Tensionless Hitch on any vertical member anchor point that can withstand the stress of the forces in the system and that will not damage the rope, such as:

- Trees.
- Columns.
- Structural Beams.

WARNING NOTE

If the anchor point has sharp edges, as sometimes found on structural beams, then you should pad the rope appropriately. (See Chapter 4, "Care and Use of Rope and Related Equipment.")

There is another knot that you can use in rope for anchoring. But it gives less flexibility for making changes than the Tensionless Hitch and may result in greater stress on the knot. However, where there will be lower loads on the anchor system, it may be preferable since it may be quicker to tie. This is the Figure 8 Follow Through Knot. As with the Tensionless Hitch, it is used where a loop of rope cannot be laid over the top of an anchor point, such as a tree or column, but has to be tied around it. (See Chapter 6, "Knots," for the procedure of tying a Figure 8 Follow Through.)

If the vertical member to be used as an anchor point is short enough so that you can easily get a loop of rope over it, then you can tie a Figure 8 on a Bight in the end of the rope and place the loop over the anchor. (See Chapter 6, "Knots," for the procedure of tying a Figure 8 on a Bight.)

ALTERNATIVE APPROACH

If local policy dictates the use of a Bowline for tying a loop in a rope, such as for a simple anchor, then you must use a Bowline in the place of a Figure 8 Follow Through or a Figure 8 on a Bight. However, you should take the following precautions:

- Since the Bowline is easily tied incorrectly, carefully inspect the Bowline to be certain that it is tied correctly. Figure 6.8, in Chapter 6 shows a simple Bowline correctly tied.
- Back up the Bowline with a "keeper knot" such as the Barrel Knot.
- See that the Bowline is not subjected to loading from a direction different from the one intended.

Using Webbing for Anchors

Webbing is a convenient material for anchoring. Among some groups whose members have a background in climbing/mountaineering, the use of webbing is very common.

Advantage:

- Less expensive than rope.

Disadvantage:

- Cannot be tied into as many different knots as rope. For example, you cannot use webbing for the Tensionless Hitch or the Figure 8 Follow Through.

Webbing is convenient for making continuous loops known as **runners.** Properly sewn runners are often used as a quick and convenient means of setting anchors. If you do not use sewn runners, then you can create a runner of a piece of webbing by tying it into a loop using a Water Knot also known as a "Overhand Bend" or "Tape Knot." (See Chapter 6, "Knots," for how to tie a Water Knot).

A good alternative to a runner would be one of the heavy duty commercial slings that are designed as lifting slings for cranes.

Fig. 7.4 Tying Webbing Around Anchor

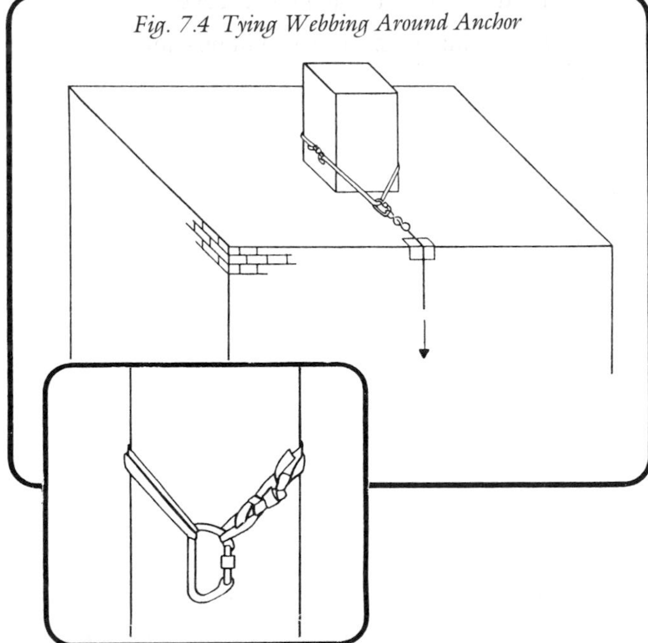

Fig. 7.5 Wrapping Webbing Around Anchor

Placement of Webbing Around Anchor Points

a) One secure method of placing the webbing around an anchor point is to tie it in a loop around the point using a Water Knot as shown in Figure 7.4. If the object being used as the anchor point is very large, this procedure can be awkward and time-consuming for one person.

b) An alternative is to have the runner tied beforehand, wrap it around the anchor point, and clip the two ends together with a carabiner as shown in Figure 7.5.

WARNING NOTE

Though it might be more convenient, it is not wise to leave knots tied in webbing unless each runner is carefully inspected before being returned to the equipment cache and again before being used:

- **You may not know *who* tied the knot and *what kind* of knot it is.**
- **Knots in webbing may eventually work their way out.**

Keeping Anchors in Place

It is generally good practice to tie the anchor on a vertical member as low down as possible to reduce the stress on it.

However, if the anchor point is strong enough, there may be certain conditions where it would be an advantage to tie the anchor webbing higher up:

- It might create a better angle for a rappeler to get over the difficult edge of a drop.
- It might reduce rope abrasion on an edge.
- In rescue, it might improve conditions for lowering a stretcher over an edge.
- It could reduce the severe friction that is stressing a hauling system.

The problem is that a simple loop of webbing around a vertical member will tend to slip down on the member. One conventional method for holding the webbing in place is to tie it around the anchor point in a girth hitch as shown in Figure 7.6(a). The drawback to this technique is the temptation to cinch the webbing back on itself as in Figure 7.6(b). This **should not** be done since it creates potentially dangerous stresses on the webbing.

One alternative to the Girth Hitch is to tie an interior loop in the webbing as shown in Figure 7.6(c). This will hold the webbing up on the anchor point, and will not stress the webbing as will a Girth Hitch.

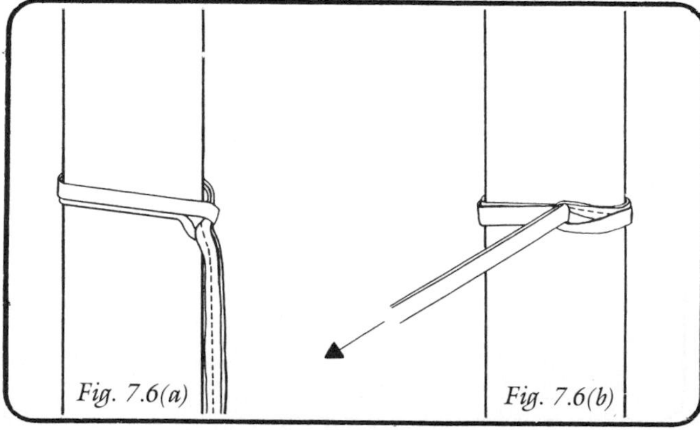

Fig. 7.6(a) *Fig. 7.6(b)*

Girth Hitch Webbing

Less Obvious Anchors

On some buildings, particularly those of recent construction, there may at first appear to be no anchor points. But after some practice at this type of problem, riggers may find some unexpected, but good, anchors. Among these might be:

- Elevator and machine housings.

  These are generally very large compared to what is often expected as anchor points. But by taking a length of rope, running it around the housing several times, and tying the ends together with a Figure Eight Bend, you may be able to create a secure anchor point for several lines.

- Scuppers (roof drain holes)

  Many buildings have low parapets with drain holes set in them at roof level. By running rope or webbing through the drain hole and back over the top of the parapet, you can create an anchor point.

  If possible, you should set the scupper on the side of the building opposite to the one over which you will run the main line. This will allow more room in a safe area on top of the building for rappelers to rig into the rope, and to set other rigging, such as lowering/raising systems.

  The more substantial parapets are those constructed of reinforced concrete. If the parapet is constructed of brick or block, the riggers should be certain that several brick or block courses are involved and the mortar is in good condition. Even under the best of conditions for a brick or block parapet, it would be wise to rig with at least two anchor points. Be certain to pad all sharp edges.

- Wall sections between windows and/or doors.

  If a set of windows and/or doors are close enough together, then you can create a substantial anchor point by passing the anchor rope or webbing through an open window or door, around the intervening wall, and back through a close-by window or door to tie it off. The anchor wall should be on the opposite side from where the main line will run out the building. This will give more safe space for rappelers to rig into the rope, or to set rigging such as lowering/raising systems.

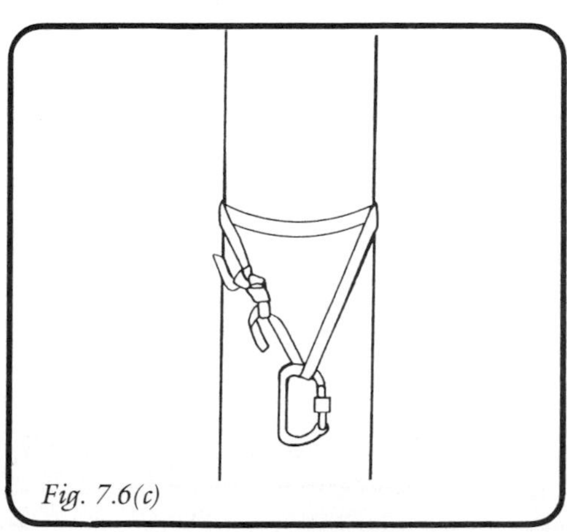

Fig. 7.6(c)

Interior Load Alternative

What To Do When There Are "No Anchors"

Extending Anchors

While you sometimes may not find anchors nearby, you may be able to establish them by running lengths of rope, sometimes for a few hundred feet, to where there are anchors. ***This should be done only with static rope.*** **Attempting to extend anchors in such a manner with dynamic rope could potentially create a dangerous situation due to the large amount of stretch.** One example of where extended anchors might work would be on the roof of a building where (absolutely, positively) no anchor points exist. Often, static rope can be run through the stairwell or through top floor windows to lower floors where anchors do exist.

Using Vehicles for Anchors

A sort of "portable anchor" that is usually available is an emergency vehicle. However, there are some safety guidelines that you should follow when using vehicles for anchors:

- Begin by setting the parking brake.
- But forces created in a high angle system can move a vehicle with its brakes set, so chock wheels. If no chocks are available, use spare tire(s).
- Idiot proof your portable anchor by removing ignition key.
- There are portions of a vehicle that are structurally weak **and should not be used:**
  a. Bumpers.
  b. Tow Hooks (these have often been subjected to intense and unrecorded stresses and contain the potential for failure).

Potential anchor points on a vehicle include structural parts, such as axles and cross members. But protect rope and webbing from oil and grease. As noted in Chapter 4, "Care and Use of Rope and Related Equipment," rope and webbing must be protected from destructive substances such as battery acids, which are often found around vehicles.

Complex Anchors

Complex, or multiple, anchors involve two or more anchor points. You use multiple anchors when one anchor point is insufficient to withstand the anticipated forces, or when one anchor point is inconveniently placed.

Multi-loop Slings on Single Anchor Point

By adding additional anchors to an appropriate anchor point, you can ensure against the collapse of an anchor system due to the failure of a single item, such as webbing, carabiner, or improperly tied knots. Figure 7.7 shows the use of two slings of equal length on a single anchor point. This kind of system is used only where the single anchor point is absolutely secure ("bomb-proof"). There is **no insurance at all** in the additional webbing and carabiners if the anchor point fails.

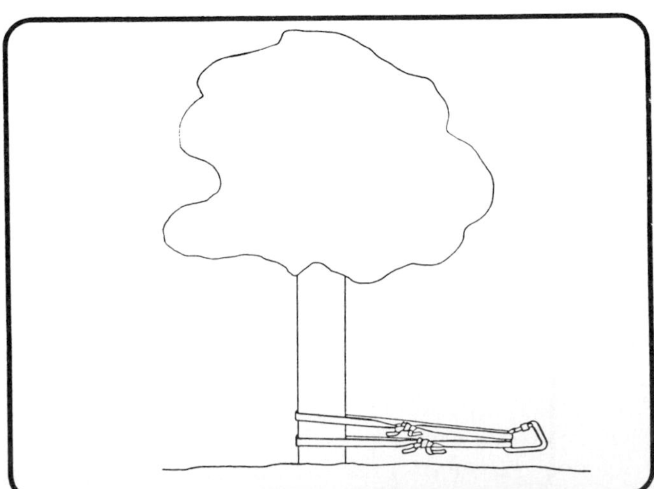

Fig. 7.7 Backing Up Single Anchor Point

Load Sharing Anchors

If anchor points are at all questionable, or inconveniently placed, then load-sharing anchors may be a solution. The simplest way to create a multiple anchor system for load sharing is to use two anchor ropes or slings of equal length. Run them from different anchor points and clip them together into a single point using one or two large locking carabiners *(see Figure 7.8)*.

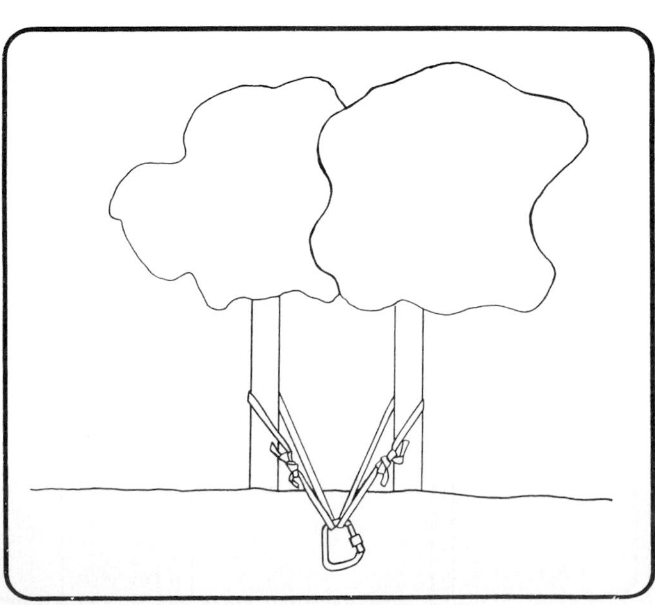

Fig. 7.8 Load Sharing Anchor

WARNING NOTE

A primary concern for rigging any type of complex anchor is not to create too wide an angle between the legs of the anchor system. Ideally, this angle should not exceed 90 degrees, and must never exceed 120 degrees. Beyond this point, the forces *on each anchor* and other elements of the system will be greater than the total load itself.

You should remember that any angle in an anchor system will increase the loading on anchors and other elements of the system. Only when the angle between the legs of the anchor system is 0 degrees will each leg carry half the load. (See Figure 7.9 for diagrams of how angles affect the forces on anchor points and other elements of the system.)

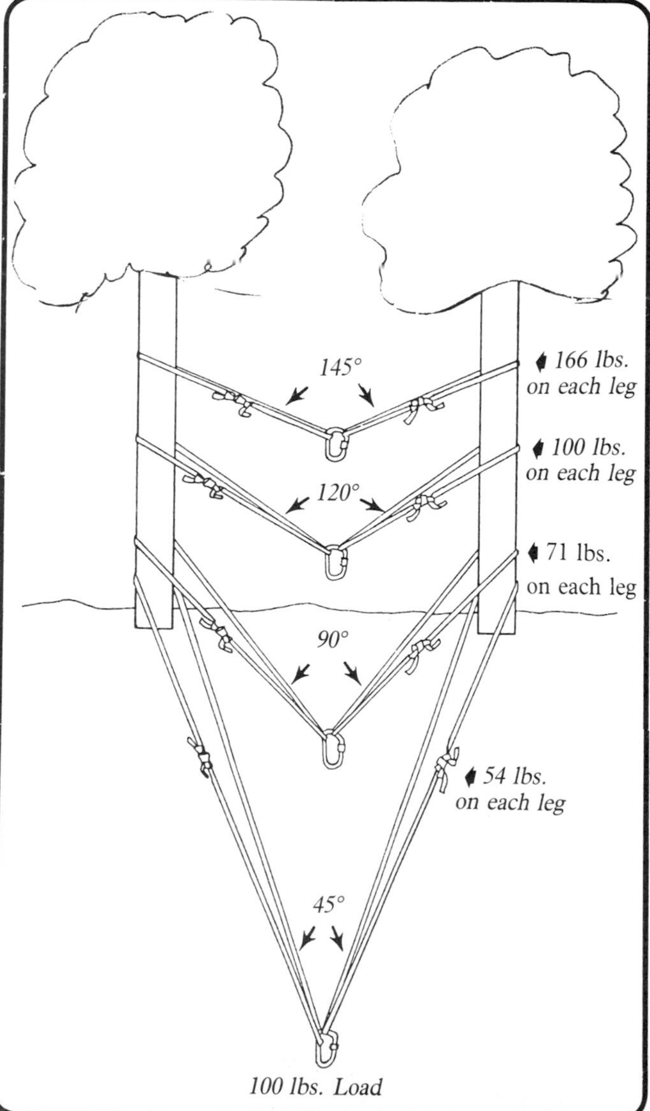

Fig. 7.9 Relationship Between Anchor Sling Tension and a 100 lb. Load at Different Angles

Fig. 7.10 Forces on Load-Sharing Anchors

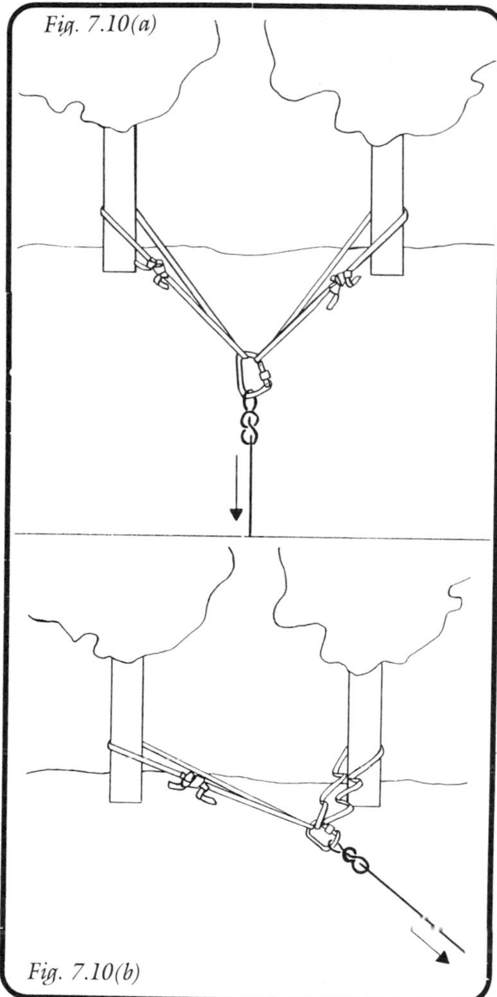

Self-Equalizing Anchors

A major problem with a fixed, load-sharing anchor system as in Figure 7.10 [a] is that the stress on the anchor points will be equal only when the force from the main line pulls directly in the center of the angle. This is actually a rare condition in high angle operations. Most of the time, the force will be pulling to one side or another, as in Figure 7.10 [b]. Often, it will be moving back and forth from one side to another. Obviously, if there is a possibility that either one of the anchor points cannot sustain these side-to-side forces, then the entire anchor system may fail.

One possible solution to this problem is to create a **self-equalizing anchor system** (also known as a **Load Distributing Anchor System**). When correctly constructed, this type of system can have some important advantages:

a) The forces on anchor points should remain close to equal, whatever the direction of pull.
b) Should any anchor point fail, the system should readjust itself to where once again, there is close to equal loading on the remaining anchor point(s).

It should be pointed out that no anchor system can be made to be completely "self-equalizing." How well the system equalizes and survives the shock loading of one anchor point failing depends on several factors:

- Keep the angles small, both to reduce magnification of forces on anchors, and to help the system readjust to the new loading.

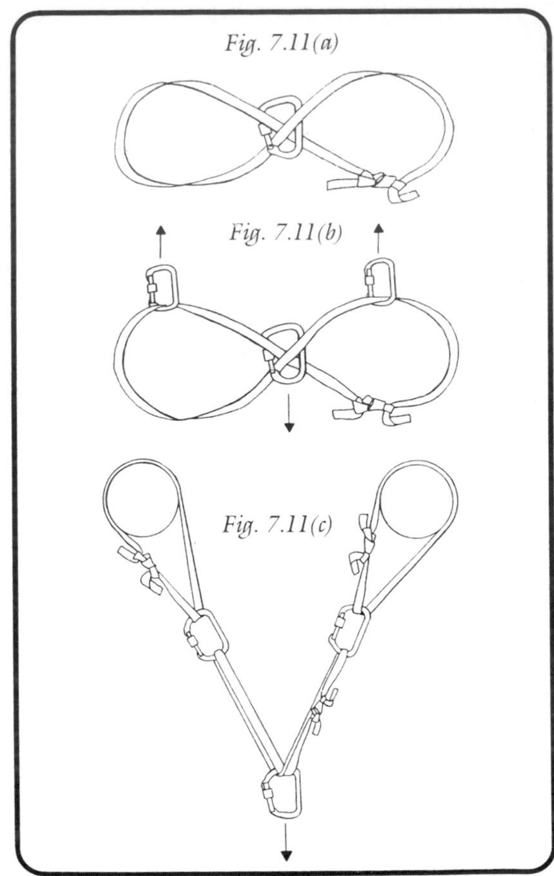

Fig. 7.11 Simple Self-Equalizing Anchor

- Avoid bulky rope or webbing, and adjust the system so that knots are less likely to run through carabiners when the system readjusts.
- Design the systems so that there will be as little drop as possible should any anchor fail.
- Avoid rope or webbing made of materials such as Kevlar™ or Spectra™ since they do not have the shock absorbing qualities of materials such as nylon.
- Make all of the anchor points in a self-equalizing system as bombproof as possible.

A Simple, Two-Point Self-Equalizing System

One of the simplest of the self-equalizing anchors involves two anchor points and uses a sling and a carabiner.

CREATING A SIMPLE TWO POINT SELF-EQUALIZING ANCHOR

(See Figure 7. 11)

1. Configure a loop of webbing or rope in the shape of an "8."
2. Clip a large locking carabiner across the inside loop.
3. Take each end of an outside loop and clip it into an anchor point.
4. Clip the carabiner on the inside loop into the main line.
5. Make certain that the angles made by the sling do not exceed the critical ones described in the **WARNING NOTE** on page 73.

Whatever the direction of pull, the central carabiner will slip along the sling to equalize the forces. Should one anchor point fail, the webbing will automatically set itself to pull on the other anchor point.

More Complex Self-Equalizing Systems

If the situation is such that three or more anchor points are needed, then you should use a more complex self-equalizing system.

TYING A SELF-EQUALIZING ANCHOR SYSTEM USING A TWO-LOOP FIGURE EIGHT

(See Figure 7.12)

1. Create a large loop by using a length of rope and tying the two ends together using a Figure Eight Bend (or Double Fisherman's Knot).
2. For ease of operation, set the circle of rope on the ground or on the floor.
3. So that the knot remains out of the way, place it at about 3 o'clock or 9 o'clock on the circle.
4. Take a large bight of rope and flip it back inside the circle, about 2/3 of the way up.
5. The lower section of the circle is now doubled, so that there are four strands of rope.
6. Gather together these four strands of rope and tie a Figure 8 Knot with them.
7. At the top of the circle, there is now a large loop and a smaller loop inside that. At the bottom of the circle is a much smaller loop created by tying the Figure 8 from the four strands.
8. Take the largest loop at the top of the circle and clip it into all the anchor points using a locking carabiner at each point.
9. Take the smaller loop at the bottom of the large loop. Using locking carabiners, clip this together to the larger loop between each anchor point.
10. Take the small loop below the knot that was created from the four strands. Clip this into the main line using one or two large locking carabiners.
11. If tied correctly, should any one point fail, the system should automatically equalize itself by spreading the load among the remaining anchor points. The system should also equalize the load among all points whatever direction the load is coming from.
12. The riggers should inspect the system to make certain that it will indeed perform in this manner before the system is loaded by people. If there are any problems, such as a knot jamming, then the system must be adjusted so that it will work as intended.

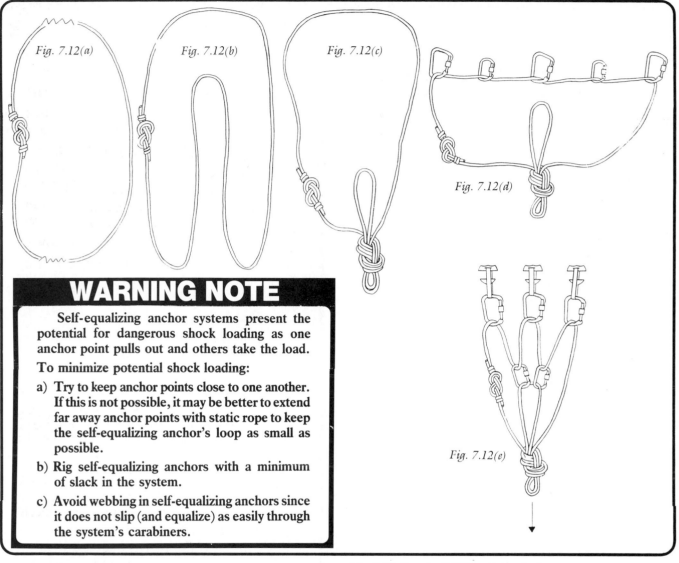

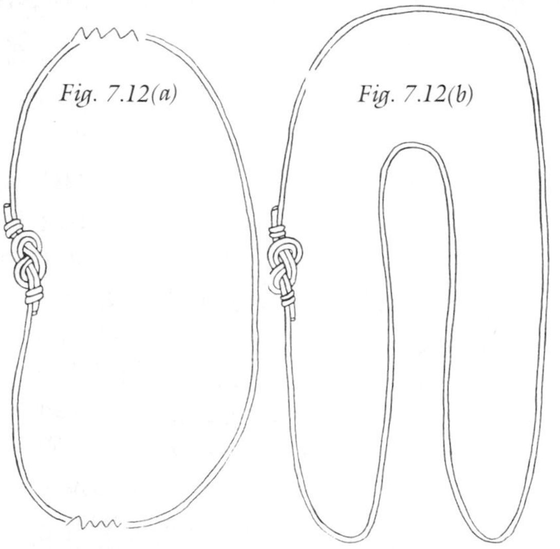

Fig. 7.12(a)

Fig. 7.12(b)

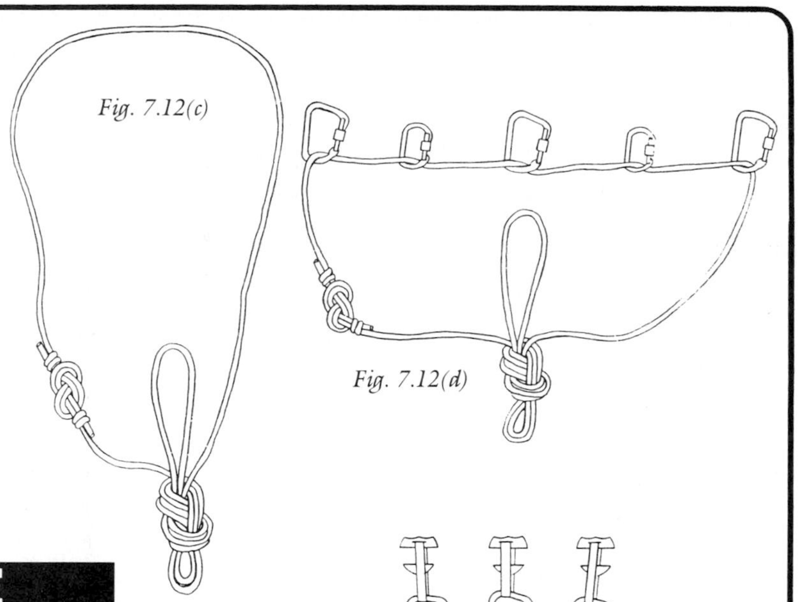

Fig. 7.12(c)

Fig. 7.12(d)

Fig. 7.12(e)

WARNING NOTE

Self-equalizing anchor systems present the potential for dangerous shock loading as one anchor point pulls out and others take the load.

To minimize potential shock loading:

a) Try to keep anchor points close to one another. If this is not possible, it may be better to extend far away anchor points with static rope to keep the self-equalizing anchor's loop as small as possible.

b) Rig self-equalizing anchors with a minimum of slack in the system.

c) Avoid webbing in self-equalizing anchors since it does not slip (and equalize) as easily through the system's carabiners.

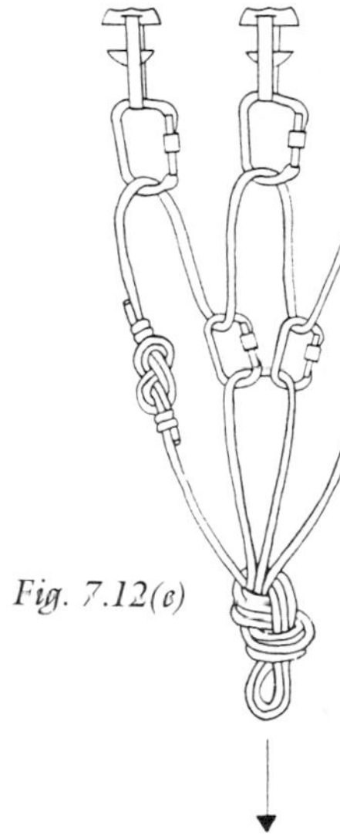

Fig. 7.12 Complex Self-Equalizing Anchor

Establishing a Picket System *(See Figure 7.13)*

Pickets

One alternative in a natural area where there are no anchors is the picket system. Though it can work very well when properly rigged, it usually takes a great deal of time to properly establish a picket system.

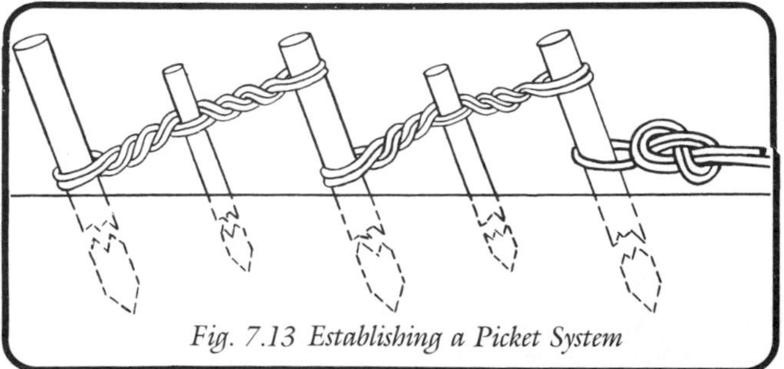

Fig. 7.13 Establishing a Picket System

1. The pickets should have a minimum length of five feet, so that there will be a minimum of three feet in the ground and a maximum of two feet above ground.
2. Drive the pickets at an angle of 15 degrees away from the force to be anchored.
3. Connect the pickets in each row together by lashing from the top of the first picket (the one closest to the load) to the bottom of the next picket. Continue in this manner until all rows of pickets are lashed together.
4. Tension the lashings by twisting with a stick four to six turns. Drive this stick into the ground to secure it.
5. Connect the main line by clipping it to the front picket in each row with a self-equalizing anchor system as described above.

QUESTIONS for REVIEW

1. Name three factors that affect the ability of an anchor point to withstand the forces of high angle activity.
2. Name three conditions where anchors should not be set directly above the high angle activity, but off to the side.
3. What is the purpose of a "directional" in an anchor system?
4. What would be the possible disadvantage in tying the main line directly into an anchor point?
5. What are three advantages in using the Tensionless Hitch or "no-knot knot" in an anchor system?
6. In examining a tree for use as a potential anchor point, attention should be paid not only to the soundness of the wood, but also to the nature of the __________ .
7. Why is it not prudent to leave knots tied in webbing (unless each runner is carefully inspected before use)?
8. How should webbing be tied to hold it in place on a vertical anchor point?
9. Name two typical reasons for anchor points on buildings being inadequate.
10. Name three potential anchor points on buildings that are an inherent part of the structure.
11. Describe parts of a vehicle that:
    a) should not be used as anchor points, and
    b) might be used as anchor points.
12. Name two situations in which complex anchors are used.
13. Ideally, the angle between the legs of an anchor system should not be more than ____ degrees and never more than ____ degrees.
14. What are two reasons for using self-equalizing anchors?

★ ACTIVITIES ★

1. Create a directional on an anchor system that was initially set at an inconvenient angle.
2. Set a Tensionless Hitch ("no-knot knot") on a tree or column for a downward pull.
3. Correctly:
    a) tie a runner from webbing in a simple loop around a vertical anchor point, and
    b) correctly tie a loop from a rope around the same anchor point.
4. Using an internal loop, set a runner tied from webbing around a tree or column so that it will not slip down.
5. Correctly set an anchor on a building using an anchor point that is an integral part of the structure. NOTE: if a training tower or similar structure is used, the student **will not** use any anchor points previously prepared, such as ring bolts.
6. Using webbing or rope on an anchor system with two anchor points, demonstrate the following:
    a) the angle that the two legs of the anchor **should not** exceed, and
    b) the angle that the two legs **must not** exceed.
7. Correctly tie a load-sharing anchor using two anchor points.
8. Correctly tie a self-equalizing anchor using a webbing loop on two anchor points.
9. Correctly set a self-equalizing anchor system using three or more anchor points.

Chapter 8

Belaying

OBJECTIVES–

At the completion of this chapter, you should be able to:

1. Define belaying.
2. Describe the elements of a belay system.
3. Describe situations that might require a belay.
4. Repeat from memory and in sequence the belay voice communications.
5. Belay using a Munter Hitch.
6. Belay using a belay plate.
7. Discuss why it is necessary to have both complete knowledge and thorough practice at belaying before attempting to belay a person in an actual high angle situation.

TERMS– *pertaining to belaying that a High Angle Rope Technician should know:*

Belay—The securing of a person with a rope to keep him from falling a long enough distance to cause harm.

Belayer—The one who performs the belay.

Belay Plate—A simple, metal plate containing one or more slots for rope, and used to create rope friction with a carabiner. It is commonly used in belaying.

Munter Hitch—A type of running knot that slips around a carabiner to create friction against itself. It is commonly used in belaying.

Belaying

The word belay comes from the days of sailing vessels. On those ships, belaying pins were set in the rails of the vessel. When sailors raised heavy objects, such as sails, they would attach a rope to the object, and then take a turn of the rope around the belaying pin to prevent the line from slipping away from them.

In the high angle environment, the principle is the same except that the purpose of modern belaying is to keep a person from falling. The person is attached to a rope and the rope is managed in such a way, or belayed, to keep the person from falling far enough to harm himself.

The ability to belay is a critical skill for anyone operating in the high angle environment. If you accept the assignment as a belayer, you have made a very serious commitment. It means that the well-being, perhaps even the life, of the person at the end of the rope is in your hands. To say that you are able to belay when you cannot, or to allow your attention to lapse from the job of belaying, could possibly mean severe injury or death for the person at the end of the rope.

The Belay System

The elements of a belay system include the following *(see Figure 8.1)*:

- A person tied to the rope and who is at risk of falling. (Often referred to as the "climber").
- The rope that is attached to the person.
- A belay device. It is in essence a braking mechanism. The rope is run through it and controlled in such a way that should the climber fall, the belay device, under the control of the belayer, holds the rope.

Fig. 8.1 Elements of A Belay System

- The belayer. He controls the belay device and the rope. His main duty is to brake the rope should the climber fall. But he must also manage slack in the rope so that slack does not add to shock loading the rope if the climber falls.

Situations Requiring a Belay

A belay may be called for any time there is "exposure" danger of falling. Some situations requiring a belay would be:

- When a person is rock climbing or mountaineering. Should he slip, a belay might be able to hold him.
- In a rescue situation where there is danger of falling. One example would be an attempt to rescue a "jumper" from a bridge.
- When a person is crossing an area not generally dangerous, but there is a small area of exposure.
- When a person is unsure of himself in attempting a new skill, such as rappeling for the first time.
- When a person's physical or mental capabilities are diminished, such as when he has been injured or is suffering vertigo.
- When environmental factors, such as potential rockfalls or areas slick with ice, increase the danger of falling.
- When one or more persons are being lowered by rope, such as in a rescue.
- When one or more persons are being raised by rope, such as in a rescue.

Judgements About When to Belay

There will be times when the need for a belay is not completely clear cut. Some examples are:

- A person who is very experienced in the high angle environment is rappeling. He may feel that a belay would only be a hinderance.
- When a belay might cause a greater problem than not having one. An example would be a situation in which there are already several rope lines involved. An additional line from a belay could cause an entanglement.
- In free drops where the load may spin. A belay line could entangle the main line and stop everything.

Judgements about these situations have to be made locally by those people who are experienced and well-trained.

A BELAY *IS NOT* THE SAME THING AS LOWERING. *Do not confuse the two.*

Belays and lowerings (see Chapter 14, "High Angle Lowering Systems"), involve different techniques, different equipment and are for different purposes (although they may be used together).

A Belay: is a safety to catch persons should they fall. A belay rope can be run either way (up or down) while it is being used for a belay. A belay rope is kept with some slack in it, and does not have weight on it except when there is a fall on it or "tension" is called for in special cases. A belay uses specialized equipment such as a belay plate or special knots such as the Munter Hitch.

A Lowering: is the controlled lowering of persons/equipment using rope through a lowering device/hardware such as the large ring of a Figure Eight Descender or a Brake Bar Rack (which cannot be used for a belay). A lowering rope goes one way: down. A lowering rope has weight on it all during the lowering operation.

However, should a fall occur, a good belay system should have the ability to lower the load a short distance to where it can be stabilized.

Attitudes about Belaying

There have been numerous accidents due to belayer failure. These are often caused by one of, or a combination of, two factors:

a) The belayer having a momentary lapse of attention just as the person on the rope fell

b) The belayer not automatically performing the correct actions due to lack of adequate training in belaying.

One basic principle about human nature is this: *In a sudden emergency, humans respond with what is instinctive.* Such emergency situations could relate to driving experience during the threat of a vehicle accident or to weapons training in a law enforcement confrontation.

Belaying is a similar activity in that it must be thoroughly learned under realistic conditions and followed by constant practice until it becomes instinctive. Otherwise, the belayer may fail to take the correct action when the sudden emergency occurs, and be responsible for severe injury or death.

It is not enough to have *intellectually* learned belaying. That is, it is not enough to have read about belaying, or even to have practiced the hand positions.

You must have combined both the *mental* and *physical* experience of the actual belay situation.

Belay Practice System

Figure 8.2 shows an overall view of a belay practice system. It consists of the following elements:

- A weight or a dummy of at least 200 pounds to simulate the weight of a falling person.

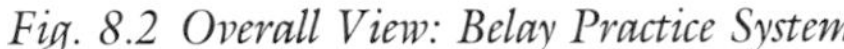

Fig. 8.2 Overall View: Belay Practice System

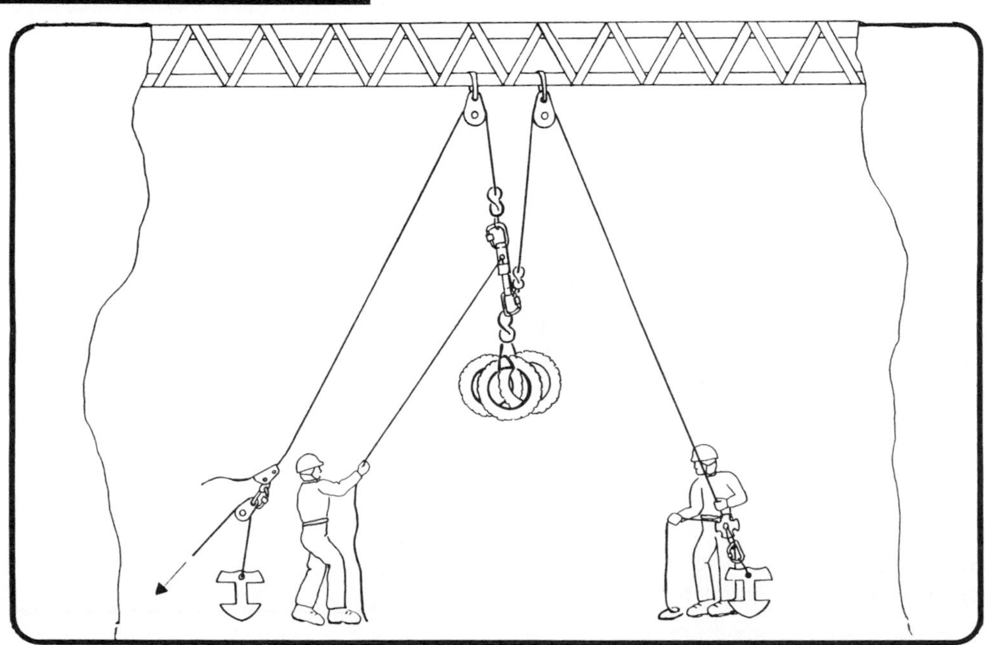

SUGGESTION:

Discarded truck tires work well as the weights for a belay practice system. They are inexpensive and tend not to cause damage to concrete/asphalt floors when they drop.

- A method of raising the weights. A winch may be the easiest and most convenient way of doing this. However, a mechanical advantage hauling system *(as in Figure 8.3)* can be used for raising the practice weights.
- A belay "station" where a belayer attempts to catch the fall with a rope.
- An instructor's station.
- The instructor has a rope that triggers the weight's fall. The triggering mechanism can be created from a seatbelt buckle, a parachute harness release, or a military helicopter harness release *(see Figure 8.4)*.

The procedure is as follows:

1. The belayer and the instructor take their stations.
2. An assistant begins to haul the weight up.
3. At some point, without warning, the instructor pulls the line that triggers the fall of the weight.
4. The student belayer attempts to arrest the fall of the weight.

WARNING NOTE

When the weight falls and the belay catches, the belay rope will come taut with great force. To avoid injury, the belayer, and others around the belayer must take the following precautions:

- **Be aware of the position the rope and belay device will take when in catches and stay out of its path. Otherwise, the impact of the rope and/or belay device can cause serious injury.**
- **Keep fingers, hair, and clothing free of the rope near the belay device. Otherwise, these could be swept into the belay device possibly causing injury and impeding the belay.**

Such a belay practice system that simulates the forces of a falling climber is essential to the instruction of belaying. This system is very simple and inexpensive to construct. Details on the system for hauling the weight and on the triggering mechanism are shown in Figures 8.3 and 8.4.

Fig. 8.3 Hauling: Belay Practice System

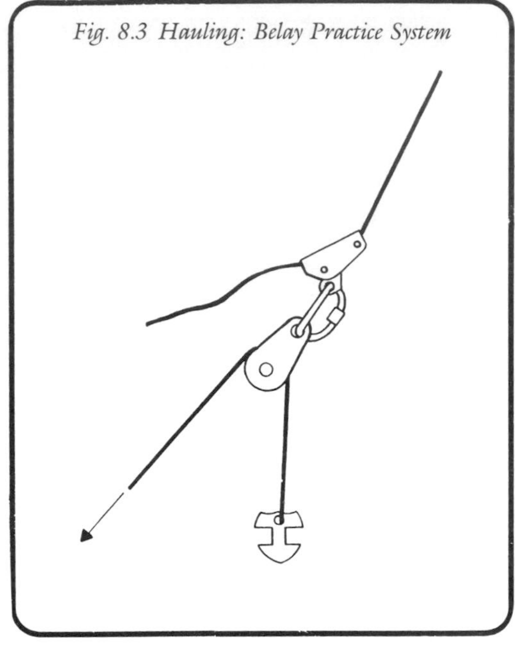

Fig. 8.4 Trigger: Belay Practice System

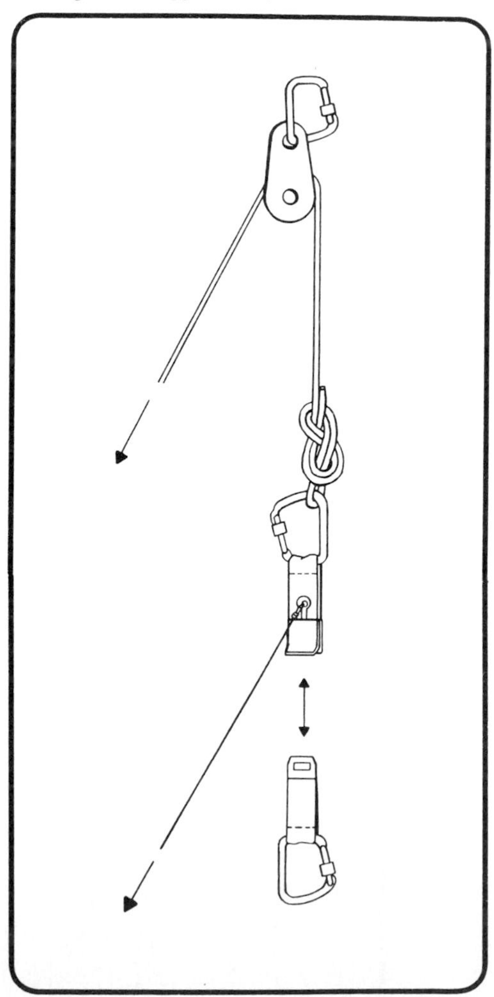

Belaying Signals

When you are belaying, it is essential that you use the standard voice signals (also called "commands" or "calls"). Otherwise, even momentary confusion could cause an accident.

In climbing and mountaineering, there are a group of voice signals that have been standard in belaying for years.

They take place as an exchange of voice signals between a climber (or rappeler) and the belayer to ensure that both are ready for any possibility.

The standard belay signals in sequence are as follows:

Climber	Belayer	Phrase Meaning
1. **"On Belay."**		*A question: "I am about to climb (or rappel), are you ready to catch me if I fall?*
2.	**"Belay On."**	*A statement: "I am ready to catch you if you fall."*
3. **"Climbing."** (**or: "Rappeling"**)		*"I am starting to climb" (or: "to rappel").*
4.	**"Climb."** (**or "Rappel"**)	*"Go ahead"*

Once the climber is at the place where he no longer needs the belay, then he initiates an exchange to end the belay:

Climber	Belayer	Phrase Meaning
5. **"Off Belay"**		*"I am in a secure place now. I no longer need the belay."*
6.	**"Belay Off."**	*"I am no longer belaying you."*

There are some additional signals that can assist communication between climber and belayer during the belay cycle. One of these is in response to a situation where the belayer is holding the rope too tightly:

Climber	Phrase Meaning
"Slack."	*"There is too much tension on the rope; I cannot move as well as I would like."*

(This requires no verbal response from the belayer, only the action of letting an appropriate amount of slack into the rope.)

The other situation is the opposite. There is too much slack in the rope. It may be that the climber is at a particularly tricky point and he needs the support of the rope to make his move.

Climber	Phrase Meaning
"Tension."	*"Hold the rope tightly for a bit; this might be a difficult move."*

(Requires no verbal response, only the action of taking slack out of the rope.)

Consistency

It must be emphasized that once these voice communications have been agreed upon, there must be no change in them without prior agreement. Otherwise, there could be some confusion in communications. If this occurs, even briefly, it can be dangerous.

Make Yourself Heard

For the belay voice communications to work, they must be heard by those involved. Do not be timid when you use them. Wind, falling water, or other conditions in the outdoors often interfere with voice communications being heard. The belay communications will take at least shouting, perhaps yelling, to be effective. If the required response to a command is not received, repeat it **louder.**

Hand Protection

In belaying, the belayer **must wear gloves** to protect his hands from rope friction. This must be done for two reasons:

- To protect his own hands from possibly severe rope burns from a running rope.
- To prevent the potential pain caused by grasping a running rope, and enable the belayer to hold the rope firmly and stop the fall of the person on the rope.

Belaying Techniques

There are a number of belaying techniques used for the high angle environment. They all work essentially the same way: they create a braking action on the rope to prevent the person at the end of the rope from falling far enough to injure himself.

We will examine two of these techniques in detail. These techniques are among those considered to be simple to learn, to be easy to use, and to pose less of a danger to the belayer and the person on the rope.

The Munter Hitch

*(also known as the **Half Ring Bend** or the **Italian Hitch.**)*

WARNING NOTE

The Munter Hitch belay must be practiced on level ground and with a weight or dummy drop before attempting it with a person in the high angle environment. This practice must be under the guidance of a qualified instructor.

PROCEDURE FOR BELAYING WITH THE MUNTER HITCH

A. **Setting the Belay on Level Ground** *(see Figure 8.5)*

A1. Tie a piece of webbing or rope sling into a secure anchor *(see Chapter 7, "Anchoring")*

A2. Clip a large locking carabiner into the anchor sling.

A3. Tie a Munter Hitch near the end of the rope where the person being belayed will be. While there are several ways of tying the Munter Hitch, *Figure 8.7* shows a simple, easy-to-remember way of doing it.

A4. Clip the large locking carabiner across the portion of the Munter Hitch that has two sections of rope parallel to one another. Lock the carabiner.

A5. It is important to remember that the Munter Hitch "runs" in both directions. This means that if the person on the rope is moving away from you, the loop in the bight will be on the side of the carabiner towards him. If you are pulling up rope (the climber is moving towards you), the loop will be on the side of the carabiner facing you.

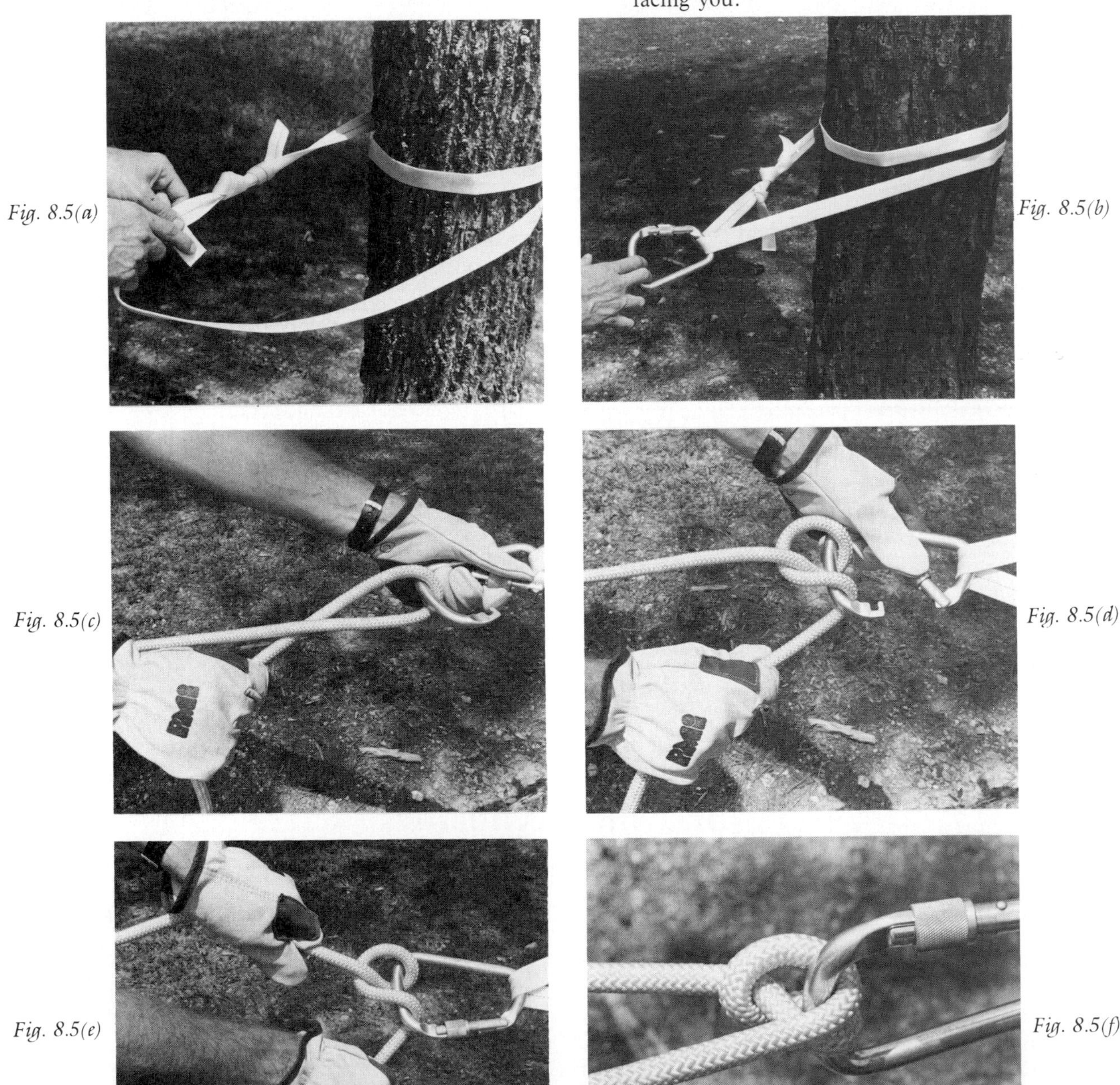

Fig. 8.5(a) *Fig. 8.5(b)* *Fig. 8.5(c)* *Fig. 8.5(d)* *Fig. 8.5(e)* *Fig. 8.5(f)*

Fig. 8.5 Setting Belay/Level Ground

The carabiner used with the Munter Hitch must be a **large** locking carabiner. A narrow carabiner can cause jamming of the Munter Hitch.

Some manufacturers sell special pear-shaped carabiners (called HMS) that are designed for use with the Munter Hitch. The gentle curves in these pear-shaped carabiners allow the Munter Hitch to move freely back and forth through the carabiner when the direction of the belay is reversed.

Fig. 8.6 Practicing Munter/Level Ground

Fig. 8.6(a)

Fig. 8.6(b)

B. Practicing the Munter Hitch Belay on Level Ground *(see Figure 8.6).*

B1. Face the carabiner. Take the rope on the side of the carabiner away from the climber in your dominant hand (the right hand for right-handed people). This is your ***braking hand.*** **YOU MUST NOT TAKE THIS HAND OFF THE ROPE WHEN THE PERSON IS "ON BELAY" AND THE BELAY IS NOT TIED OFF.**

B2. Take the rope on the side of the carabiner near the climber in your weaker hand (the left hand for right-handed people). This is your ***guide hand.*** You will use it to help manipulate the rope.

B3. Now have a partner wearing a seat harness clip into the end of the rope.

B4. Remove slack in the rope between the carabiner and the climber by pulling it out with the brake hand.

B5. Begin the belay voice communications (the person on the rope says, "On Belay." If the belayer is ready, he says, "Belay On."

B6. Have the person on the rope begin slowly walking backwards. With the guide hand, pull the rope out from the carabiner and feed it to the climber. Allow the "climber" to set the rate of rope run. As your brake hand approaches within a foot of the carabiner, slide it back ***while still holding the rope.*** **NEVER TAKE THE BRAKE HAND OFF THE ROPE.**

B7. Now, to give a feel of the control you have, hold the rope firmly with the brake hand, stopping the person on the rope from going farther.

Fig. 8.6(c)

Fig. 8.6(d)

B8. In an area where he will not slip, have the "climber" firmly plant his feet on the ground and lean back from the rope. Hold the rope firmly with the brake hand.

B9. To get a feel of the control, slowly let a few inches of rope out with the "climber" leaning against the rope.

Fig. 8.6(e)

Fig. 8.6(f)

Fig. 8.6(g)

Fig. 8.6(h)

Fig. 8.6 Practicing Munter/Level Ground

Reversing Direction

B10. Now have the partner walk slowly towards you. The problem now is to take in rope while keeping the person safely on belay.

B11. Have your brake hand on the rope about one foot from the carabiner.

B12. Place your guide hand about three feet from the carabiner and on the rope.

B13. As slack comes into the rope, pull it out by pulling with the brake hand. Also pull the rope with your guide hand to reduce friction on the Munter Hitch and help ease the rope through the carabiner. Just as your guide hand approaches your brake hand, grasp both sides of the rope with your guide hand. Do not cross hands.

B14. Holding both lines taut with the guide hand, move your brake hand up the rope towards the carabiner **WITHOUT TAKING IT OFF THE ROPE.**

WARNING NOTE

The moving of hand positions on the rope is one of the most critical operations in belaying. Remember: the person at the end of the rope may fall *at any time* while you are belaying. If you do not have your brake hand on the rope at all times, ready to grasp it in an instant, you may drop the person and cause severe injury or death.

B15. Continue the procedure until the person on the rope reaches your position.

B16. Exchange the voice signals that conclude the belaying cycle. (*Person on rope: "Off Belay."* Belayer: **"Belay Off."**)

Practice With a Belay Weight/Dummy

Now practice the same operation using a belay dummy or weight as shown in Figure 8.2. Do a procedure first with the weight falling as it is being raised, and then with it falling as it is being lowered.

> **WARNING NOTE**
>
> **When the weight falls, and the belay catches, the belay rope will come taut with great force. To avoid injury, the belayer, and others around the belayer must take the following precautions:**
>
> - **Be aware of the position the rope and belay device will take when it catches and stay out of its path. Otherwise, the impact of the rope and/or belay device can cause serious injury.**
> - **Keep fingers, hair, and clothing free of the rope near the belay device. Otherwise, these could be swept into the belay device possibly causing injury and impeding the belay.**
> - **Belay plates are designed for catching single-person loads and may not be the most appropriate device for catching rescue loads.**

Fig. 8.7 Tying the Munter Hitch

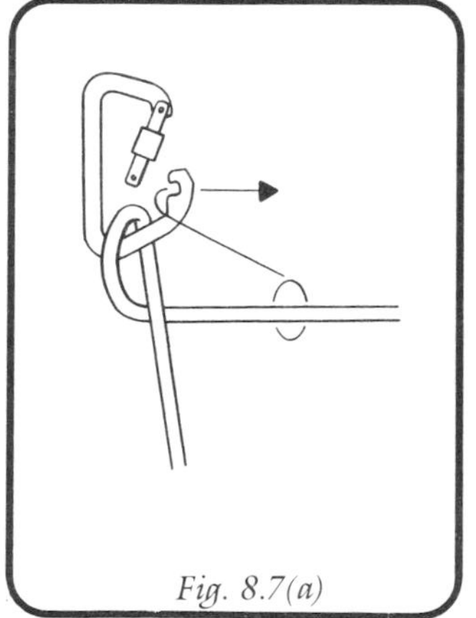

Fig. 8.7(a)

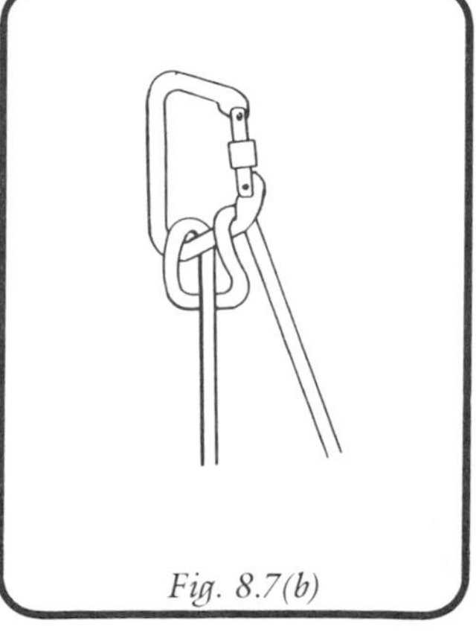

Fig. 8.7(b)

In an actual belay situation, you would, if possible, keep the "climber" in view. But for practice, you will face away from the weight/dummy so you cannot visually anticipate the fall. This will sharpen your skills at catching an unexpected fall.

As each fall of the dummy/weight comes onto your rope, and you catch it, lower it to the ground using the Munter Hitch.

Using a Belay Plate

Another method of belaying is the use of a belay plate. Provided the belayer has practiced until he is skilled with it, and is alert to falls, it is a relatively secure technique for single person loads.

Shown in Figure 8.8 are various brands of belay plates. All operate in essentially the same manner. Note the elongated slot through which the rope passes. Some belay plates have two slots of the same size. They are designed for possible belaying with a double rope system. Some plates have two slots for different sizes. They are designed for people who might be using a 11mm rope on one occasion, and a 9mm rope another time.

The Sticht Plate also comes in a model with a spring on one side. This spring is for use with a soft rope, such as a dynamic climbing rope, to keep it from jamming in the plate.

Note that all the plates have one very small hole near the edge. The "Keeper Loop," a loop of small diameter cord, approximately eight inches long, can be attached through this hole. This will attach to a convenient place to keep the device from sliding down the rope out of reach of the belayer.

Belay plates are designed only for use with kernmantle ropes (either static or dynamic). **They must not be used with laid ropes.**

Fig. 8.8 Various Belay Plates

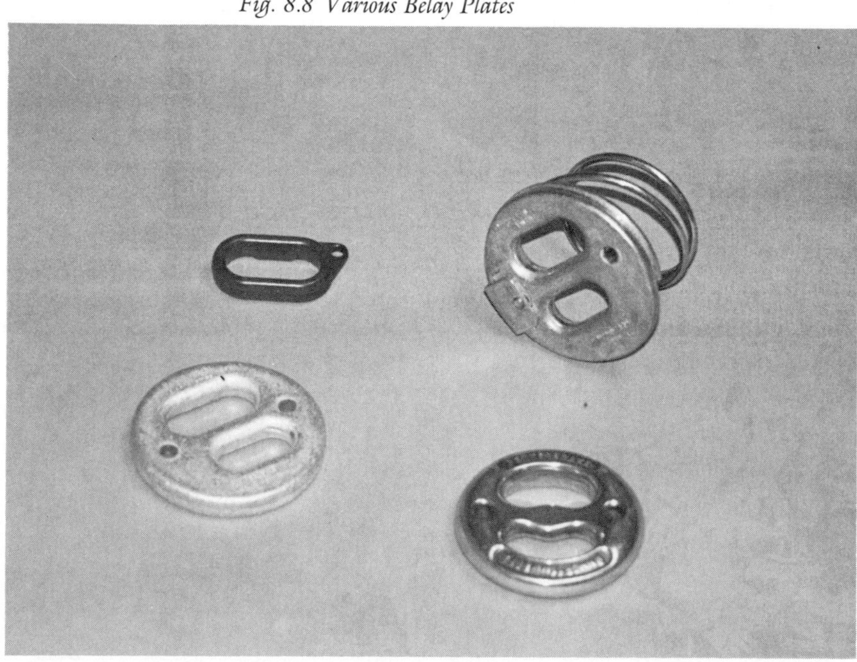

PROCEDURE FOR BELAYING WITH A BELAY PLATE

WARNING NOTE

Belaying with a belay plate must be practiced on level ground and with a weight or dummy drop before attempting it with a person in the high angle environment. This practice must be under the guidance of a qualified instructor.

Fig. 8.9 Rigging the Belay Plate

Fig. 8.9(a)

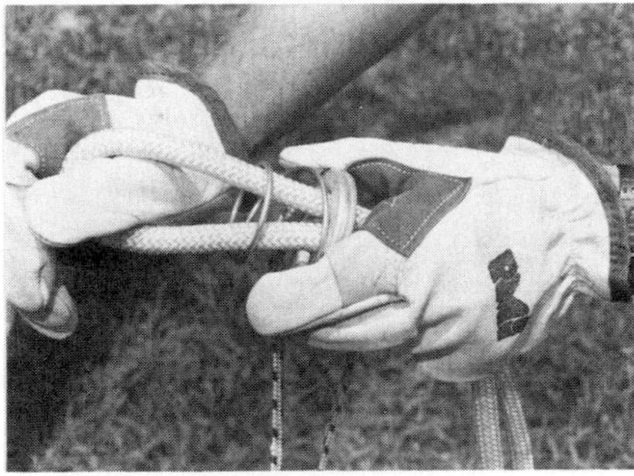

Fig. 8.9(b)

Fig. 8.9(c)

C. **Rigging The Belay Plate** (see Figure 8.9)

C1. Tie a webbing or rope sling into a secure anchor *(see Chapter 7, "Anchoring").*

C2. Clip a locking carabiner into the end of the loop.

C3. Take a bight of rope near the end of the rope where the person to be belayed will be.

C4. Push a few inches of the bight through the hole in the belay plate and clip the carabiner onto the bight so that the belay plate is secured onto the rope.

NOTE: Some belay plates are designed to be facing only one way.

a) Sticht Plate with a spring: the spring should be next to the carabiner.

b) Clog Cosmic: Groove should be next to the carabiner.

C5. Secure the "keeper loop" by clipping any style of carabiner through it and clipping the carabiner directly into the anchor sling.

D. **Practicing The Belay on Level Ground** *(see Figure 8.10).*

D1. Face the belay plate. Take the rope that is on the side of the belay plate away from the "climber" in your dominant hand (the right hand for right-handed people). This is your ***braking hand.*** **You must not take this hand off the rope when the person is "on belay" and the belay is not tied off.**

D2. Take the rope on the side of the brake plate that is nearest the climber in your weaker hand (the left hand for right-handed people). This is your ***guide hand.*** You will use it to help manipulate the rope.

D3. To stop the rope from running out, grasp the rope with both hands and spread the hands apart until the rope forms a 180° angle. Note that this forces the brake plate against the carabiner, causing friction on the rope. At the same time, hold the rope tight with your brake hand.

D4. Now, have a partner wearing a seat harness clip into the end of the rope.

D5. Remove slack in the rope between the brake plate and the climber by holding the rope strands at a small angle and pulling the slack out with the brake hand.

D6. Begin the belay voice communications. *(Person on the rope: "On Belay."* If the belayer is ready, he says, **"Belay On."**)

Fig. 8.10 Belay Plate/Level Ground

Fig. 8.10(a)

Fig. 8.10(b)

Fig. 8.10(c)

Fig. 8.10(d)

Fig. 8.10(e)

Fig. 8.10(f)

D7. Have the person attached to the rope begin slowly walking backwards. Keep the rope strands at the brake plate in a small angle. This will allow the rope to feed through the brake plate. With the brake hand, control the rope's run into the brake plate. If the plate jams against the carabiner and slows the rope when you do not want it to, take the guide hand and pull the brake plate away from the carabiner. **ALWAYS KEEP THE BRAKE HAND ON THE ROPE.**

D8. Now, initiate a braking action by spreading the angle of the rope strands apart to 180° and holding the rope firmly in the brake hand.

D9. In an area where he will not slip, have the "climber" firmly plant his feet on the ground and lean back from the rope. Hold the rope firmly with the brake hand.

D10. To get a feel of control, slowly let out a few inches of rope with the "climber" leaning against the rope.

Fig. 8.10(g)

Reversing Direction

D11. Now have the partner walk slowly towards you. The problem now is to take in rope while keeping the person safely on belay. To practice control of rope tension, do not allow the rope between the climber and the belay plate to touch the ground.

D12. Have your brake hand on the rope about 1 foot from the brake plate.

D13. Place your guide hand about 3 feet from the brake plate.

D14. As slack appears in the rope, pull the slack out by pulling the rope with the brake hand. Your guide hand can help move rope towards the brake plate.

D15. Just as your guide hand approaches your brake hand, grasp both sides of the rope with your guide hand.

D16. Holding both lines taut with the guide hand, move your brake hand quickly up the rope toward the brake plate **WITHOUT TAKING THE HAND OFF THE ROPE.**

WARNING NOTE

The moving of hand positions on the rope is one of the most critical operations in belaying. Remember, the person at the end of the rope may fall *at any time* while you are belaying. If you do not have your brake hand on the rope at all times, ready to grasp it in an instant, you may drop the person and cause severe injury or death.

D17. Continue the procedure until the person on the rope reaches your position.

D18. Exchange the voice signals that conclude the belay cycle. (*Person on rope: "Off Belay."* Belayer: **"Belay Off."**)

Practice with a Belay Weight/Dummy

Now practice the same operation using a belay weight or dummy as shown in Figure 8.2. Do a procedure first with the weight or dummy falling as it is being raised and then with the weight or dummy falling as it is being lowered. In each case, as the weight comes onto your belay line, lower it to the ground with the brake plate.

WARNING NOTE

When the weight falls, and the belay catches, the belay rope will come taut with great force. To avoid injury, the belayer, and others around the belayer must take the following precautions:

- **Be aware of the position the rope and belay device will take when it catches and stay out of its path. Otherwise, the impact of the rope and/or belay device can cause serious injury;**
- **Keep fingers, hair, and clothing free of the rope near the belay device. Otherwise, these could be swept into the belay device possibly causing injury and impeding the belay.**

In an actual belay situation, you would try to keep the "climber" in view as you belay him. But for practice, you will face away from the weight/dummy so you cannot visually anticipate the fall. This will sharpen your skills at catching an unexpected fall.

Belay Plates in Figure 8 Descenders

Some Figure 8 descenders are designed with a belay plate either in the small ring or between the two rings. However, you should use 8 belay plates with caution for the following reasons:

- The slot in the Figure 8 may not be the correct size for the rope you are using.
- Some Figure 8s have slots that are not well-designed for use as a belay plate.
- A Figure 8 is not as well-balanced and may not be as easy to use as a belay plate.

WARNING NOTE

***DO NOT* use the Figure 8 descender as a belay device with the rope wrapped in the large ring as for rappeling or lowering. This large ring is not designed with enough friction to stop a rope that is shock-loaded from a fall.**

Additional Cautions for Belayers

Belay Direction

The main elements of the belay system, the anchor, belay device, and climber must be in as direct a line as possible, so the instant the climber falls, the force will come directly onto the belay device and the anchor. If these elements are not in a direct line, any or all of the following could happen:

- The belay device could fail to work properly.
- The belayer could be thrown off position.
- The anchor could fail.
- The system could be shock-loaded.

Also, the belay rope must not be around or against the belayer or any other person. They could be injured by the rope's suddenly coming taut.

MAINTAINING PROPER SLACK IN THE BELAY ROPE

It is critical that the belayer maintain a proper amount of slack in the belay rope. If the rope is too taut, it can interfere with the climber (if there is a climber on the rope) or with the brakeman (if the belay is a safety for a separate lowering system). Good judgement on belay rope slack is another skill that develops with practice at belaying. There should be at least some **visible** slack in the rope, but not so much slack that there would be intense shock loading of the rope during a fall.

Securing the Belayer

If the belayer is near a place where he could fall, he must be secured via a safety line to an anchor.

This safety line should not be interconnected to the belay system, but via a separate safety line to an anchor. This is to ensure that the belayer himself is not endangered by a climber's fall and that whatever happens, he remains in a stable situation to continue the belay or otherwise assist the climber.

Bottom Belay *(Only for Rappeling)*

A bottom belay is a pull on the rappel rope from the bottom *(see Figure 8.11)*. A common use of the bottom belay is to assist a rappeler who is in danger of losing control. It is, in essence, a substitute for the rappeler's control hand. This pull from the bottom increases friction on the rappeler's descender.

Fig. 8.11 Bottom Belay

A bottom belay is not a substitute for the belays previously described. It should be used when a top belay is not available and in an emergency. Bottom belays have the following drawbacks:

- The rope can easily slip out of the grip of the person at the bottom.
- The belayer can only exert as much pressure as his body weight. This is often sufficient if the belayer is directly below the rappeler and the rappeler has not gained too much momentum out of control. But it is often not effective if applied from an angle.
- A bottom belay is not effective with all rappel devices.
- A person doing a bottom belay is in danger of being hit by objects, such as rocks, being dislodged by the person or rope above.
- A bottom belay does not provide backup for the failure of the main line rope, anchor, or rappel device, only for an out-of-control rappel.

Body Belays

One additional technique for belaying is known as the body belay. With this technique, the belayer creates friction by running the rope around his body, usually around the waist. Except in emergencies, the technique is not recommended for the following reasons:

- It is not as easy to stop a fall as with a belay device.
- It can injure the belayer.
- The belayer's ability to hold a fall is only as high as his pain threshold.
- If the belayer is at the top of a drop, it can pull him over.
- It can entangle the belayer in the rope. If the climber falls, it may entrap the belayer and put him in a position where he is unable to assist the climber.

QUESTIONS for REVIEW

1. List the four elements of a belay system.
2. Describe six situations that might require a belay.
3. Describe how a belay differs from a lowering in the following: a) purpose, b) technique, and c) equipment.
4. With you playing the role of a belayer and another person playing the role of a climber, recite from memory the cycle of belay voice communications. Reverse roles.
5. List three reasons why belay plates on Figure 8 Decenders may pose problems.
6. List the conditions in which a bottom belay should be used.

★ ACTIVITIES ★

1. On flat ground, rig a Munter Hitch belay system and, using a partner tied into the rope, have him walk away from you in a belay cycle.
2. Repeat the above with the person approaching you.
3. Repeat 1 and 2 using a belay plate.
4. Using a belay dummy/weight fall, rig a Munter Hitch belay and simulate a situation where a climber falls as the rope is being let out.
5. Using a belay dummy/weight fall, rig a Munter Hitch belay and simulate a situation where a climber falls as rope is being taken in.
6. Using a belay dummy/weight fall, repeat 4 and 5 using a belay plate system.

WARNING NOTE

When the weight falls, and the belay catches, the belay rope will come taut with great force. To avoid injury, the belayer, and others around the belayer must take the following precautions:

- **Be aware of the position the rope and belay device will take when it catches and stay out of its path. Otherwise, the impact of the rope and/or belay device can cause serious injury.**
- **Keep fingers, hair, and clothing free of the rope near the belay device. Otherwise, these could be swept into the belay device, possibly causing injury and impeding the belay.**

Chapter 9

Rappeling

PREREQUISITES

Before attempting the activities described in this chapter, you must have demonstrated that you can properly.

1) Use and care for rope.
2) Use and care for other equipment employed in the high angle environment.
3) Tie correctly, confidently, and without hesitation the eight knots described in Chapter 6.
4) Apply the principles of anchoring and rig a safe and secure anchor.
5) Apply the principles of belaying and can safely and confidently belay another person using either a Munter Hitch or belay plate.

OBJECTIVES–

At the completion of this chapter, you should be able to:

1. Describe the purposes of rappeling.
2. Describe the necessity for maintaining control during a rappel.
3. Discuss the principles behind the body rappel, along with its dangers and limitations.
4. Discuss the principles behind the arm rappel, along with its dangers and limitations.
5. Set a safe and secure anchor for rappeling.
6. Safely belay a person who is rappeling and interact with a person belaying you while you are rappeling.
7. Discuss the principles behind rappeling with Figure 8 Descenders.
8. Describe the advantages and disadvantages of using a Figure 8 with Ears.
9. Safely and under control, rappel with a Figure 8 descender and be able to securely lock off the device while on rope, and then return to a rappel.
10. Discuss the principles behind the use of a Brake Bar Rack.
11. Safely and under control, rappel with a Brake Bar Rack, securely lock off the device while on rope, and then return to a rappel.
12. Discuss the principles behind self-belay techniques.

TERMS—*that a High Angle Rope Technician should know:*

Arm Rappel—(Guide's Rappel) A type of rappel in which the rope wraps around both outstretched arms and across the person's back. The technique is better suited for sloping terrain than for vertical situations.

Body Rappel—(Dulfersitz Rappel)—A type of rappel that uses the body as friction by running the rope through the legs, across one hip, over the opposite shoulder, and to a braking hand. Because of the discomfort involved and the potential damage to body parts, the technique has largely been supplanted by other techniques.

Brake Bar Rack—A rappel device that consists of a series of short metal bars fixed to, and sliding along, a "U"-shaped metal rack with an eye at one end for attachment.

Brake Hand—The hand, usually the dominant one, that grasps the rope to help control the speed of descent during a rappel.

Carabiner Wrap—A rappel technique that uses several rope wraps around a seat harness carabiner to create friction and control the descent. It is generally not considered a safe and secure technique for rappeling.

Descender—A rappel device that creates friction by a rope running through it and is attached to a rappeler to control descent on a rope. Most descenders can also be used as a fixed brake lowering device.

Dulfersitz—See **Body Rappel.**

Figure 8 Descender—A commonly used descender that is made roughly in the shape of an 8.

Guide Hand—The hand, usually not the dominant one, that cradles the rope to help in balancing the rappeler.

Locking Off—The technique of jamming a rope into a descender or tying off securely so that the rappeler can stop the descent and operate hands free of the rope.

Rappeling—The controlled descent of a rope using the friction of the rope against one's body or through a descender.

Rappeling is controlled descent on a rope by using the friction of the rope against one's body or through a descender. It is a skill that is essential for a person to operate comfortably in the vertical environment. The learning of safe and controlled rappeling skills is a step toward developing vertical competency. Rappeling may appear to the inexperienced to be a spectacular act. However, just to be able to rappel does not necessarily mean that a person is skilled and knowledgeable in the vertical environment. Rappeling is not the ultimate goal for a high angle technician, but one personal skill to be used in combination WITH OTHER skills for activities in the vertical environment. Rappeling, for example, may be the means of travel in a controlled descent of a vertical face. And it may be used in combination with other essential vertical skills for performing a rescue.

One important thing that indicates a person's competence in rappeling is **control**:

- Being able to control the descent with a minimum amount of physical effort.
- Rappeling in a controlled manner so that the rope is not damaged by heat buildup in the rappel device and the anchors are not damaged by shock loading.
- Being able to stop the rappel at any time.
- Being able to tie off securely and operate hands free of the rope and rappel device.
- Being able to operate in any body position including upside down.

How Rappeling Works

Any technique for rappeling uses friction with the rope to slow the rate of descent. Both the **body rappel** and the **arm rappel** use friction of the rope on the body to slow the descent (both of these techniques are described below). This friction and the resulting heat and discomfort are what make these techniques unpleasant to use. Because of this discomfort (and because of the potential dangers associated with them) these techniques are no longer widely used.

Most modern techniques use rappel **devices** or **descenders.** They are attached to the rappeler, usually by means of a seat harness carabiner. The rope runs through the device to create friction, so the heat and discomfort go into the device and not the rappeler's body. Also, a well-designed rappel device offers more control of the descent than does a body rappel.

With most rappel devices, the rate of descent is controlled by pulling down on the portion of the rope that is below the device. This increases friction by increasing rope tension and pressure on metal parts of the device. This controlling action is usually done with the rappeler's dominant hand (usually the right hand on a right-handed person). This hand is known as the **brake hand.**

The other hand cradles the rope above the device or (with some descenders) the device itself. THIS HAND DOES NOT SUPPORT BODY WEIGHT BY GRASPING THE ROPE. In most rappel devices, this hand, known as the **guide hand,** helps to balance the rappeler. It may also assist in controlling the descender (in some rappel devices).

WARNING NOTE

It is essential that from the beginning, anyone learning to rappel maintain absolute control of the descent. Among the ways of ensuring this control are:

- **Avoiding rapid, bouncing rappels that can lead to loss of control, damaged rope, overloaded anchors, and injuries.**
- **Using a top belay.**
- **Learning under the guidance of a qualified and experienced instructor.**

TYPES OF RAPPELS

Arm Rappel

WARNING NOTE

The arm rappel is *only* for use on short, low-angle slopes. It does not produce enough friction to adequately control full body weight in a completely vertical situation. Because of the potential of rope abrasion injuries on arms and hands, it must be used only when wearing long-sleeved shirts and gloves.

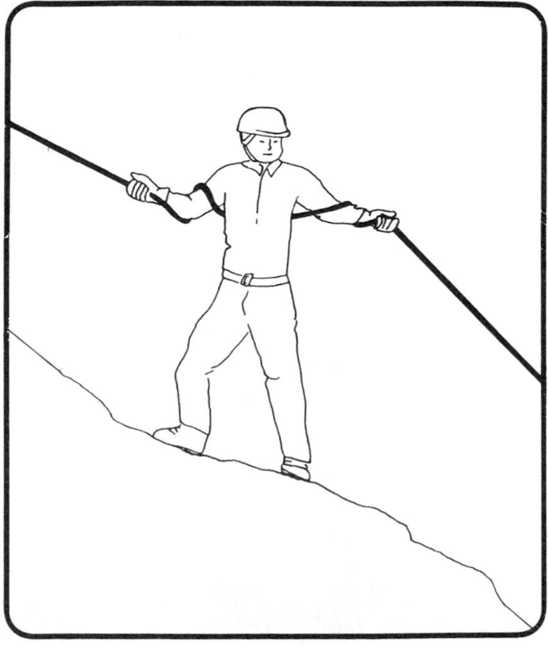

Fig. 9.1 Arm Rappel

Figure 9.1 shows the **arm rappel,** which is sometimes used for short distances on low angle slopes.

The user sets up the rappel by having his upper back to the anchored rope while he faces uphill. He then wraps both extended arms around the rope. The friction, and rate of descent, is controlled by varying hand grip.

Body Rappel *("Dulfersitz")*

WARNING NOTE

The body rappel must be attempted on a low-angle slope before it is attempted on a steep slope. The body rappel should not be used in general practice, but only in case of emergency, when no hardware is available to use as a rappel device.

The body rappel technique presents two potential dangers:

- **The rope could become unwrapped from the leg. As a result, the rappel would lose friction with the rope and possibly fall free to the ground. This is a particular danger on a vertical face.**
- **Rope abrasion and pressure damage body parts, particularly at the crotch and shoulder. Thick padding in these areas is recommended.**

Figures 9.2(a) through 9.2(c) illustrate a procedure for wrapping the rope for a body rappel.

1. The rappeler straddles an anchored rope facing the anchor.
2. He brings the strand of rope that is below him around one hip (the side with the dominant arm).
3. He then brings the rope across the chest and over the opposite shoulder.
4. He now brings the rope down across the back to the braking hand (the dominant hand), which is on the same side as the hip where the rope runs.
5. Control of the descent is maintained by (dominant) hand strength and by bringing the rope across the chest with the brake hand.

As mentioned earlier, one of the dangers of the body rappel is on high angle rappels where the rope can become unwrapped from the leg. For this reason, the wrapped leg must be kept lower than the unwrapped leg and the upper body.

Fig. 9.2 Body Rappel

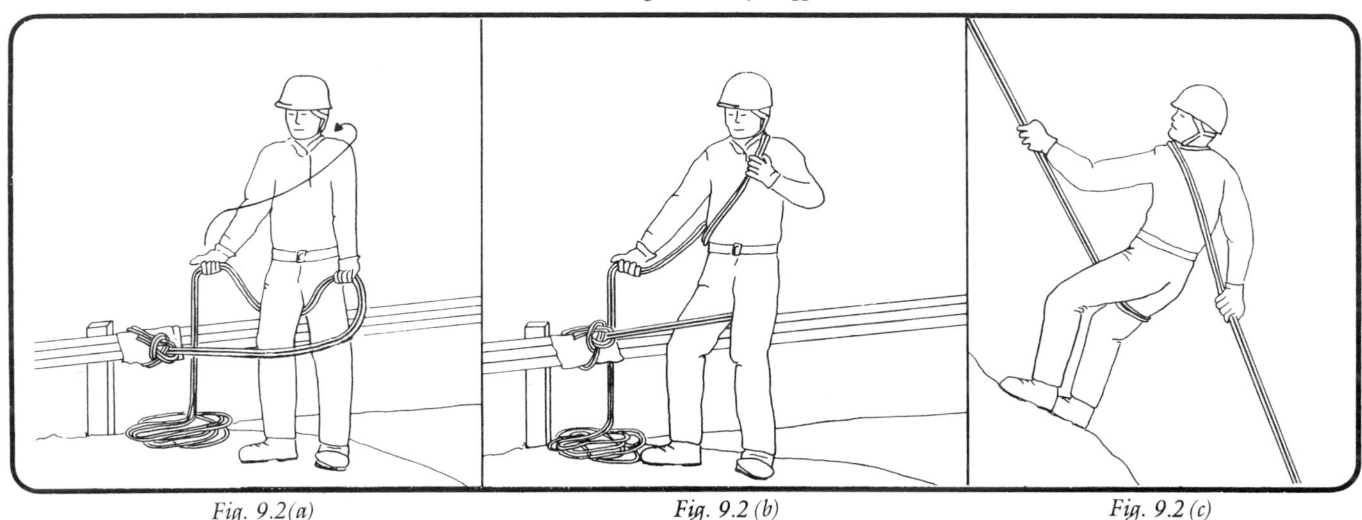

Fig. 9.2(a) *Fig. 9.2 (b)* *Fig. 9.2 (c)*

FIGURE 8s

One design of a conventional Figure 8 descender is shown in Figure 9.3. (For more details on materials and the various designs, see Chapter 5, "Basic Vertical Hardware.") Depending on the specific design, the larger ring of the Figure 8 may be rounded or squared off. The larger ring is the portion of the descender that creates friction on the rope, while the smaller ring is attached to a seat harness carabiner.

Most of the conventional Figure 8 descenders are found in smaller sizes. Because of their compactness and light weight, these types are often used by climbers or other persons involved in recreational activities.

While it is a strong device for rappeling, the conventional Figure 8 has two main drawbacks:

- You cannot use the smaller version with larger diameter rope, and you may find it difficult to use the smaller Figure 8 descender "double wrapped" ("double wrapping" of a Figure 8 descender is explained on page 103 below).
- When you are rappeling with the conventional Figure 8, it is possible for the rope wraps to slip up and around the larger ring to form a girth hitch *(see Figure 9.4).*

Fig. 9.3 Regular Figure 8 Descender

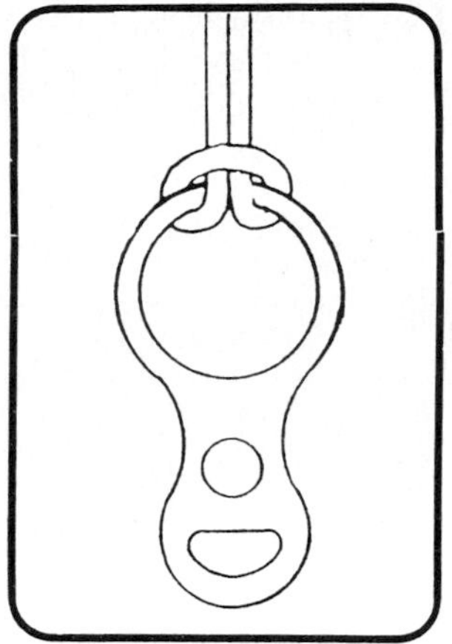

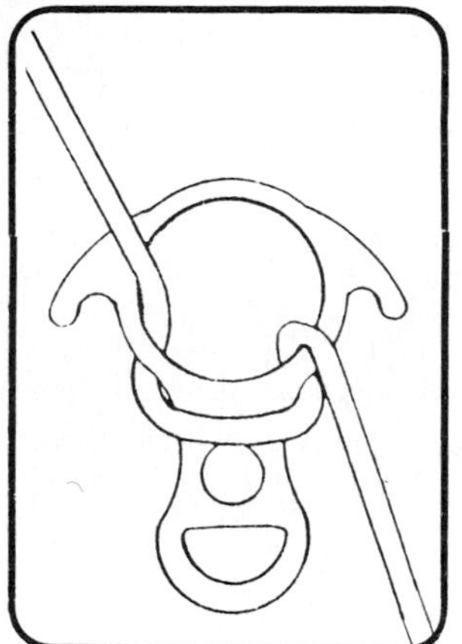

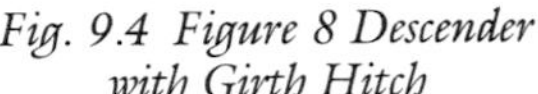

Fig. 9.4 Figure 8 Descender with Girth Hitch

Fig. 9.5 Lacing Rope Into Figure 8 From Top

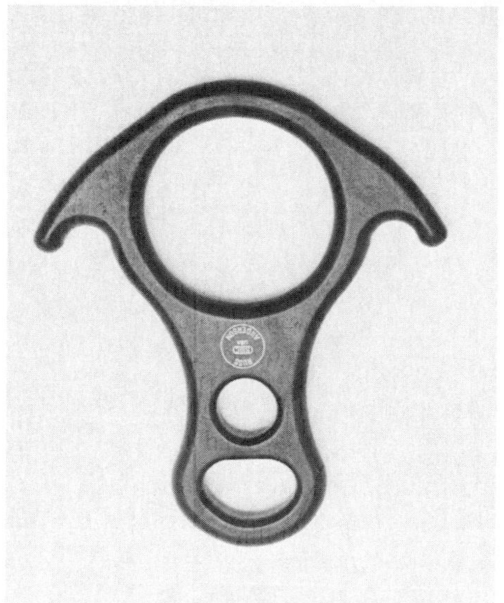

Fig. 9.6 Figure 8 Descender with Ears

The Girth Hitch Problem

Accidental girth hitching of the Figure 8 ring tends to occur in two circumstances:

- When you attempt to ease over a difficult edge or ledge and the edge catches a wrap of rope on the bottom of the Figure 8. The weight of the rappeler forces the wrap over the top of the ring and into a girth hitch. To help prevent this kind of occurrence, it is best to lace the rope onto the descender by first bringing the bight of rope into the large ring from the top, as shown in Figure 9.5.
- When you cause momentary slack in the rope, such as rappeling over ledges, over uneven faces, or in bouncing rappels. These situations cause a momentary slack in the rope wrap, allowing it to slip over the large ring.

The girth hitching of the Figure 8 immediately stops the rappel and prevents you from descending further. You will begin to move again only when you can remove your weight from the girth hitch long enough to slide it back over the large ring. This is a difficult position for you to get out of unless your possess the right skills and equipment. If you have good upper body strength, you might be able to forcefully extricate yourself from this situation. But it would be easier and usually safer for you to set an ascender or Prusik knot on the rope above the Figure 8 and step into an attached sling. (see Chapter 10, "Ascending," for possible ways of extricating yourself from this kind of situation.)

But probably the best solution to the girth hitch problem on a Figure 8 ascender is to prevent it from happening. One method of prevention is the use of a Figure 8 with Ears *(see Figure 9.6)*. The "ears," which are protrusions on the larger ring, primarily serve to prevent the rope from slipping over the larger ring to form a girth hitch. But the ears also serve other functions, such as helping to contour the rope around the ring and holding the rope in place when it is locked off.

Consequently, unless space and weight are strong considerations, the Figure 8 with Ears is preferable for most rappelers.

Procedure for Using the Figure 8 With Ears

WARNING NOTE

1. The learning and practice of rappeling techniques must be under the guidance of a qualified instructor.

2. Rappeling techniques must be practiced first on level ground and then on short and moderate slopes before using them on a steep face.

3. All experienced personnel using rappel techniques on a steep face or any other area where a severe fall could result must use a top belay.

A. **On Level Ground** *(see Figure 9.7)*

A1. Don a sewn, manufactured seat harness with thigh supports and a front tie-in point.

A2. Clip a large locking carabiner or a quick link into the seat harness tie-in point. If the carabiner or quick link is in a vertical plane after being clipped in, make certain that the gate is toward your body. (This is to help prevent friction against a vertical face from opening the gate.)

A3. Establish a secure anchor point.

A4. Firmly attach a main line rappel rope to the anchor point.

A5. Take a Figure 8 with Ears in your guide hand *(9.7a)*.

Fig. 9.7 Using Figure 8 on Level Ground

Fig. 9.7(a)

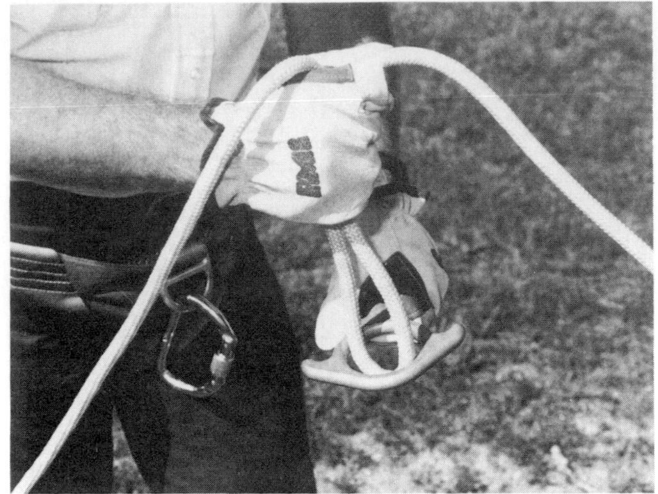

Fig. 9.7(b)

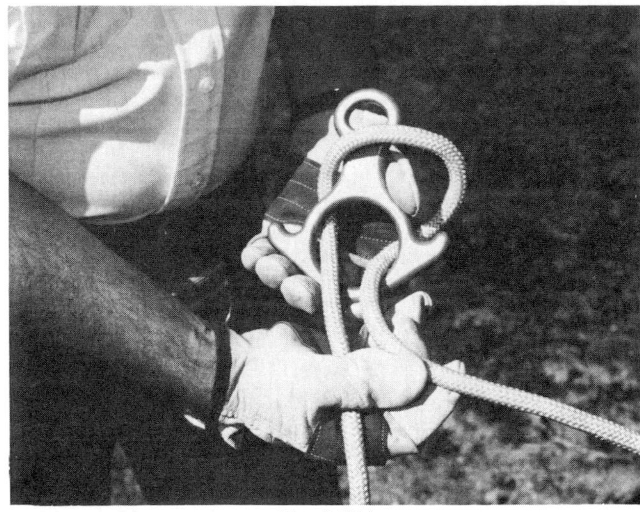

Fig. 9.7(c)

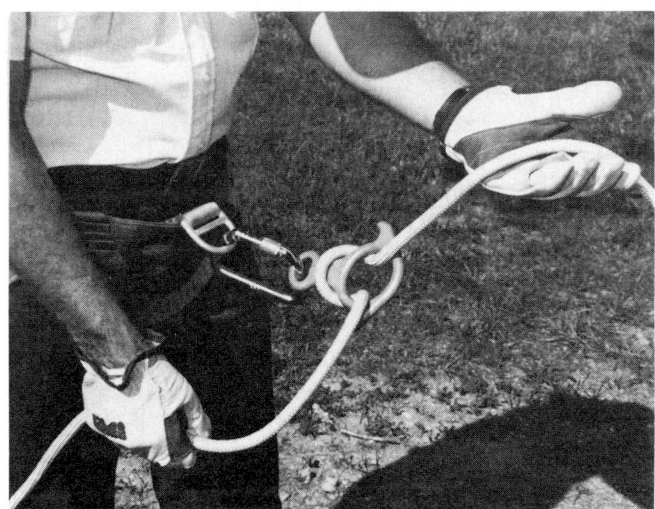

Fig. 9.7(d)

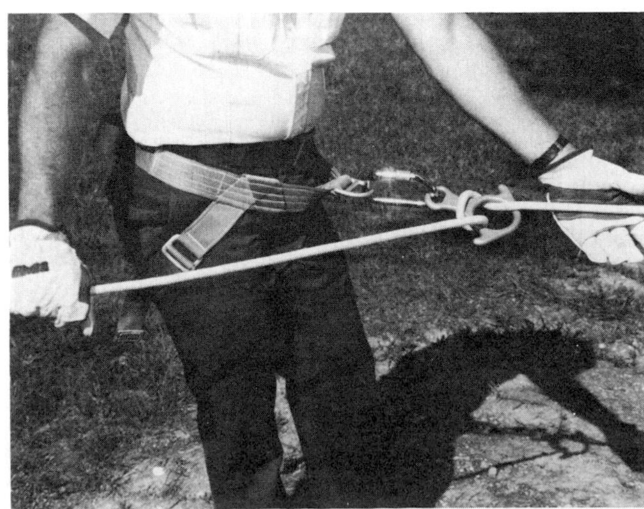

Fig. 9.7(e)

Fig. 9.7(f)

A6. Face the anchor with the rope running past you on the brake hand side.

A7. On the rappel main line near the anchor, take a bight of rope in your brake hand and push it through the large ring of the Figure 8 descender **from the the top** *(9.7b).*

A8. Bring the bight of rope around the end of the small ring of the Figure 8 descender and across the waist of the rappel device. Pull the rope snugly around the Figure 8 descender by pulling on the strand of rope that is on the side of the descender away from the anchor (running end). If the Figure 8 descender will be in a horizontal plane when it is clipped into your seat harness, the rope should run across the **top** of the waist of the Figure 8 descender. If the Figure 8 descender will be in a vertical plane once it is clipped into your seat harness, the rope should lay across the side of the descender that is near your brake hand *(9.7e).*

NOTE: When you have to slide over a difficult edge to rappel, and if you get the Figure 8 descender caught on the edge, the rope is on the side of the descender away from the edge, and will be less likely to jam on the edge. Also if you are using a conventional Figure 8 descender and you keep the rope on the side away from the edge, the edge is not as likely to push the rope up over the larger ring to form a girth hitch.

A9. Clip the small ring of the Figure 8 descender into your seat harness carabiner. Lock the carabiner gate *(9.7d).*

A10. Take the rope that is on the lower side of the Figure 8 descender in your dominant hand (the right hand on right-handed people). This is your **brake hand.** When you are rappeling and the descender is not **securely** tied off, you must NEVER TAKE YOUR BRAKE HAND OFF THE ROPE.

A11. With your less dominant hand (the left hand for right-handed people), lightly cradle the rope above the Figure 8 descender. This is your **guide hand.** This hand is **not** for supporting your weight, but to help balance yourself. YOU MUST NOT SUPPORT YOUR WEIGHT WITH THE GUIDE HAND, or it could throw you off balance when you try to rappel.

A12. Now, by pulling down on the rope with your brake hand, pull the slack out of the rope between the Figure 8 descender and the anchor. If this is difficult, you can assist the process with your guide hand *(9.7e).*

A13. Grasp the rope below the Figure 8 descender with your brake hand and pull it taut against your hip with your hand about six inches below your hip. This is the position when you need extra stopping power by using the friction of the rope against your hip. However, you should not constantly keep the rope against your hip since it may abrade your seat harness webbing. So, while keeping the rope taut with your brake hand, swing the rope out about 2 feet from your hip at an angle that is comfortable for your arm.

A14. Lean back away from the anchor so that the rope between the Figure 8 descender and the anchor becomes taut. Always remove the rope slack between your descender and the anchor before beginning a rappel.

NOW IS THE TIME FOR A SAFETY CHECK. Have your instructor do a visual check of the rigging, carabiners, your descender, gloves, etc.

A15. As you lean back against the rope, begin to walk backwards, letting rope slowly through the Figure 8 descender with the brake hand, and holding your guide hand lightly on the rope above the rappel device. As you let the rope slip through your brake hand, keep the same distance on the rope between your hand and the Figure 8 descender *(9.7f).*

Locking Off *(see Figure 9.8)*

A16. To lock off the Figure 8 descender, have the brake hand allow rope to slide through the descender until the brake hand is about 1 foot from the rappel device *(9.8a).*

A17. Now, **hold the rope taut with the brake hand.** In a continuous, smooth motion, use the brake hand to pull the rope in an arc from the rappel position, straight out in front of you, passing below the main line, and then to where the rope and your hand are 180 degrees from the rappel position *(9.8b).*

Fig. 9.8 Locking Off Figure 8 Descender

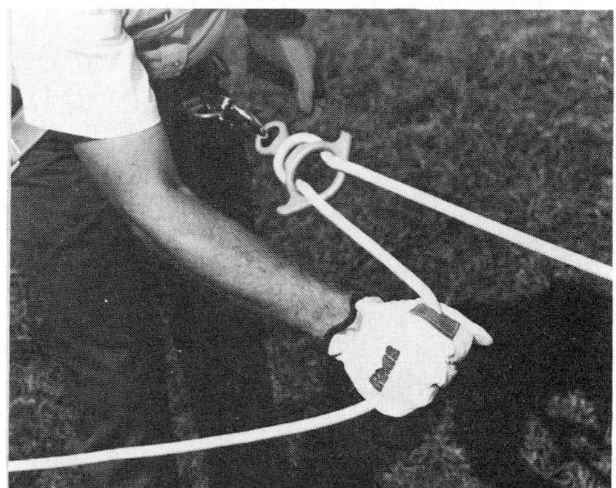

Fig. 9.8(a)

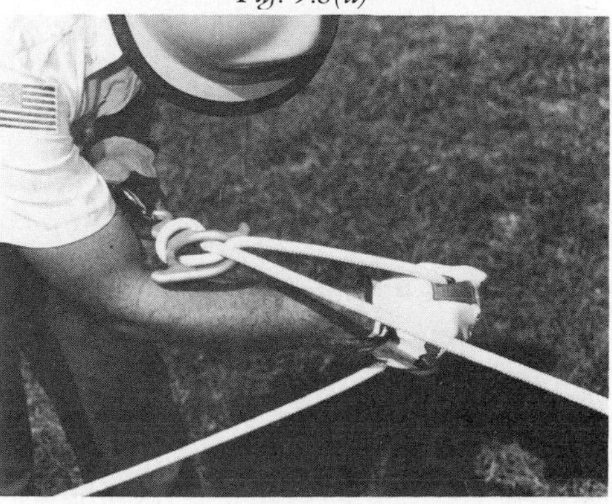

Fig. 9.8(b)

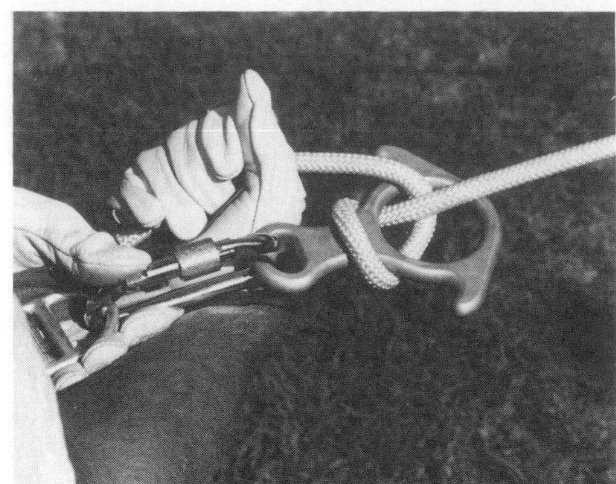

Fig. 9.8(c)

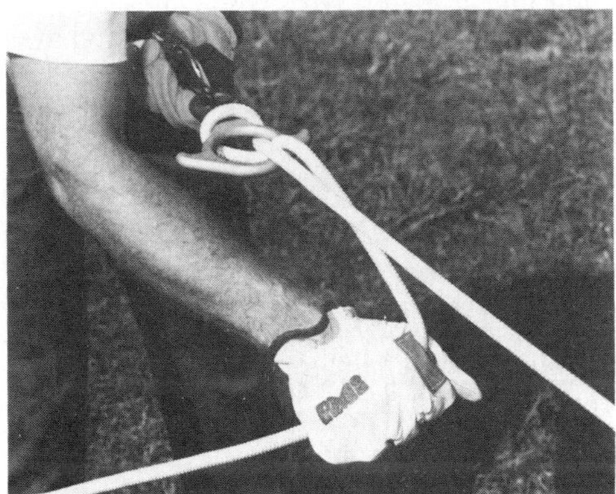

Fig. 9.8 (d)

NOTE: This may be a little difficult to do while standing on flat ground. Once you are suspended by the rope on a steep slope or on a vertical face, this movement is much less awkward to perform. This has to do with the the angle of the rope and anchor in relation to the Figure 8 and the rappeler.

A18. **Maintain a firm grip on the rope with your brake hand.** With the brake hand, take the strand of rope it is holding and pull it down farther towards you to trap it between the strand of rope that goes to the anchor and the large ring of the Figure 8 descender *(9.8c).*

CAUTION: The braking side of the rope must be firmly trapped between the rope that goes out of the descender to the anchor and the large ring of the Figure 8. For this to happen, there must be tension on the rope between the Figure 8 descender and the anchor.

Unlocking

A19. Take the rope **firmly in your brake hand.** In a smooth, continuous motion, pull the rope first straight towards the anchor and in an arc back to the rappel position. You will hear a slight "pop" and feel a slight bump as the rope unlocks from the rappel device. **Keep the rope firmly in your brake hand.** Continue rappeling as before *(9.8d).*

B. Rappeling with the Figure 8 Descender on a Slope *(see Figure 9.9.)*

B1. Establish a secure anchor point at the top of a short slope of about 45 degrees (if a slope is not available, then use a stairway). Attach the rappel rope securely to the anchor point.

B2. Establish an anchor point for a belay. Attach an anchor sling securely into the anchor point. Clip a large locking carabiner into the end of the anchor sling.

B3. Have a belayer take position. If a Munter Hitch is being used for belay, have the belayer tie it into the belay rope and clip it into the belay carabiner. If the belayer is using a belay plate, attach it to the rope and clip it into the belay anchor carabiner. Be sure the carabiner gate is locked.

NOTE: (Keeping a proper distance between the belay anchor rope and the main line rappel rope.) As noted elsewhere in this manual, it is good practice to keep rope strands, such as the belay line and the main line rappel rope, apart, for the following reasons:

- To prevent tangling.
- To prevent damage to the rope from heat fusion as a result of rope cross.

However, the distance between the anchors should not be too far. If there were a great distance between the anchors and the main line rappel anchor failed, but the belay caught, there would be a possibility of a **pendulum fall** (a sudden swing on the line that could result in injury to the rappeler).

B4. Clip into the belay rope. Initiate the belay cycle with the belay voice communications. (Rappeler: **"On Belay."** Belayer: ***"Belay On."*** *[9.9a]*).

B5. At a secure point where you are not in danger of falling, follow steps A5 to A14 above to lace the Figure 8 descender onto the rope *(9.9b & c).*

DO A SAFETY CHECK ON ALL EQUIPMENT AND RIGGING. In particular, inspect carabiners for sideloaded gates and unlocked gates and the anchors for slipped anchor slings. Make certain that loose clothing is tucked in, hair is not in danger of being caught, and the helmet chin strap is secure.

B6. Begin to back down the slope, controlling your descent with the brake hand and following the steps outlined in A15 through A18 above. Keep the following principles in mind *(9.9d)*:

a) Keep your body generally perpendicular to the slope.

b) Keep your feet apart about the width of your shoulders.

c) Keep your knees relaxed and slightly flexed.

d) Take slow and deliberate steps backwards.

e) Keep your body slightly turned in the direction of the brake hand, looking downslope to select a path of travel.

f) Use your guide hand for balance. DO NOT SUPPORT YOUR WEIGHT WITH THE GUIDE HAND.

g) NEVER TAKE THE BRAKE HAND OFF THE ROPE UNLESS IT IS SECURELY LOCKED OFF.

B7. When part way down the slope, stop the rappel and lock off the Figure 8 descender using the principles described in steps A16 through A19 above *(9.9e, f & g)*.

B8. Unlock and continue the rappel to the end of the slope *(9.9h)*.

B9. Conclude the belay cycle with the belay voice communications. (Rappeler: **"Off Belay."** Belayer: ***"Belay Off."***)

Fig. 9.9 Using Figure 8 on Slope

Fig. 9.9(a)

Fig. 9.9(d)

Fig. 9.9(b)

Fig. 9.9(e)

Fig. 9.9(c)

Fig. 9.9(f)

Fig. 9.9(a)

Fig. 9.9(b)

C. **Rappeling Down A Vertical Face with Figure 8 Descender** *(see Figure 9.10)*

WARNING NOTE

1. The learning and practice of rappeling techniques must be under the guidance of a qualified instructor.

2. Rappeling techniques must be practiced first on level ground and then on short and moderate slopes before using them on a steep face.

3. All personnel learning rappel techniques on a steep face or any other area where a severe fall could result must use a top belay.

C1. Choose a short vertical face (approximately 20 feet) where the top breaks over gradually into a steep face. On the first try, do not choose a face with a sharp edge.

C2. Establish a secure anchor point safely back from the edge. If possible, have the anchor point high above the edge. This will assist any rappeler going over the edge. Attach the main line rappel rope securely to the anchor point.

C3. Establish a separate anchor point for a belay. Attach a sling securely to the anchor point. The anchor point and sling should be established so that when the belayer takes position, he has a good field of view of the top and face, but is not in danger of falling over the edge. In the end of the sling, clip a large locking carabiner.

C4. Have a belayer tie into a safety line and take position. If a Munter Hitch is being used for belay, have the belayer tie the Munter Hitch into the belay rope and clip it into the belay carabiner. If the belayer is using a belay plate, attach it to the rope and clip it into the belay anchor carabiner.

C5. Clip into the belay rope. Initiate the belay cycle with the belay voice communications. (Rappeler: **"On Belay."** Belayer: ***"Belay On."*** *[9.10a]*).

C6. At a secure point, where you are in no danger of falling, follow steps A7 through A14 above for attaching the Figure 8 descender to the rope *(9.10b, c & d)*.

C7. Make certain that the slack is out of the rope between the Figure 8 descender and the anchor.

Do a SAFETY CHECK on all connectors such as carabiners, on the seat harness, on the anchor, and other rigging. Make certain that no loose clothing or hair is going to be sucked into the descender, and make certain your helmet is secure.

Fig. 9.10 Using Figure 8 on Vertical

Fig. 9.10(a)

Fig. 9.10(b)

C8. Slowly begin backing to the edge. Keep the following principles in mind:

a. **Keep the body generally perpendicular to the slope.** This means that you will have to deliberately lean out as the slope becomes vertical. This may seem to be an unnatural stance at first, but is necessary to keep your feet from slipping out from under you *(9.10e & f)*.

b. Keep the feet apart about shoulder width for balance. This will help keep you from being pulled over to one side.

c. Keep the knees relaxed and slightly flexed.

d. Take slow and deliberate steps backwards.

e. Keep the body slightly turned in the direction of the brake hand, looking down slope to pick a path for your descent.

f. Use the guide hand for balance. DO NOT SUPPORT YOUR WEIGHT WITH YOUR GUIDE HAND.

g. NEVER TAKE THE BRAKE HAND OFF THE ROPE UNLESS YOU ARE **SECURELY** LOCKED OFF.

Fig. 9.10(c)

Fig.9.10(d)

As it is normally used, the belay is for the safety of the rappeler and IS NOT to be used by the belayer to control the rate of descent of the rappeler nor to share the load when the rappeler is in a controlled descent. For a student to learn the proper control of a rappel device, he must be controlling his full weight with no control coming from the belayer. Therefore, it is important that the belay line have a small amount of slack as the rappeler descends.

The belayer controls the rappeler's weight and/or rate of descent only if the rappeler loses control or requests assistance (such as "tension").

Fig. 9. 10(e)

Fig. 9.10(f)

Fig. 9.10(g)

Fig. 9.10(h)

Fig. 9.10(i)

C9. As you move over the edge, you will note that your weight comes more onto the rope. This means you need greater effort for control with the brake hand. As your weight comes onto the rope, if you feel pulled in a direction right or left, step slightly in that direction until you feel a better balance *(9.10g).*

C10. If you slip, fall over, or even turn upside down, KEEP CALM. HOLD TIGHT WITH YOUR BRAKE HAND UNTIL YOU ORIENT YOURSELF. Then, slowly, place your feet against the face and rebalance yourself. DO NOT HOLD YOUR WEIGHT WITH YOUR GUIDE HAND, but use it to balance yourself.

C11. When you are midway down the face, stop the rappel and lock off. On a vertical face, with your weight on the rope and descender, you will find it more difficult to trap the rope between the main line and the large ring of the Figure 8 descender. But hold the brake line steady and pull it across and down with deliberate force until it is securely trapped *(9.10h & i).*

A More Secure Lockoff for The Figure 8 Descender

In certain vertical situations a more secure lockoff of the Figure 8 descender may be desireable. Such situations include when the rappeler may have to be locked off for long periods of time, when he has to do a great deal of moving about in one position to manipulate equipment or a rescue subject, or any other circumstance where he feels he needs greater security.

Figure 9.11 illustrates one technique for more securely locking off a large Figure 8 descender.

a. Trap the brake side of the rope between the line going to the anchor and the large ring on the descender.

b. Pull the brake side of the rope firmly down toward the seat harness carabiner, across the surface of the Figure 8, and around BEHIND the ears. DO NOT BRING THE ROPE THROUGH THE LARGE RING OF THE FIGURE 8. It should be between the line going to the anchor and the large ring, and above the line first locked off. Make certain that the rope lays firmly around the device and there is no slack.

c. Bring the brake side of the rope down and around the Figure 8 again as in (b.), and then behind one ear, but do not place it between the line going to the anchor and the large ring. Instead, form a large bight of rope from the brake side of the rope.

d. Bring the bight up parallel with the rope going to the anchor.

e. Tie an overhand knot with the bight onto the rope going to the anchor.

f. Be certain that the overhand knot is contoured well and there is no slack in the knot.

To unlock, untie the overhand knot and unwrap the brake side of the rope from around the Figure 8. ALWAYS KEEP THE ROPE FIRMLY IN YOUR BRAKE HAND WITH NO SLACK IN THE ROPE BETWEEN YOUR BRAKE HAND AND THE FIGURE 8.

C12. Unlock. Untie the Overhand knot. You will find it will be more difficult to pull the brake side of the rope out of its trap between the large ring and the standing part of the rope because in a vertical situation your full weight is involved. But **grasp the rope tightly** and pull it slowly away from you until you feel the slight jolt indicating that it has come unlocked Maintain tension on the rope with your brake hand. NEVER TAKE YOUR BRAKE HAND OFF THE ROPE.

C13. Rappel to the bottom and complete the belay cycle. (Rappeler: **"Off Belay."** Belayer: ***"Belay Off."***)

Gaining Extra Friction from the Figure 8 Descender

One of the advantages of the Figure 8 with Ears is the ease of creating increased friction and therefore greater control of the descent. This is because those Figure 8 descenders that have ears are larger devices with greater surface area to create friction. Also, the ears help to contour the rope and hold it in place.

Fig. 9.11 Secure Lock-Off for Figure 8

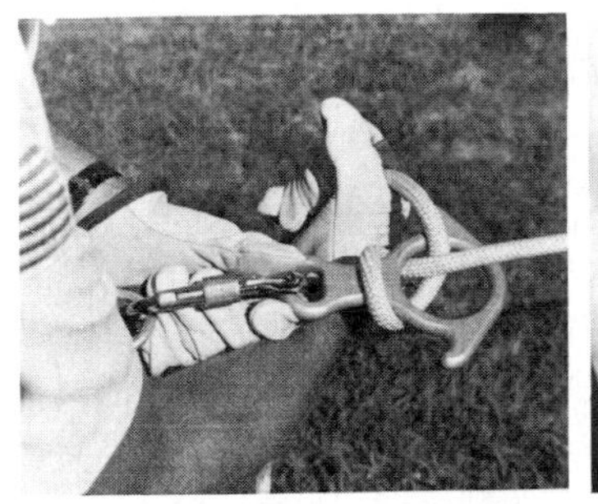

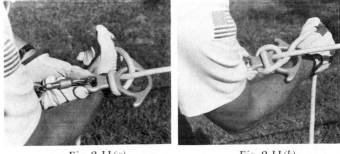

Fig. 9.11(a)

Fig. 9.11(b)

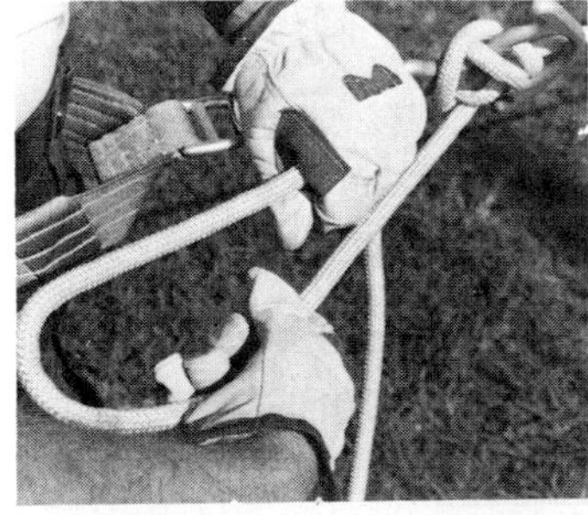

Fig. 9.11(c)

Fig. 9.11(d)

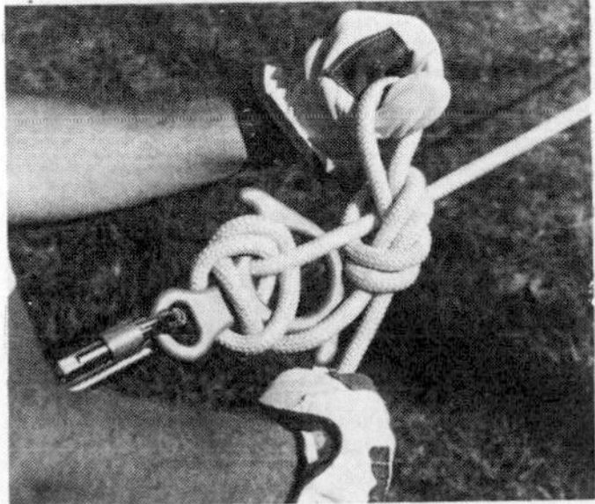

Fig. 9.11(e)

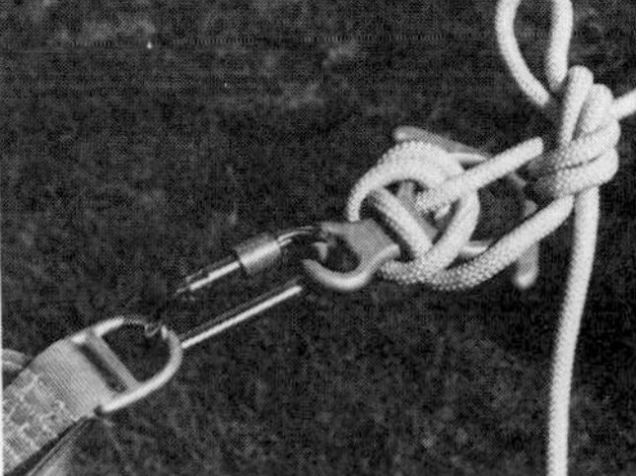

Fig. 9.11(f)

Fig. 9.12 Double Wrapping the Figure 8

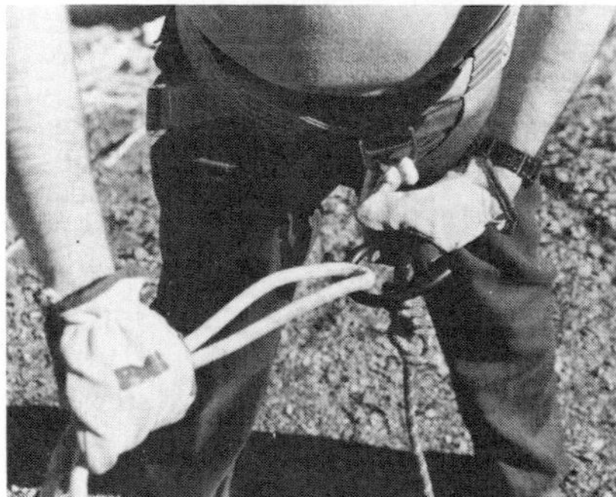

Fig. 9.12(a)

Fig. 9.12(b)

Fig. 9.12(c)

Fig. 9.12(d)

Fig. 9.12(e)

SUGGESTION:

Getting Off the Rope

With any rappel device, it is easier to get off rope if you have a small amount of slack in the rappel rope to work with. A quick trick to gain this slack is to keep your weight on the rope and rappel device as your feet touch the bottom, and do a deep knee bend before stopping your rappel. As you return to standing straight up, you will have a foot or so of slack in the rappel rope that will assist you in unlacing the rappel device.

Double Wrapping a Figure 8 Descender

(see Figure 9.12)

NOTE: This cannot be done once the rappeler is on rope. It must be done before attaching the Figure 8 descender to the seat harness carabiner.

1. Face the anchor with the rappel rope running past you on your brake hand side.
2. At the place on the rappel rope where you want to attach yourself for the rappel, take a bight of rope in your brake hand. Push it through the large ring of the Figure 8 down through the top (if the ascender will be in a horizontal plane) or from the side with the brake hand (if the descender will be in a vertical plane).
3. Bring the bight of rope around the small ring of the Figure 8 and over the waist of the device.
4. Push the bight of rope through the large ring again, between the two strands already there. If you need more rope, pull on the bight.
5. Now bring the center of the bight back over the waist and pull the rope strands snug.
6. Attach the Figure 8 descender to your seat harness carabiner and lock the carabiner.

Rappeling with a Double Wrapped Figure 8 Descender

While a double-wrapped Figure 8 descender can give a rappeler added control through greater friction, it does require some increased attention to technique.

The double-wrapped Figure 8 descender works smoother with the brake hand out to the side as shown in Figure 9.13. This position will help to guide the rope better around the Figure 8 descender. If the hand is closer to the body, the following may occur:

1) The strand of rope running around the Figure 8 descender on the side by the brake hand will begin to cross itself.
2) There is no danger in this, but the rappeler will feel himself slowing down due to increased friction, and the descent will possibly be a little rougher.
3) To uncross the strands, simply bring the brake hand back away from the body. You will feel a slight bump as the rope strands uncross.

Fig. 9.13 Hand Position with Double Wrapped 8

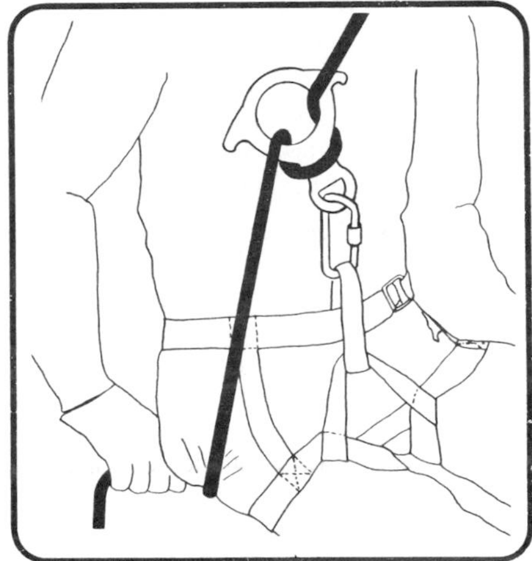

Fig. 9.14 Gaining Hip Friction on Rappel

Fig. 9.15 Butt Thrust

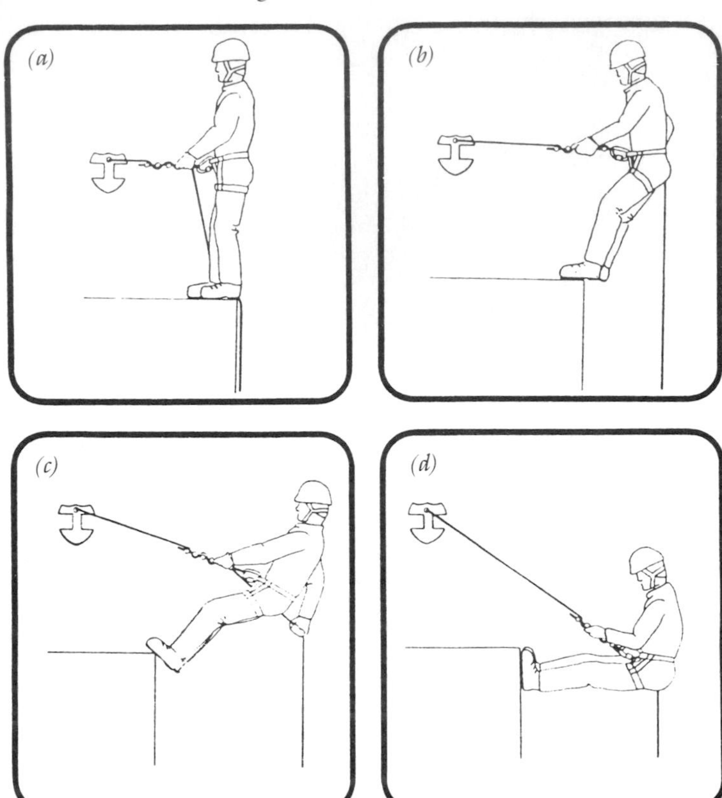

Variation Number One (The Butt Thrust)

(see Figure 9.15)

1. Facing the anchor, back up, slowly letting slack through your descender until you are standing with the balls of your feet on the edge.
2. Imagine that something is pushing you at your waist, so that your butt is slowly being thrust back out over the drop and opposite the anchor. **Keep your feet in place.** This should get the weight off of your toes and onto the insteps of your feet against the face of the wall (this serves to have your weight pressing against the wall).
3. As this is happening, the rope should be coming downward to meet the edge of the drop. When the rope does reach the edge, it will create greater stability for you by making a three-legged tripod (the rope on the edge plus your two feet kept a shoulder's width apart).
4. If you need to assist this process of getting the rope down to the edge, you can quickly shuffle your feet down the face of the wall. Taking small steps increases your stability.

Increasing Friction with the Body

A rappel should only be done when the rappeler has friction (and, therefore, **control**) to spare. If during the rappel there is not enough friction, one technique is to bring the rope sharply against the thigh as shown in Figure 9.14. It is not a good rappel technique to continually have to do this because:

- The rope may run across seat harness webbing and damage it.
- This leaves no extra friction to spare in case it is needed.

Rappel Stance

When learning to rappel, most people feel uncomfortable backing over a sharp edge while standing. However, for most circumstances, this is the most effective stance for rappeling. When they are intimidated by a difficult edge, some rappelers may try to "sneaky Pete" their way over the edge by rolling over it on their side or stomach. This action should not be done for the following reasons:

- On cliffs it will brush rocks and other debris over the side and endanger those people who are below.
- It can easily trap feet or hands between the rappel rope and the edge.
- It can snag rappel devices on the edge, stranding the rappeler.
- It can damage equipment, such as carabiners and seat harnesses, that gets snagged on the edge.

The preferred stance is on both feet (or alternatively, on both knees). On particularly difficult edges, there are variations that can assist in getting over. **DO NOT ATTEMPT ANY OF THE FOLLOWING FOR THE FIRST TIME WITHOUT A TOP BELAY.**

Fig. 9.16 Knees Over Edge Rappel

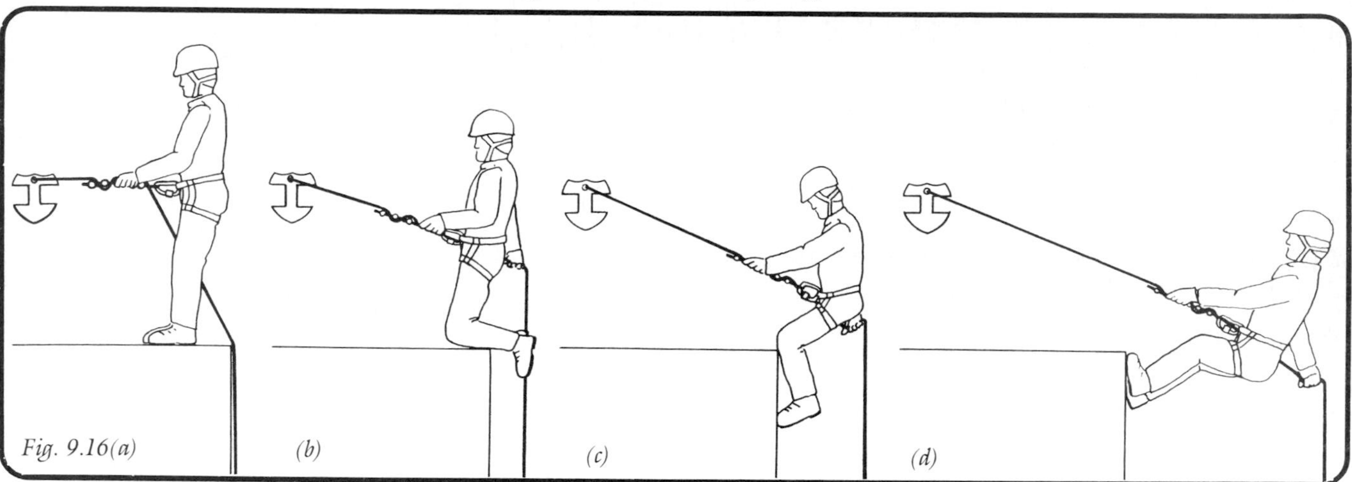

Fig. 9.16(a) *(b)* *(c)* *(d)*

Variation Number Two (Knees) *(see Figure 9.16)*

1. Walk back to the edge with slack out of the rope.
2. Get down on your knees right at the edge of the drop.
3. Lean back, getting your butt back away from the edge.
4. Slide over the edge on your knees. Your toes will hit the wall and you will stabilize as the rope comes down on the edge. Continue to rappel backwards and stand against the wall on your feet.

Clearing the Descender

It is very important that as you clear the edge on a rappel, the descender also clears and does not catch on the edge. Otherwise, the descender may become jammed and you will be stranded in a precarious position.

Thus, when backing over the edge in a rappel, ALWAYS OBSERVE THE POSITION OF THE DESCENDER AND MAKE CERTAIN THAT IT IS GOING TO CLEAR THE EDGE.

Undercut Edges

Undercut edges are those where the edge is overhung so far back that your legs cannot reach the wall as you start your rappel over the edge. Undercut edges present the rappeler with a special problem, and require an advanced technique, to be attempted only after you have developed full confidence and skill to control the descender.

Rappeling from an undercut edge is similar to rappeling from a helicopter skid. The procedure involves maintaining your feet on the edge, while lowering the rest of the body until your head is well below the feet and the edge of the overhang. Only after you are certain that the rappel device and your torso are far enough below the edge to clear it, do you step off the edge. **This results in a forward pendulum.** Your feet must absorb the shock of the forward motion against the vertical face.

Effect of Rope Angle on Rappeling

One factor that will significantly affect the degree of difficulty in rappeling over an edge is the angle the rope makes from the rappeler to the anchor point *(see Figure 9.17)*. This will range from the most difficult for a horizontal angle (the anchor on the same level as the rappeler) to the easiest for a vertical angle (the anchor above the rappeler). It is rare that a vertical angle will be found. Most of the time it will be a compromise: getting the rope angle as high up as possible while maintaining a safe and secure anchor point.

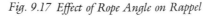

Fig. 9.17 Effect of Rope Angle on Rappel

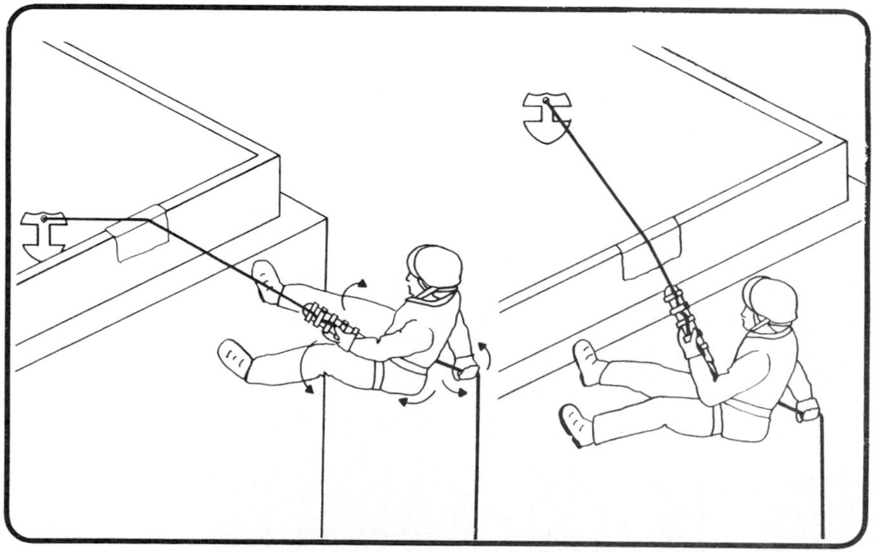

Fig. 9.17(a) Lower Anchor Point *Fig. 9.17(b) High Anchor Point*

THE BRAKE BAR RACK

The Brake Bar Rack descender offers several advantages for rappeling, including the following:

- It offers greater friction; therefore, greater control than most descenders.
- It provides the ability to change friction once the person has begun to rappel.
- It provides the ability to easily rappel longer drops than most descenders.

Among the potential disadvantages are:

- The Brake Bar rack is somewhat more complex than descenders such as the Figure 8. Consequently:

  It takes a bit longer to put it on the rope, and

  It is somewhat bulkier and heavier.

Rappeling with The Brake Bar Descender

(see Appendix II for guidelines on attaching the bars to the rack.)

(For more specific details on the Brake Bar Rack itself, see Chapter 5, "Basic Hardware.")

Attaching the Rack to Yourself and the Rope

A1. Decide on how many bars to begin the rappel with. When learning to use the Brake Bar Rack, always begin a rappel with all six bars engaged. As you become more experienced, you can learn how many bars are needed according to the specific situations. Figure 9.18 shows the Brake Bar Rack in position on a seat harness ready to be put on the rope (this figure shows the rack in position for a right-handed person).

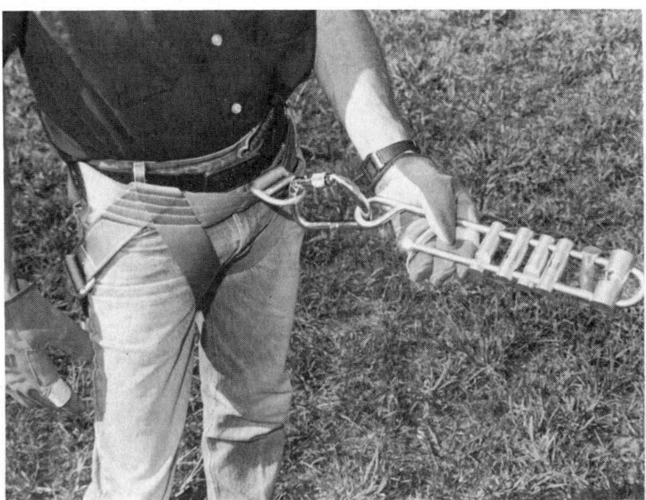

Fig. 9.18 Brake Bar Rack on Harness

A2. If your seat harness carabiner is in a horizontal plane, attach the rack to it with the short leg of the rack down. If you have a seat harness carabiner in a vertical position, then have the short leg of the rack toward the brake hand (the right hand on a right-handed person).

Attaching the Brake Bar Rack to the Rope

(see Figure 9.19)

A3. Establish a secure anchor point.

A4. Attach the rappel rope securely to the anchor point.

A5. Clip the rack into your seat harness carabiner and lock the carabiner (with the gate toward your body).

A6. Stand facing the anchor with the rappel rope on your brake-hand side.

A7. Hold the rack out in front of you in your guide hand.

A8. Disengage all bars except the top one on the rack. Do this by sliding them one at a time toward the bottom of the rack (towards the eye). Squeeze the two legs of the rack together with one hand and, with the other hand, flip back each bar.

A9. Pick up the rope with your brake hand. Drop the rope between the two legs of the rack and across the top bar.

DO NOT PASS THE ROPE BETWEEN THE TOP BAR AND THE BEND OF THE RACK *(Figure 9.20)*. **THIS WILL PINCH THE ROPE, MAKE THE DESCENT HARDER TO CONTROL, AND CAUSE EXCESSIVE WEAR ON THE RACK** *(9.19a)*.

A10. Reach down below the rack, grab the rope, and pull it across the top bar away from you (toward the anchor) pulling the slack out of it.

A11. With the other hand, clip in the second bar at the bottom of the rack, and slide it up to trap the rope between it and the top bar *(9.19b)*.

A12. Now bring the free end of the rope back across the second bar, pulling it toward the anchor so that the second bar is snugged in by the force of the rope pulling against it. Note that the rope must be on the side of the bar opposite the notch to hold the bar in place on the rack frame *(9.19c)*.

A13. Repeat the process with the remainder of the bars, until all six are clipped in *(1.19d & e)*.

A14. In an area with good footing, so that you will not slide down, lean back against the rope. The preferred position for the brake hand is for it to be below the rack and off to the side. This position for the brake hand is similar to other rappel devices *(9.19f)*.

Fig. 9.19 Attaching Rack to Rope

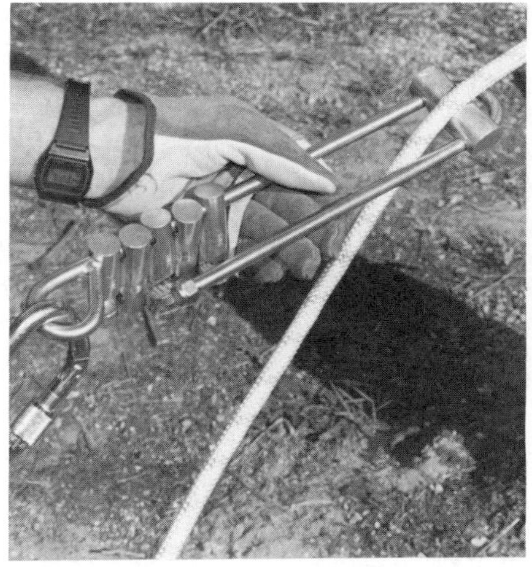

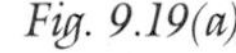

Fig. 9.19(a)

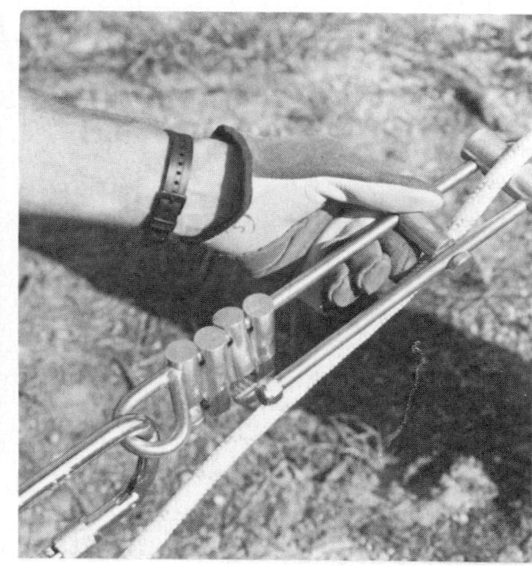

Fig. 9.19(b)

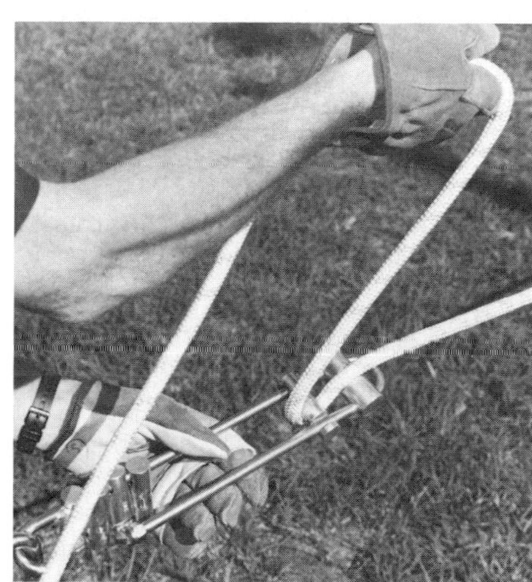

Fig. 9.19(c)

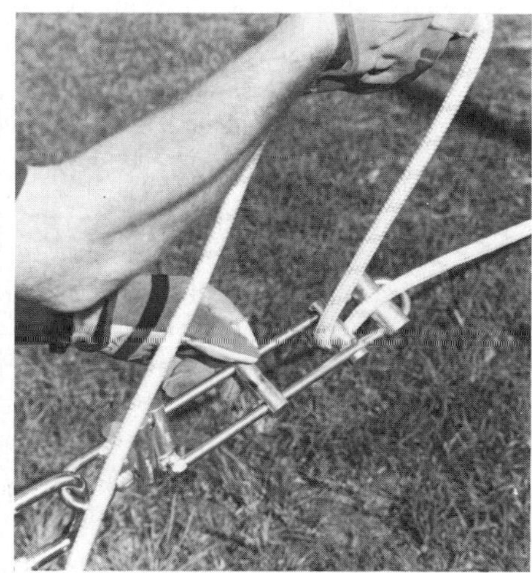

Fig. 9.19 (d)

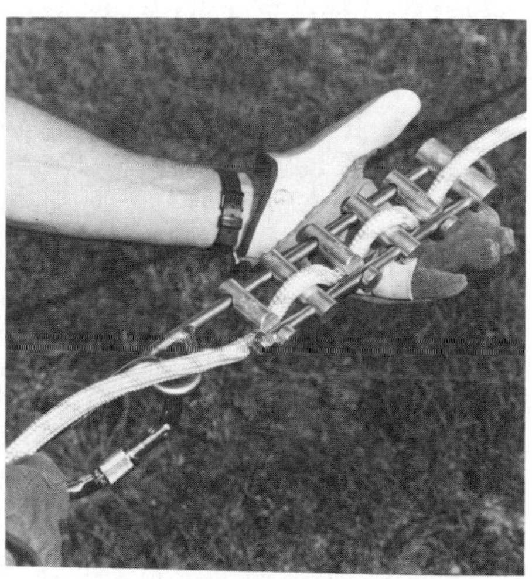

Fig. 9.19(e)

Fig. 9.19(f)

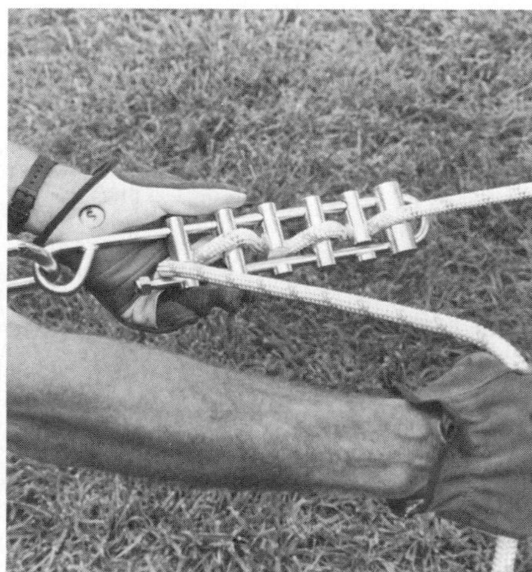

Fig. 9.19(g)

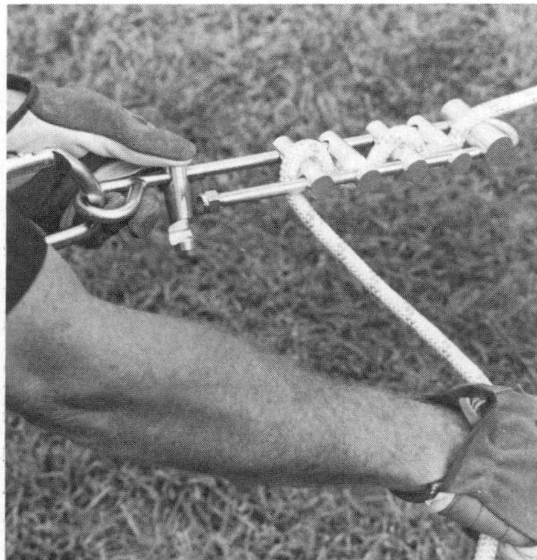

Fig. 9.19(h)

A15. The difference is in the position for the guide hand. Instead of being on the rope above the device, as with other descenders, the guide hand should be resting on the bars of the rack, holding the bar ends between the thumb and fingertips.

A16. Take your brake hand and pull the rope way from you (toward the anchor). If they are laced correctly, this should pull all of the bars together toward the top of the rack. This is known as the "quick stop" position when you are rappeling *(9.19g).*

A17. Now bring the rope back to the normal rappel position.

A18. With your guide hand grasp the bottom bar on either side of the rack and push it—along with the other bars—toward the top of the rack. This is the "stop" position for the guide hand. By jamming the bars together in this manner toward the top of the rack, you increase the friction on the rope and add another element of control.

A19. Using the guide hand, pull the bars, one by one, back towards you. As you are doing this, ease your grip on the rope with the brake hand. This is increasing the "go" mode of the rack by reducing the friction between the bars and the rope. As you are leaning back against the rope, you may feel the rope begin to move a bit through the rack and through the brake hand.

A20. If you have not moved, disengage the bottom bar. Do this by first swinging the rope with the brake hand in an arc to the opposite side of the rack to uncover the bottom bar *(9.19h).* Then squeeze the two legs of the rack together at the open end of the rack that is near you. Unclip the bottom bar. Let it slide down the rack toward the eye and out of the way. Now spread the remaining bars apart along the length of the rack. This is lessening the friction even more.

Fig. 9.20 Incorrect Lacing of Brake Bar Rack

A21. If you still have not moved, remove the fifth bar (which is now on the bottom) in the same manner that you disengaged the sixth bar. Spread the remaining bars along the length of the rack.

A22. Now, reverse the process by clipping bars back in to gain friction. Do this by using your guide hand to squeeze the legs of the rack together and clip the bars in one at a time at the bottom and lacing the rope back between them. This is the same process you used when you initially laced up the rack on the rope.

Tying Off *(see Figure 9.21)*

A23. Lean back against the rack so that the rope between the rack and the anchor is taut. Start the tie-off process by taking the rope with your brake hand and pulling it away from you, to the top of the rack and towards the anchor *(9.21a).*

A24. With the brake hand, pull the rope over to the side of the rack by your guide hand so that the rope runs across the top bar between the curve of the rack and the section of the rope going to the anchor *(9.21b).*

A25. Bring the rope back toward you, pulling it taut so that it locks all the bars together. Bring the rope through the two legs of the rack and across the bottom bar.

A26. Pull the rope away from you, towards the anchor, in the same path you did before. Pull it firmly so that all the rope sections are taut and the bars locked together *(9.21c).*

A27. The rack should now be locked in a "stop" position. With your brake hand extended parallel with the strand that runs to the anchor, hold the rope out away from you. At the point the trailing end of the rope crosses your brake hand, form a large bight with the rope using the assistance of your guide hand *(9.21d).*

A28. Treating this bight as one rope, use it to tie an overhand knot in the line that is going to the anchor. Cinch the overhand knot firmly against the top bar of the rack. THERE MUST BE NO SLACK IN THE ROPE RUNNING OVER THE BAR, NOR SPACE BETWEEN THE BARS. The rack is now locked off *(9.21e & f).*

Unlocking

A29. When unlocking, ALWAYS KEEP A FIRM GRIP ON THE ROPE AND ALLOW NO SLACK IN THE BRAKE END OF THE ROPE. To unlock, reverse the locking process. To untie the overhand knot, pull slowly towards you on the brake end of the rope, holding your guide hand at the center of the bight of rope so that it comes out slowly.

Fig. 9.21 Tying Off Brake Bar Rack

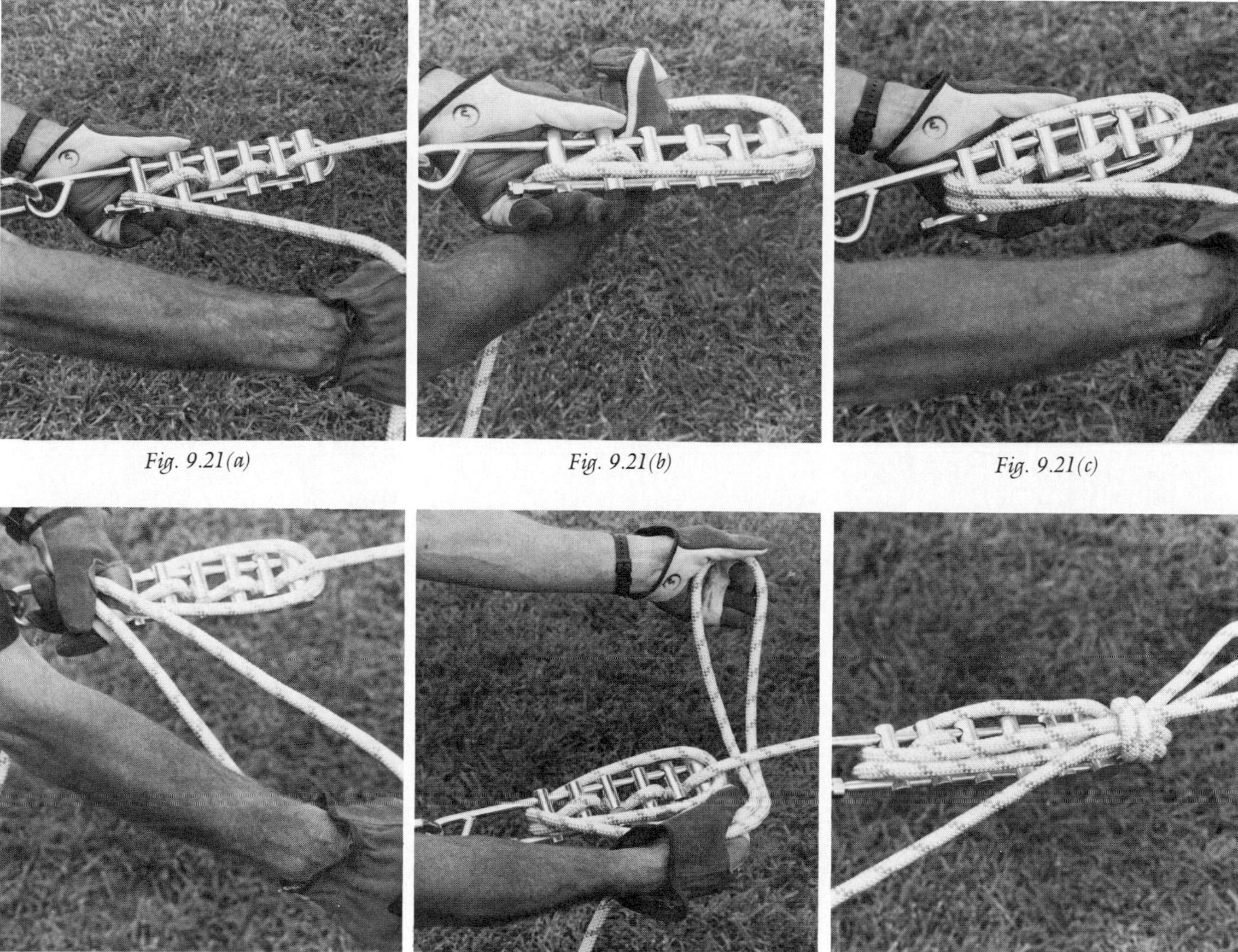

Fig. 9.21(a) *Fig. 9.21(b)* *Fig. 9.21(c)*

Fig. 9.21(d) *Fig. 9.21(e)* *Fig. 9.21(f)*

A30. With your brake hand firmly on the rope, pull the brake end of the rope in a 180 degree arc until it is straight out in front of you.

A31. Now, still grasping the rope firmly with the brake hand, pull the rope straight out towards the side by the guide hand, then through a 180 degree arc and back to the normal rappel position.

A32. Place the guide hand back in its normal position of cradling the bars. If you have not begun to move again, pull the bars apart with the guide hand and proceed to rappel again.

Getting Off Rope

A33. Getting the rack off the rope is a reversal of the process of putting it on. You may leave the rack attached to your seat harness carabiner while doing this. With your brake hand, pull the rope back in the direction of the anchor so that it uncovers the bottom bar completely.

A34. Using your guide hand, squeeze the legs of the rack together and clip off the bottom bar. Let the bar slide to the bottom of the rack.

A35. With the brake hand, move the rope back through the legs of the rack, uncovering the next bar up and pulling the rope back toward the anchor. Unclip the next bar up with the guide hand. Continue this procedure until all bars are disengaged.

Fig. 9.22 Using Brake Bar Rack on Slope

Fig. 9.22(a)

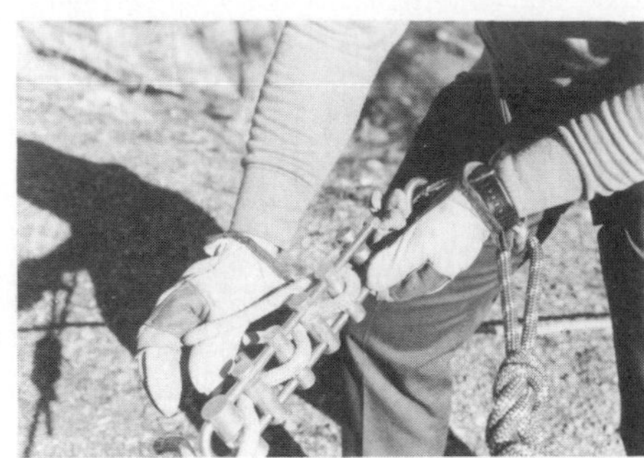

Fig. 9.22(b)

Fig. 9.22(c)

Fig. 9.22(d)

Fig. 9.22(e)

Fig. 9.22(f)

Fig. 9.22 (g)

Rappeling with the Brake Bar Rack on a Slope

(see Figure 9.22)

B1. Establish a secure anchor point at the top of a short slope of about 45 degrees. (If a slope is not available, then use a stairway.) Securely attach the main line rappel rope to the anchor point.

B2. Establish an anchor point for a belay. Attach a sling securely to the belay anchor point. Clip a large, locking carabiner into the end of the anchor sling.

B3. Have a belayer take position. If a Munter Hitch is being used for belay, have the belayer tie it into the belay rope and clip the rope into the belay carabiner. If the belayer is using a belay plate, attach it to the rope and clip it into the belay anchor carabiner.

B4. Clip yourself into the belay rope. Initiate the belay cycle with the belay voice communications (Rappeler: **"On Belay."** Belayer: ***"Belay On."*** *[9.22a, b & c]*).

B5. In a secure position, where you are not in danger of falling, follow steps A5 through A15 above to lace the rope onto the rack. Be certain that there is no slack between the rack and the anchor *(9.22d)*.

Have the instructor do a SAFETY CHECK. Among the other critical elements in the belay system, make certain that the bars are laced up correctly. Check all connectors such as carabiners to make certain they are locked and in position of function. Be certain the seat harness is buckled correctly and the anchors are secure. Be sure that no loose clothing or hair will be drawn into the rappel device. Be certain that the helmet is secure.

B6. Begin backing down the slope, controlling your decent with your brake hand and, if necessary, with the guide hand on the bars. If you are unable to move, use the guide hand to pull the bars down toward you, as in step A20 above *(9.22e)*.

B7. If, after spreading the bars along the length of the rack, you still have not moved, disengage bars as described in steps A21 and A22 above. But NEVER HAVE LESS THAN FOUR BARS ON THE RACK. *(9.22f)*.

B8. Rappel until you are about midway down the slope. With your brake hand, do a "quick stop" as described in step A16 above *(9.22g)*.

B9. Rappel a short distance farther. Now, using your guide hand on the bars, attempt to stop yourself as described in step A18 above.

B10. Relax your guide hand and tie off the rack as described in steps A24 through A29 above.

C. Rappeling with the Brake Bar Rack Down a Vertical Face *(see Figure 9.23)*

C1. Choose a short vertical face (approximately 20 feet) where the top breaks over gradually into a steep face. On the first try, do not choose a face with a sharp edge.

C2. Establish a secure anchor point safely back from the edge. If possible, have the anchor point high up. This will assist the rappeler in going over the edge. Attach the main line rappel rope securely to the anchor point.

C3. Establish a separate anchor point for a belay. Securely attach a sling into the belay anchor point. The anchor point and sling should be established so that when the belayer takes position, he has a good field of view of the top and face, but is not in danger of falling over the edge. Clip a large, locking carabiner into the end of the belay sling.

C4. Have the belayer tie into a safety line and take position. If a Munter Hitch is being used for belay, have the belayer tie it into the belay rope and clip the rope into the belay carabiner. If the belayer is using a belay plate, attach it to the rope and clip the rope into the belay carabiner.

Fig. 9.23(a)

Fig. 9.23(b)

Fig. 9.23(c)

Fig. 9.23(d)

Fig. 9.23(e)

Fig. 9.23(f)

Fig. 9.23(g)

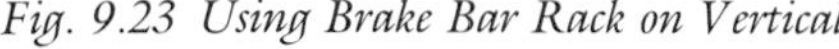

Fig. 9.23 Using Brake Bar Rack on Vertical

C5. Clip yourself into the belay rope. Initiate the belay cycle with the initial belay voice communications (Rappeler: **"On Belay."** Belayer: ***"Belay On."*** *[9.23a]*).

C6. In a secure position, where you are in no danger of falling, follow steps A5 through A15 above to lace the rope onto the rack. Be certain that there is no slack between the rack and the anchor.

Have the instructor do a SAFETY CHECK. Among other elements of the rappel system, make certain the the bars are laced correctly. Be certain that the carabiners are locked and in correct manner of function. Be certain that the seat harness is on correctly and buckled securely. Check the anchors for security. Be certain that no loose clothing or hair will be drawn into the descender, and that the helmet is secure *(9.23b)*.

C7. Begin backing to the edge. Because you are on top, there is a great deal of friction in the rack, but little weight to pull the rope through. Consequently, you may have to feed the rope through the rack by letting slack with the brake hand. And you may have to reduce friction with your guide hand by spreading the bars apart or perhaps disengaging one or two bars. **BUT REMEMBER: AS SOON AS YOU START OVER THE EDGE, YOU MAY NEED THE FRICTION WHEN YOUR FULL WEIGHT COMES ONTO THE ROPE. SO BE PREPARED TO REENGAGE THE BARS WITH YOUR GUIDE HAND AND TO ESTABLISH CONTROL WITH YOUR BRAKE HAND** *(9.23c)*.

WARNING NOTE

One of the major concerns in going over the edge with a rack is the possibility of catching the device on the edge. This is a possibility with any rappel device. But, because of the rack's length, you need to take special care when you use it. If you do catch the rack on the edge, any of the following things might happen:

1) You could jam the rack on the edge, preventing the rope from running through the device, thereby stranding yourself in that position.

2) The pressure of your body weight could bend the rack, causing it to malfunction in the future.

The solution to these problems is to avoid edge catch. As you go over the edge, make certain that you lean out enough and push back with your feet before you step down so that the rappel device clears the edge before the rope lays across the edge *(see Figure 9.24)*.

Fig. 9.24 Clearing Rack from Edge

Fig. 9.24(a)

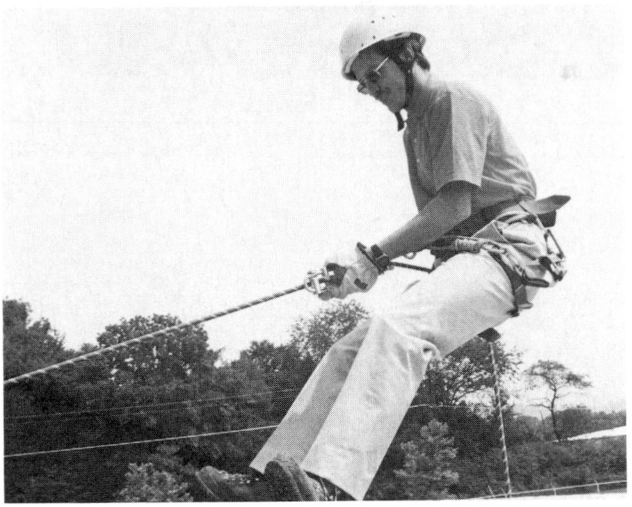

Fig. 9.24(b)

C8. As you go over the edge, keep the following principles in mind:

a) Keep the body generally perpendicular to the wall. This means that you will have to deliberately lean out as the wall becomes vertical. This may seem unnatural at first, but it is necessary to keep your feet from slipping out from under you.

b) Keep the feet apart a shoulder's width for balance. This will help prevent you from being pulled over to one side.

c) Keep the knees relaxed and slightly flexed.

d) Keep the body slightly turned in the direction of the brake hand, looking down slope, picking a path.

Fig. 9.24(c)

Fig. 9.24(d)

e) Use the guide hand for balance and for control of the bars. DO NOT SUPPORT YOUR WEIGHT WITH IT.

f) NEVER TAKE THE BRAKE HAND OFF THE ROPE UNLESS YOU ARE SECURELY LOCKED OFF *(9.23d)*.

C9. After you are over the edge, rappel a few feet, then bring the brake side of the rope up in a "quick stop" by pushing it away from you toward the anchor *(9.23e)*.

C10. Tie off the rack so you can be hands free of the rope *(9.23f & g)*.

C11. Unlock and rappel a few feet farther.

C12. Attempt to stop your descent with the balance hand by jamming the bars up together.

C13. Pull the bars apart and rappel to the bottom.

C14. Complete the belay cycle with the appropriate voice commands. (Rappeler: **"Off Belay."** Belayer: ***"Belay Off."***)

C15. Remove the rack from the rope.

C16. Immediately move away from the "drop zone" to lessen your chances of being hit by falling objects and to clear the rope for others.

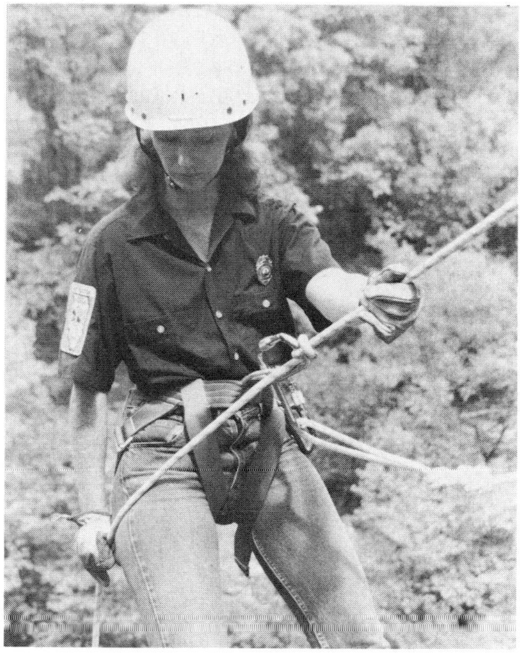

Fig. 9.25 Munter Hitch Rappel

Emergency Descent Systems

There may be emergency situations when a rappel is necessary, but the person has no descender with him. There are possible solutions to this problem.

One solution might be the **Body Rappel**, described earlier in this chapter. But, as noted, it has some distinct disadvantages and dangers.

One system that has been used in the past is the **Carabiner Wrap.** This consists of wrapping a seat harness carabiner with several turns of the rappel rope to create friction. However, THE CARABINER WRAP RAPPEL IS NOT CONSIDERED A SATISFACTORY AND SAFE TECHNIQUE FOR RAPPELING for the following reasons:

1) If the rope wraps are not correctly put onto the carabiner, they can spiral out of the carabiner gate, resulting in a free fall.

2) The wraps can bear on the carabiner gate and break it. One alternative to the carabiner wrap might be the Munter Hitch rappel (see Chapter 6, "Knots," and *Figure 9.25*).

WARNING NOTE

As with all other rappel techniques, the Munter Hitch rappel must be practiced on level ground, on a moderate slope with a rappel, and on a short drop with a rappel before adopting it for an emergency rappel technique.

Self-Belay Techniques

Where the belay of a rappeler by another person is not possible or practical, self-belay techniques may be possible. Most self-belay techniques are based on the use of some type of rope grab device on the main rappel line **above** the descender. Usually the rappeler is attached to the device via a short sling to a chest harness. The self-belay mechanism is triggered by a definitive action by the rappeler, such as leaning over backwards.

WARNING NOTE

Self-Belay Techniques are NOT completely automatic safeties, but require some positive action on the part of the rappeler. Remember: *in an emergency one reacts with an instinctive action.* Whether one responds with the correct action in a self-belay emergency may depend on how well-trained and disciplined he is to do so.

One example of a self-belay device is the **Spelean Shunt** as shown in Figure 9.26. The Spelean Shunt consists simply of a Gibbs cam, an oval carabiner, and a short piece of webbing or rope that attaches to a chest harness.

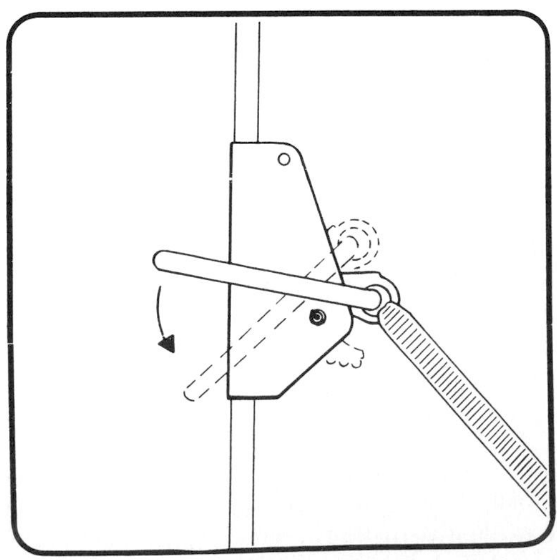

Fig. 9.26 Spelean Shunt

WARNING NOTE

Whenever you use a self-belay device, such as the Spelean Shunt, you must be certain that the sling connecting the device to the person is not too long. Otherwise, it could put the device out of reach on the rope above you. The result would be that the device could lock itself out of reach and you could be stranded on the rope.

Why a Prusik "Safety" May Not Be

One traditional means of rappeling safely has been the use of what is called a "Prusik Safety," which uses a sling with a Prusik knot on the rope above the rappeler and connected to the rappeler's chest harness (see Chapter 10, "Ascending," for further information on the Prusik knot). The theory has been that should the rappeler get out of control, the Prusik knot can be used to tighten on the main rope to stop the fall. However, this practice has been falling from favor because its "safety" may be an illusion:

- In a panic, the falling rappeler may grab for the Prusik knot itself, which does not close it, but opens it up. The result is a free fall—until the rappeler hits bottom.
- If the Prusik knot does not close and stop movement immediately, the friction heat of the knot on the rope can cause the Prusik sling to melt through and fail.

One disadvantage of the Spelean Shunt is that it can catch and set at times when that is not desired. This happens particularly when the rappel is not completely free.

NOTE: Regulations of the Federal Occupational Health and Safety Administration (OSHA) require that in a workplace environment, such as in high angle window cleaning, there be a self-belay device ON A SEPARATE LINE from the rappel line.

Self-Belay Device on Second Line

One alternative for a self-belay device is the use of a **spring-loaded** cam on a line that is parallel and close by the rappeler *(see Figure 9.27)*. A sling attaches the cam to the rappeler's harness.

WARNING NOTE

When using a cam on a separate line as a belay, the cam MUST BE spring-loaded so that it will automatically catch should there be a fall. A free-running cam may not catch in an emergency, and could run to the ground with the person it was supposed to protect.

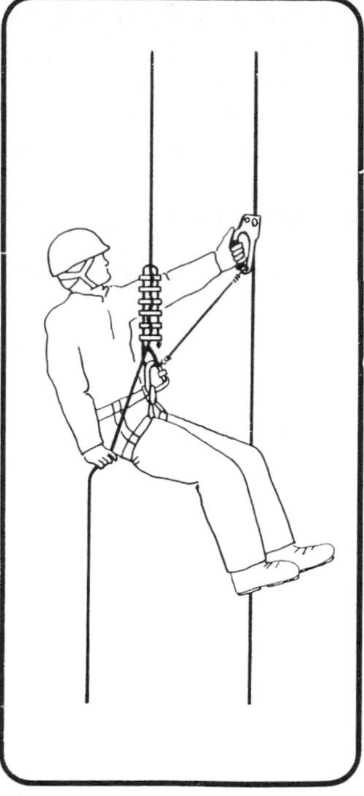

Fig. 9.27 Rappeling with Spring Loaded Can on Separate Line

Protecting the Rappel Rope That is Below You

In most cases, the rappel rope is simply dropped down the vertical face where the rappeler is about to travel. There are, however, circumstances where this might not be desirable:

- In tactical operations where there is a hostile person below you who could grab the rope (and thereby control you).
- Where there is a very frightened/unpredictable person below you.
- Where unstable rocks could be knocked loose by the rope and fall on persons below.
- Where the rope could become tangled or jammed and you could not retrieve it.

In these cases, it might be desirable for the rappeler to keep the rope with him. One way of doing this is to attach the rope bag to the rappeler as shown in Figure 9.28. There are three possible ways of attaching the bag:

- If it is very light, then the bag might be attached to a seat harness equipment sling.
- If the bag is heavy, then it might be attached with a carabiner directly into the bottom of the descender.
- Special design rope bags have straps that attach to the lower leg.

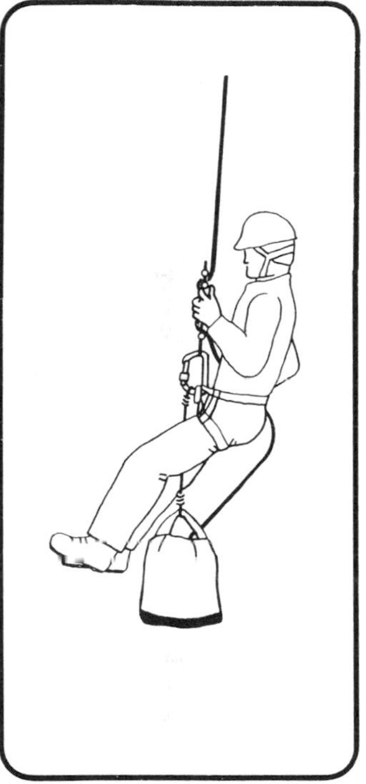

Fig. 9.28 Rappeling with Bagged Rope

WARNING NOTE

It is difficult to accurately estimate the length of a rope when it is in a bag. So there is the danger that when rappeling from a bagged rope, you could rappel off the end of the rope. Always either tie a stopper knot in the bottom end of a bagged rope or tie it into the bag to prevent rappeling off the end of the line. (see “Preventing a Rappel Off the End of a Rope” on page 116.)

Extricating Jammed Rappel Devices

Rappel devices are notorious for sucking up loose material to become jammed and, perhaps, strand the rappeler in a very difficult and painful position that might require a rescue. Among the possibilities are:

- “Tee” shirts and other loose clothing.
- Hair.
- Body parts, such as loose flesh on an underarm.

The best solution to this problem is, of course, prevention:

- Tuck in shirt tails and other loose clothing.
- Keep hair trimmed. If that is not possible, keep it tied back and tucked into helmets.
- Keep flabby body sections (underarms, stomachs, etc.) away from rappel devices.

Techniques for Extrication

DO NOT USE KNIVES. When rope yarn is stretched, as it is under the loading of one or two persons, the slightest touch with a knife blade causes it to part and the rope to catastrophically separate. If you are trying to cut jammed material from a descender, you are probably under pressure, physically unbalanced, and, possibly, also in pain. So it is extremely difficult for you to cut away the offending material without also touching the rope with the knife.

The way out of such a situation is to take your weight off the descender with the use of an ascender or Prusik knot above the descender. See Chapter 10, "Ascending," for a description of this technique.

Preventing a Rappel Off the End of a Rope

There is the potential in some situations of rappeling off the end of a rope. This usually occurs when you cannot see the bottom of the drop before you begin the rappel. As you near the bottom end of the rope, you may not be paying attention or you may lose control. One form of insurance against rappeling off the end is to use a "stopper knot." One example of this is a Figure 8 knot tied in the bottom end of the rappel line.

Even better is a Figure 8 on a Bight which forms a loop. This gives you something to stand in while you figure out what to do next.

Questions for Review

1. List five characteristics of controlled rappeling.
2. Why are the body rappel and arm rappel uncomfortable to use?
3. The rate of descent in rappeling is controlled by the _______________ hand.
4. The ____________________ hand helps to balance the rappeler, but does not support weight.
5. The arm rappel should only be used for what situations?
6. What two dangers are posed by using the body rappel?
7. What are the two main drawbacks in using the conventional Figure 8 descender?
8. What is one way of getting out of the girth hitch occurrence with the conventional Figure 8?
9. What is one way of preventing the girth hitch occurrence?
10. What is one way of lacing up a Figure 8 descender that will help prevent it from jamming if caught on an edge?
11. Why should you keep a proper distance between a rappel line anchor and a belay anchor?
12. What is the danger in having too great a distance between a rappel line anchor and a belay anchor?
13. Name seven principles to keep in mind when rappeling down a vertical face.
14. What is one technique for gaining extra friction from a Figure 8 descender?
15. What can occur when rappeling on a Figure 8 with the brake hand too close to the body?
16. Describe the preferred stance for rappeling.
17. Describe the effect of the angle the rope makes from the rappeler to the anchor point.
18. Why is it important that a rappeler not catch the rappel device on the edge of a drop?
19. Name three advantages and three disadvantages of the Brake Bar Rack.
20. In lacing the rope onto the Brake Bar Rack, one *(should) (should not)* pass the rope between the top bar and the bend of the rack.
21. What is different about the position for the guide hand when using a Brake Bar Rack in contrast to other rappel devices?
22. What is the minimum number of bars that one should use on a Brake Bar Rack?
23. What are two reasons that the carabiner wrap rappel is not considered a satisfactory and safe technique?
24. When using a self-belay device, why should one not get the connecting sling too long?
25. When using a self-belay device on a separate line, why should it be spring-loaded?
26. What is one way of preventing a rappel off the end of a rope?

Chapter 10

Basic Ascending Techniques

PREREQUISITES

Before attempting the activities described in this chapter, you must have demonstrated that you can properly:

1) Use and care for rope.
2) Use and care for other equipment employed in the high angle environment.
3) Tie correctly, confidently, and without hesitation the eight knots described in Chapter 6.
4) Apply the principles of anchoring, and rig a safe and secure anchor.
5) Apply the principles of belaying, and safely and confidently belay another person using either a Munter Hitch or belay plate.
6) Apply the principles of rappeling: rappel safely, confidently, and under control; tie off the rappel device to operate hands free of the rope, and then return to a safe and controlled rappel.

OBJECTIVES–

At the completion of this chapter, you should be able to:

1. Describe the purposes of ascending and the principles behind it.
2. Select equipment for an ascending system.
3. Tie and use a Prusik knot.
4. Safely and efficiently use handled ascenders.
5. Ascend a fixed rope safely and efficiently using an ascending system created from three ascenders.
6. Define what is meant by "changing over" from ascending to rappeling, or from rappeling to ascending, and what is involved in these procedures.
7. Safely and efficiently change over from ascending to rappeling.
8. Safely and efficiently change over from rappeling to ascending.
9. Describe what is meant by an ascending "system" and what criteria are necessary for ascending systems that are safe and efficient.
10. Describe the need for and use of the procedure known as "tying off short."
11. "Tie off short" when ascending.
12. Describe how ascenders can be used to extricate yourself from a jammed rappel device or other similar emergency.
13. Use ascenders to extricate yourself from a jammed rappel device or other similar emergency.

TERMS— *pertaining to ascending that a High Angle Technician should know:*

Ascender—A mechanical device, or a friction knot, that is used in ascending a fixed rope. They are secured to the rope, and attached via attachments (slings) to the person using them.

Ascending—A means of traveling up a fixed rope with the use of either mechanical devices or friction knots that are attached with slings to the user's body.

Ascender Sling—Attachments of webbing or rope that connect a person to his ascenders.

Cams—Mechanical rope grab devices without handles which slide in one direction on a rope and are used for ascending.

Changeover—To transfer from an ascending mode to a rappeling mode or from a rappeling mode to an ascending mode.

Chicken Loop—A safety loop that fits around the ankle to secure the ascender sling and prevent the foot from slipping out of the sling should an upper connection fail and the person ascending fall over backwards.

Handled Ascenders—Ascenders with frames large enough to accommodate built-in handles that can be comfortably gripped with the hands. Most handled ascenders have toothed cams which grip the rope.

Prusik—A type of friction knot used in ascending. It has also come to be used by some individuals as a term synonymous with ascending, even when mechanical devices are used, i.e., "to Prusik."

Prusik Loop—A continuous loop of rope in which a Prusik knot is tied.

Tying Off Short—A safety technique that creates an extra point of attachment during ascending by tying the person directly into the main line rope.

The Purpose of Ascending

Ascending a rope is in essence the opposite of rappeling. It is the use of mechanical devices (or in some cases, friction knots) to safely and efficiently ascend a fixed rope. Ascending is a further development of competency in the vertical environment. To only be able to rappel means that you can only travel one way on the rope: down. But to be able to both competently rappel **and** ascend means that you have developed the freedom to travel both down and up on the rope.

To further develop competency and enhance skills at rappeling and ascending, you must also be able to safely transfer between rappeling and ascending while on rope. This procedure is known as **changeover.** If you are skilled at changeover, you have the ability to both make the transition from rappeling to ascending and from ascending to rappeling.

To be able to both ascend and make changeover safely and efficiently requires a mix of equipment and the simultaneous use of skills. This means that the you must have an absolute knowledge of the equipment and an instinctive use of the skills. These only come through practice of the necessary techniques.

Along with the development of ascending and changeover skills and a thorough knowledge of the equipment involved comes the ability to extricate yourself from certain difficult situations. You may, for example, use these skills to extricate yourself from a jammed rappel device without the dangerous use of knives.

How Ascending Is Accomplished

Ascending is accomplished through the use of rope grab devices, called **ascenders.** When properly secured to the rope, ascenders slide only in one direction: up. There are two basic types of ascenders:

- **Friction knots.** There are several different kinds of these, but the type most commonly used is the **Prusik knot.**
- **Mechanical Ascenders.** These work by an offset camming action that presses against the rope to keep the device from sliding down the line. They are further subdivided into two types:
  - Cams, which hold by squeezing the rope. Two brands currently on the U.S. market are the Gibbs and the Rescuecender™.
  - Handled ascenders, which hold primarily by gripping the rope with teeth on a cam, which also presses the rope.

There are several brands of handled ascenders available, including:

- CMI.
- Clog.
- Petzl.
- Jumar.

In all cases, ascenders are attached to the user's body by **slings,** connectors made either of webbing or of rope. These slings are connected to a seat harness, chest harness, feet, or combinations of these.

The actual ascending process works by the alternating action by the user. While resting his weight on the first ascender as it grips the rope, the person keeps his weight off the second ascender and moves it up the rope. He then shifts his weight to the second ascender and takes weight off the first ascender as he moves it up the rope. By repeating this cycle, the person moves up the rope.

In any ascending activity, **at least two ascenders are required,** and three ascenders increase the margin of safety.

The Basics: A Prusik Knot

Friction knots were the first type of rope grab devices used in ascending. For the most part, they have been replaced in ascending by mechanical devices. But the knowledge of how to tie and use a friction knot is still important. If you know how, you can often improvise a friction with rope or cord when you do not have mechanical ascenders. On numerous occasions, the ability to improvise a friction knot has saved lives by helping people extricate themselves from difficult situations. Friction knots have also helped people perform a self-rescue after the failure of a mechanical ascender. While there are a number of different friction knots, the one most commonly used is the **Prusik knot.**

Selecting Material for A Prusik Knot

- *Diameter:*

  The Prusik knot will operate more efficiently if the rope it is made from is a smaller diameter than the main line rope to which it is attached. However, be certain that the Prusik cord is strong enough to support the intended load, with a proper safety margin.

- *Stretch:*

  The Prusik material should not be stretchy, otherwise it will be difficult to loosen once it is "set" on the rope.

- *Construction.* This is a compromise among the following:
  - A softer lay construction grips better, but will be more difficult to loosen and move up the rope, and will wear out faster.
  - Harder lay ropes are easier to loosen and move up the rope, but do not grip as well.

A general principle is that a Prusik knot holds better if it is a different material from the mainline rope, for example: polyester Prusik cord on a nylon main line.

WARNING NOTE

Material used in Prusik knots wears quickly, and should be inspected before each use.

Creating A Prusik Loop

Create a continuous loop from an approximately 6-foot length of rope which you have chosen for the loop material. Do this by tying the two ends together with a Figure 8 Bend knot.

ALTERNATIVE APPROACH

Many people prefer to form a Prusik loop using a Grapevine Knot (also known as a "Double Fisherman's knot"). This is also a secure knot for supporting body weight and may create a knot of slightly less bulk than the Figure 8 Bend. For an illustration of the Grapevine Knot, see page 59.

Attaching the Prusik Loop to a Rope (*see Figure 10.1*).

1. Securely anchor a main line static kernmantle rope vertically so that it will support a person. There should be some means of letting slack into the rope while it still supports a person. (This is so that the person practicing with a single Prusik knot can be let back to the ground when he gets up as far as he can.) One possibility is the "Ascending Practice System" described in this chapter on page 126.
2. Don a sewn, manufactured seat harness with leg/thigh supports. Clip a locking carabiner into the seat harness front tie-in point.
3. Stretch the Prusik loop out between two hands with the connecting knot about mid-point.
4. At about eye level begin to place the loop on the rope by forming a Prusik knot. Do this by holding the loop against the rope on the side of the rope near you. Have a smaller portion of the loop (about 6 inches) off to the right of the rope *(10.1a).*
5. Bring the larger side of the loop around the main line rope toward you and pull it through the smaller side of the loop. Be sure that the Figure 8 Bend (or Grapevine) knot passes well through and that the coils formed around the main rope are even *(10.1b & c).*
6. Bring the larger side of the loop through the same path again as before. Make certain that the coils of the Prusik knot around the main rope are even and parallel *(10.1d).*
7. Tighten the Prusik knot around the main rope. Do this by the following: a) with one hand, pull the end of the Prusik loop away from the main rope; b) at the same time, with the other hand, grasp the knot by placing the fingers on the coils of the knot on the side away from you, with the thumb on the bar portion of the knot on the side next to you. Grasp the knot with the thumb and pull the bar tight against the rope *(10.1e & f).*

Fig. 10.1 Attaching Prusik Loop to Rope

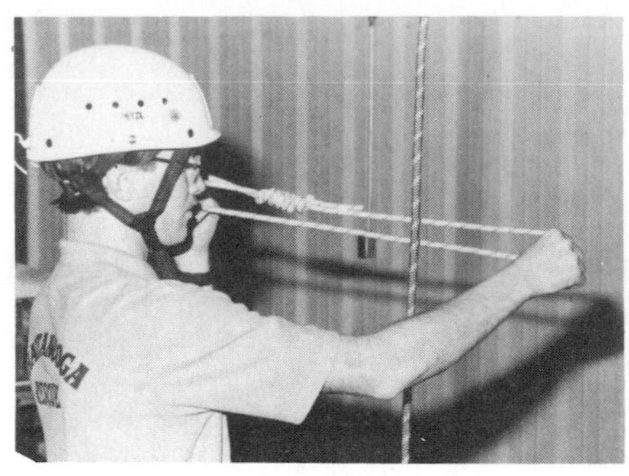
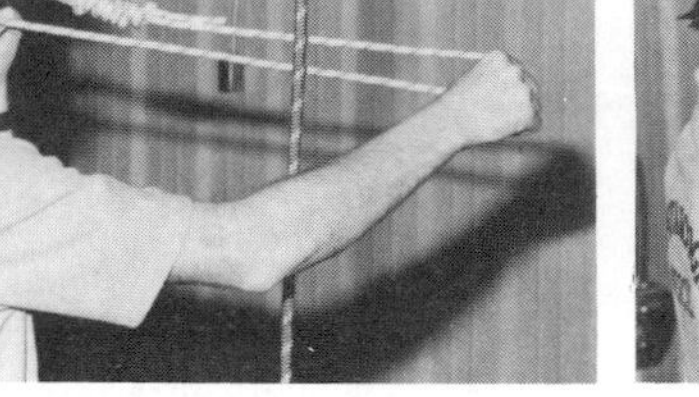

Fig. 10.1(a)

Fig. 10.1(b)

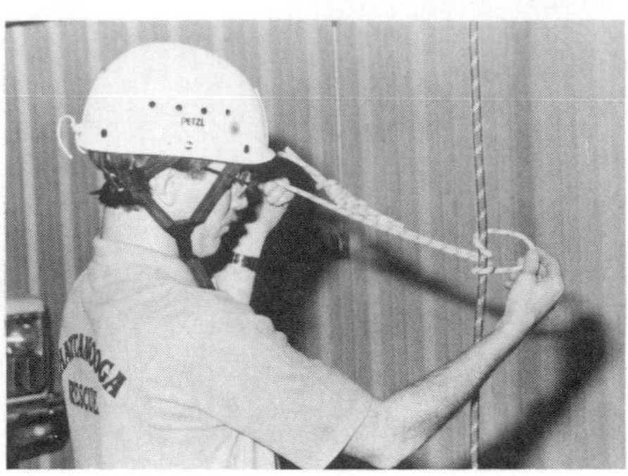

Fig. 10.1(c)

Fig. 10.1(d)

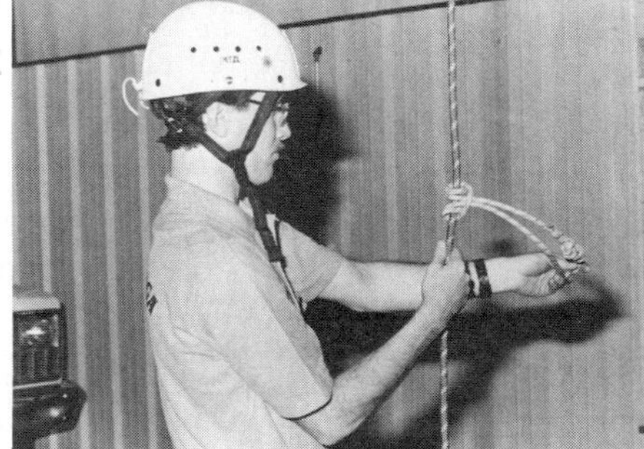

Fig. 10.1(e)

Fig. 10.1(f)

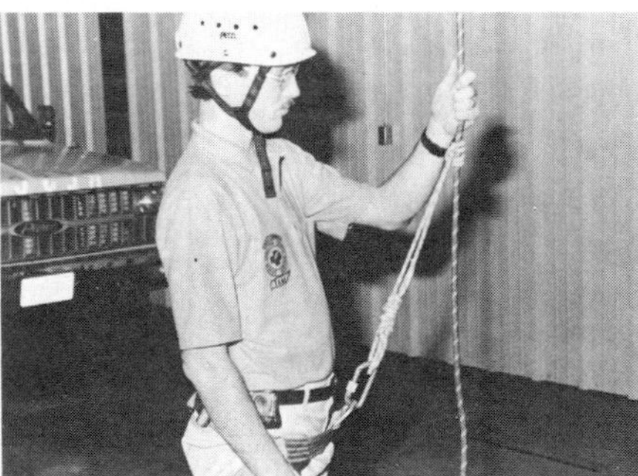

Fig. 10.1(g)

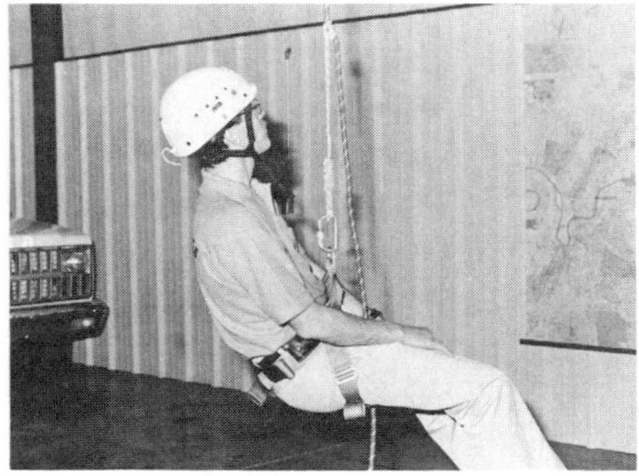

Fig. 10.1(h)

Weighting The Prusik Knot

8. Clip your seat harness carabiner into the end of the Prusik loop. Lock the carabiner *(10.1g)*.
9. Now sit down so that the Prusik knot locks on the rope and the loop supports your weight. If the knot begins to slide, lock it further by holding it in your hand and pressing it closed using the thumb as in step #7.
10. Stand up and take your weight off the knot.
11. Slide the knot 6 inches up the rope. Unlock the knot by grasping it with your hand as before. But this time pull the bar of the knot with your thumb **away** from the main rope so that it loosens. As you do this, move the knot up the rope with your hand around it. Be certain that you have taken your weight off the Prusik sling.
12. After you slide the knot up the desired length, relock it as you did before. Sit down with the knot supporting your body weight *(10.1h)*.
13. Repeat this process until you cannot slide the knot any farther up.
14. If you cannot get back down, have someone allow slack into the rope so you have your weight off the knot. Remove the knot from the rope.

You have now examined the basics of using a Prusik knot. Because you used only one knot, your movement was obviously limited. In the actual practice of ascending with Prusik knots, you would not shift your weight to the ground in order to raise the Prusik knot. Instead, you would shift your weight to another Prusik knot, which you would also be advancing up the rope.

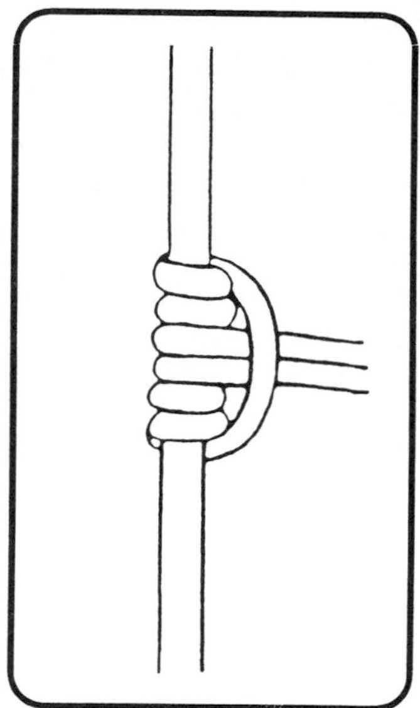

Fig. 10.2 Six Coil Prusik

Greater Holding Power: The Six Coil Prusik

The four coil Prusik knot, described above, will be adequate for most vertical applications. However, for greater holding power, such as where the rope is made slippery by mud or ice, or where greater weight is involved, the Six Coil Prusik knot may be desired *(see Figure 10.2)*. The Six Coil Prusik is created by running the Prusik sling one more time through the loop. This is known as "adding another wrap."

Though offering the potential of greater holding power, the Six Coil Prusik knot may be more difficult to manipulate than the Four Coil Prusik knot.

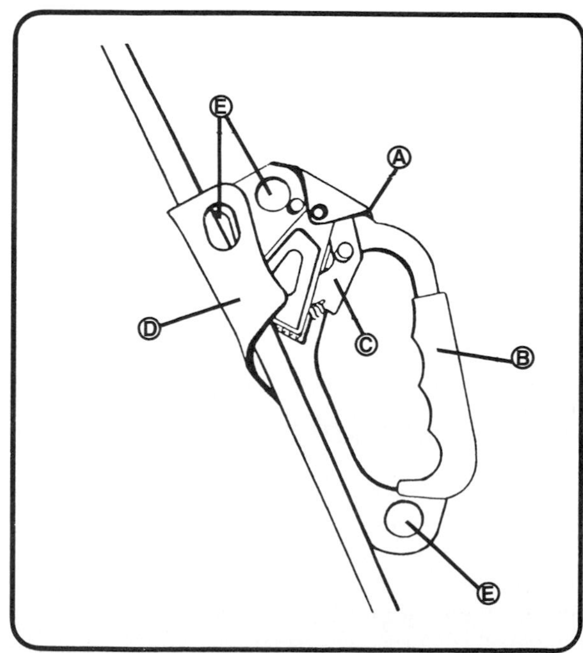

Fig. 10.3 Typical Handled Ascender (Right Hand)

Handled Ascenders

In recent years, most people have begun to use mechanical ascenders. They tend to be easier, more efficient, and more convenient to use than knots. One type of mechanical ascender is the **handled ascender.** Figure 10.3 shows a typical handled ascender. While some models, may not have t' parts exactly as shown here, they all work in essentially same manner.

Ⓐ **The Frame.**

This is what the parts are attached to and what mostly determines the ascender strength. The frame may be fabricated from extruded, stamped or plate aluminum or, in some cases, from cast aluminum.

Ⓑ **The Handle.**

This may be an integral part of the frame or it may be attached to the frame with rivets or bolts. In some designs, the handle is molded to fit the contour of the hand and can be comfortably used with gloves or mittens.

Ⓒ **The Safety Lever.**

When in the locked position, the safety lever prevents full downward movement of the cam. This is designed to keep the ascender from accidentally coming off the rope. (However, see warning below.)

Ⓓ **The Nose.**

This forms the inside channel into which the cam pushes the rope so that the ascender stays on the rope.

Ⓔ **Tie-In Points.**

These are usually an integral part of the frame and are used to fasten a sling that is attached to the user. Most ascenders have tie-in points at the bottom, so that the ascender can support a person below them. Some ascenders also have an additional tie-in point at the top. These are used in certain ascending systems where the ascender is pulled along as the user advances up the rope.

■ Right- and Left-Handled Ascenders

Most handled ascenders are manufactured in right- or left-handed models. Some manufacturers color code the ascender so you can tell the difference. However, you can tell the difference between a right- or left-handed model by doing the following:

1. Turn the ascender so that the opening for the rope between the nose and the cam is facing you.
2. The left-handed model will have its handle to your left. The right-handed model will have its handle to your right.

WARNING NOTE

Handled ascenders *can and do fail when misused.* The following are the most common modes of failure:

- ***Frame Breakage.***
  Ascender frames constructed of cast aluminum can crack or break when subjected to the high stress of being dropped. Cast aluminum frames can crack under their paint so there is no visible sign of damage. Previously-owned ascenders of any type may have been subjected to stresses that could result in failure. For this reason, a used ascender should not be purchased unless its complete history is known.
- ***Rope Damage.***
  This occurs when there is so much stress on the ascender that the teeth of the cam tears the rope sheath. Sheath tearing has been known to occur with as little weight as 800 pounds on the ascender and rope. For this reason, ascenders with toothed cams should never be used in situations where more than one person's body weight is involved. One example of where handled ascenders should not be used is in rescue hauling systems.
- ***Rope Slipping Out of an Ascender.***
  Handled ascenders are designed to operate most efficiently and safely on vertical ropes and when moved in a direct line with the rope. Handled ascenders have been known to slip off the rope when they have been pulled away from the rope. This could happen when using handled ascenders on a rope that is not completely vertical, but at an angle, such as ascending a sloping highline or traversing rope along a ledge.

  If there is a chance that the ascender might be operated in this kind of situation—which they were not designed for—then a safety carabiner should be clipped across the ascender and the rope as shown in Figure 10.4. This may not prevent the ascender from slipping from the rope, but the sling will remain connected to the rope via the carabiner.

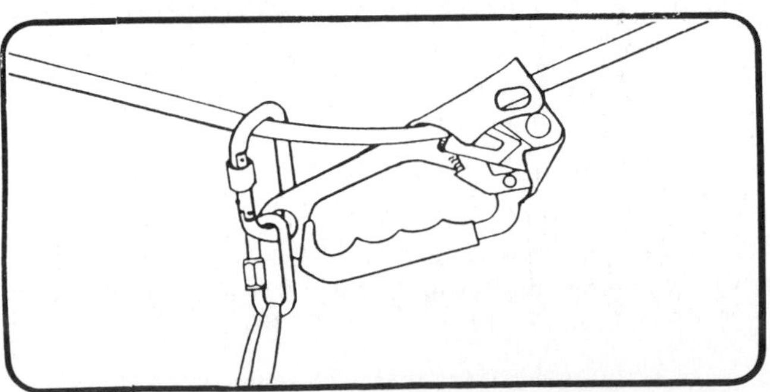

Fig. 10.4 Clipping Crab Across Ascender/Rope

Ascender Slings

To be used safely and efficiently, ascenders must be attached to the user's body with connections known as **slings.** These slings may be made either of webbing or of rope. Many people prefer rope for the following reasons:

- An appropriate design of rope may abrade less easily than webbing.
- Rope will operate better than webbing if the sling has to go through a roller device, such as used in certain ascending systems.

If you decide to construct ascender slings from rope, you may be able to use line that is a smaller diameter than what is normally used for the main line rope. A sling made of 3/8 or 5/16 rope might be appropriate as long as it has an adequate safety factor.

Most experienced ascender users prefer slings constructed of static rope. Static rope tends to stretch less than dynamic rope, so slings of static rope tends to transfer the energy involved in ascending directly to the ascenders, rather than absorbing it.

The actual length of the slings will depend on factors such as the proportion of your body and the type of ascending system you use. These factors are explored later in this chapter.

Tying the Slings to the Ascender

Because of the different designs for ascender tie-in points, the actual method of connecting the slings to the ascender will vary depending on the individual brand and design.

In connecting the sling to an ascender, remember that a sharp bend in a rope will diminish the strength of the rope (see Chapter 3, Rope, on the 4:1 rule). If the rie-in point of an ascender is wide enough, then the sling may be tied directly into it using a Figure 8 Follow Through knot. If the tie-in point is narrow so that it will create too sharp a bend in the rope, then the rope should be first clipped into a carabiner or snap link. The carabiner or snap link is then clipped directly into the carabiner tie-in point.

WARNING NOTE

Some recommend against clipping a carabiner or snaplink directly into an ascender with a *cast* aluminum frame. It is thought that under a severe impact, the carabiner or snap link might cause the cast aluminum frame to crack.

A Figure 8 Follow Through knot is tied in the end of the sling to be attached to the person. If the sling is to go into a seat harness, then the knot is connected to the carabiner, which is clipped into the seat harness front tie-in point. If the sling is to go to a foot, then a large loop in the Figure 8 knot will be needed. This should be large enough to slip through a "chicken loop" and onto a boot. (More on this in the section on "Ascending Systems" below.)

Using a Handled Ascender *(see Figure 10.5)*

1. For this exercise, you will need only one handled ascender. It may be either right- or left-handed, depending on which feels more comfortable.
2. Using either webbing or rope, create an ascender sling. For this exercise, the total length after tying should only be about two feet. Attach the sling to the lower tie-in hole of the ascender either directly or with a carabiner following the guidelines outlined above. Tie a loop in the end of the sling that is to attach to the person.
3. Don a sewn, manufactured seat harness with leg/thigh supports. Clip a locking carabiner into the seat harness tie-in point.
4. Clip the seat harness carabiner into the loop of the ascender. Lock the carabiner.

Attaching the Ascender onto the Rope

5. Securely anchor a main line, static, kernmantle rope vertically so that it will support a person's weight with an adequate safety factor. As in the exercise with the Prusik knot, there must be a means for the person on the rope of getting back down once he has pushed the ascender up as far as it will go *(10.5a & b).*
6. Take the ascender in one hand and press the safety lever so that the cam swings down and open. The literature that comes with the ascender should provide specific instructions for triggering the safety lever. The triggering is usually done with the thumb of the hand that is holding the ascender. With some ascenders, this triggering can be done as the hand grasps the handle. With other ascenders, the hand must be grasping the entire ascender, with the handle against the palm and the index and middle finger around the nose *(10.5c & d).*
7. Place the ascender on the rope at about eye level with the nose up. Do this by holding the cam down with the trigger. Place the ascender so that the main line rope runs in the channel of the nose. Release the cam so that it presses the rope into the channel of the nose and the ascender remains on the rope. Release the safety lever, making sure it now locks the cam on the rope. **WHEN YOU ARE ASCENDING, NEVER TOUCH THE SAFETY LEVER UNLESS YOU INTEND THE ASCENDER TO COME OFF THE ROPE** *(10.5e).*

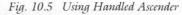

Fig. 10.5 Using Handled Ascender

Fig. 10.5(a)

Fig. 10.5(b)

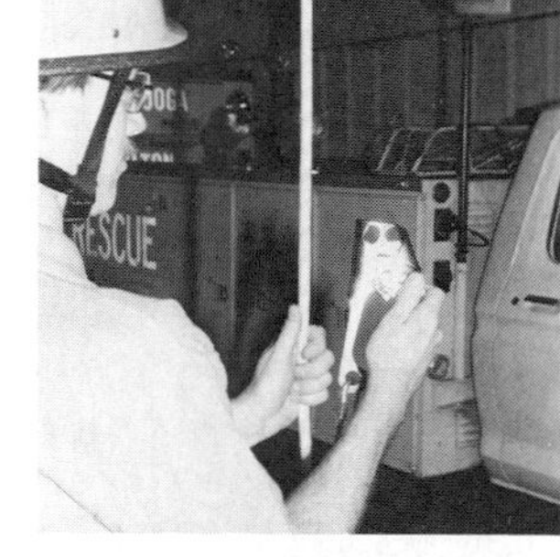

Fig. 10.5(c)

Fig. 10.5(d)

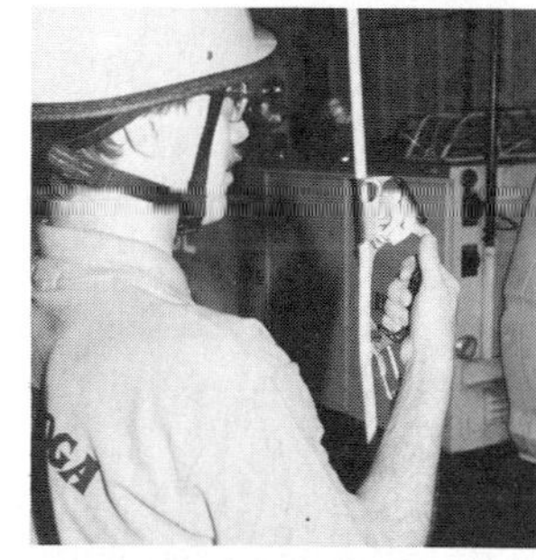

Fig. 10.5(e)

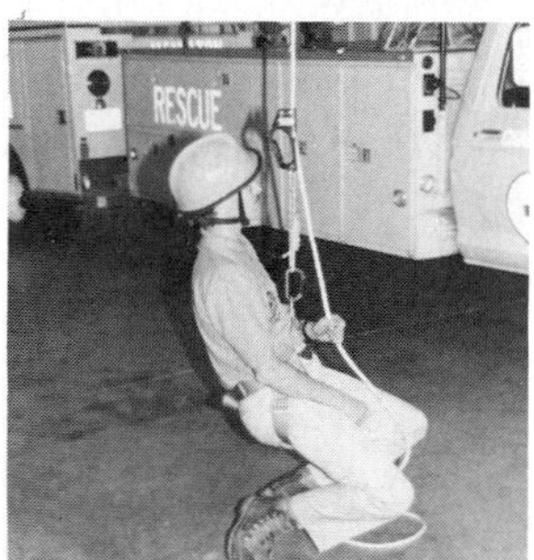

Fig. 10.5(f)

Fig. 10.5 (g)

8. Sit down so that the sling comes taut and the ascender supports your weight on the rope. (If the rope stretch is such that you end up on the ground, reset the ascender up the rope so that you are being supported *[10.5 f]*.)
9. Now, stand up so that your weight is off the sling and push the ascender up a foot. Sit down again *(10.5g).*

Backing the Ascender Down the Rope

10. Stand up so that your weight is completely off the ascender. While grasping the upper part of the ascender around the nose, take your thumb and move the cam down while slightly lifting the ascender. **DO NOT TOUCH THE SAFETY LEVER.** Move the ascender down a foot and release the cam so that it reengages the rope. Sit down again on the ascender until it supports your weight.

You have now examined the basics of using a handled ascender. Because you used only one ascender, your movement obviously was limited. In the actual practice of using ascenders, you would not shift your weight to the ground in order to raise the ascender. Instead, you would shift your weight to another ascender, which you would also be moving up the rope.

Creating an Ascending "System"

As seen from the previous exercise, one ascender can hold you on the rope. But to effectively move up the rope, you need two or more ascenders. When two or more ascenders, whether they be friction knots or mechanical ascenders, are worked together to travel up a rope, the arrangement is known as an ascending **system.**

Dozens of different ascending systems are used in rope work. They differ in terms of what kinds of ascenders they use, in what combinations of ascenders are attached to the body, what parts of the body they are attached to, and in other ways. Each ascending system may offer advantages in safety, in ease of use, in speed of movement, and in ease of movement. No one ascending system includes all these advantages. Every system has at least one drawback.

When you are beginning to work with ascending, you should initially seek ascending systems that combine safety and ease of use. Some characteristics of ascending systems that contribute to safety and ease of use are:

- The system should require more use of your legs and feet and less use of your arms and hands. Your legs and feet are stronger and have greater stamina than your hands and arms.
- The system should help hold you upright on the rope with your body weight over the legs. This requires less arm strength, encourages use of the legs, and contributes to safety.
- The system should be able to support you in a sitting position while you are on the rope. Ascending is a tiring activity and short periods of rest while sitting are essential.
- The system should have attachments to the seat harness (and in some cases to a chest harness). Some systems have attachments only to the feet and depend on arm and body strength to hold the user upright. These systems are commonly referred to as "death rigs."

It is important that whatever system you use, it be fine-tuned to your body height and build. A well-tuned system makes rope ascending no more work than climbing a ladder. A poorly fitted, untuned system will quickly exhaust even the most physically sound person.

WARNING NOTE

Maintaining Three Points of Attachment in Ascending

A commonly accepted safety guide in ascending is the "three points of attachment" rule. This means that at all times while ascending, a person is attached to the rope in at least three places. In many ascending systems, such as the ones shown in this chapter, three ascenders are used on the rope. In situations where only two ascenders are used, it may be necessary to "tie off short." (See Below)

Tying Off Short

Tying off short is the safety procedure of tying directly into the main rope to ensure an additional attachment. It is used in certain situations during ascending when there are less than three points of attachment on the rope. Examples include:

- When a person is using only two ascenders and he must take one of the ascenders from the rope for a procedure such as moving past a knot or going over an edge of a cliff or building.
- When, for any other reason, there are less than three ascenders on the rope.

To tie off short, perform the following *(see Figure 10.6):*

1. Reach down below the lowest ascender to the slack rope hanging below you *(10.6a).*
2. Take a large bight of that rope and pull it up.
3. Tie a Figure 8 on a Bight in the bight.
4. Clip the Figure 8 on a Bight into a spare carabiner that is clipped into the seat harness *(10.6b & c).*
5. Make your move past the obstacle *(10.6d).*
6. When finished with this safety, unclip the Figure 8 on a Bight from the carabiner, untie the knot, and allow the rope to drop back down below you *(10.6e & f).*

Tying Off Short *(Figure 10.6).*

If it happens that you are ascending with only two ascenders, you can create a continuous safety by tying off short:

1. Tie off short as soon as you have ascended high enough so that a fall would injure you.
2. Leave the Figure 8 on a Bight knot clipped into your seat harness carabiner as you continue to ascend.
3. Ascend until you create enough slack in the rope that it no longer offers adequate protection from a fall (this distance is usually around ten feet, or one story).
4. Untie the old knot, drop the slack out of the main line rope, and retie a new knot closer to you.
5. Repeat this procedure until you have finished the climb.

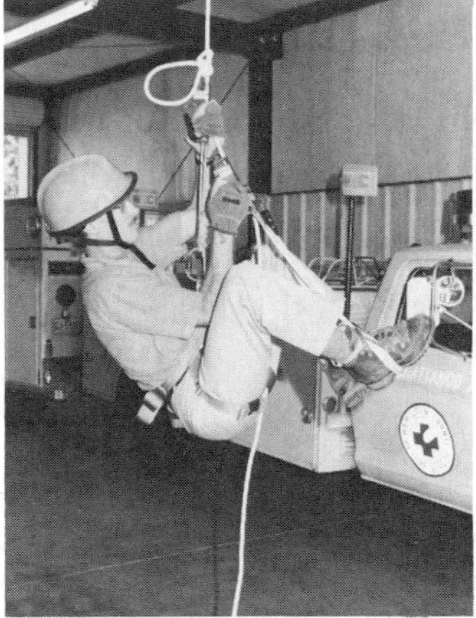

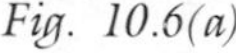

Fig. 10.6(a)

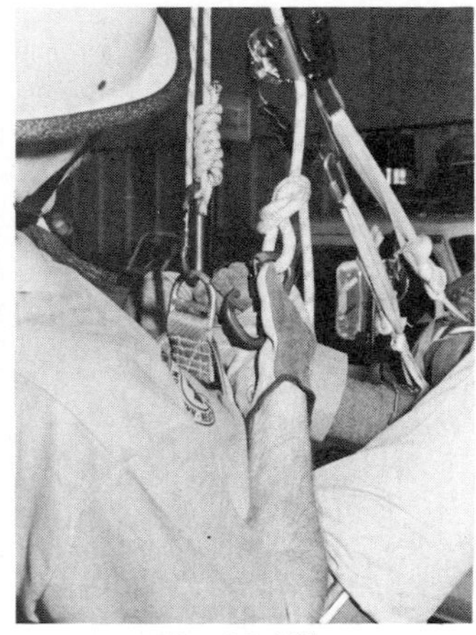

Fig. 10.6(b)

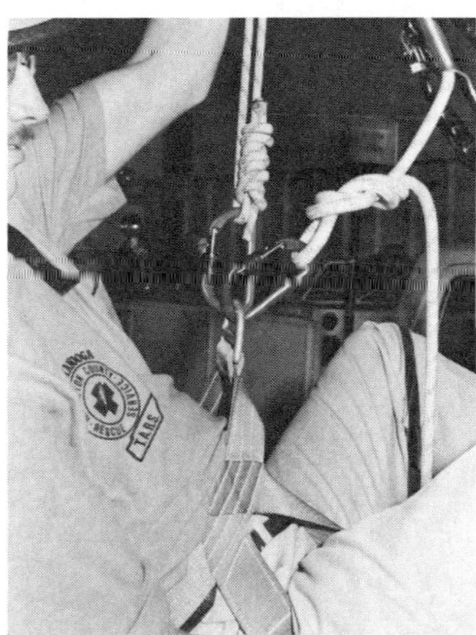

Fig. 10.6(c)

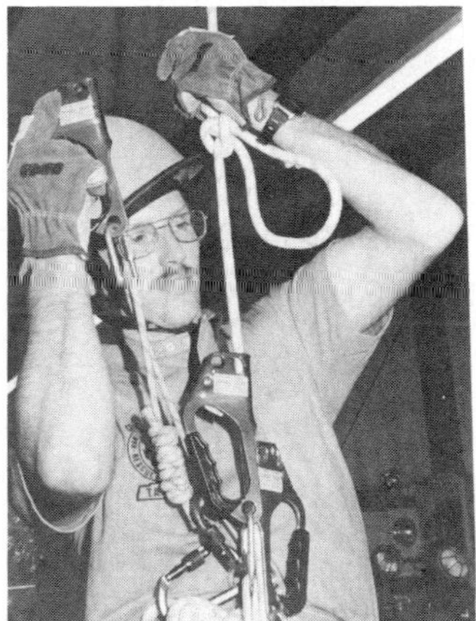

Fig. 10.6(d)

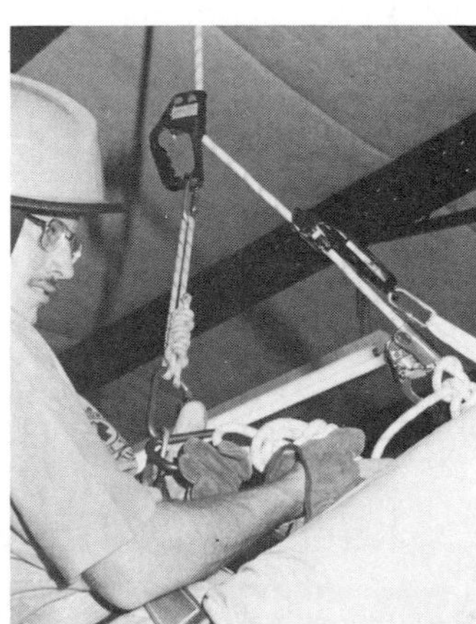

Fig. 10.6(e)

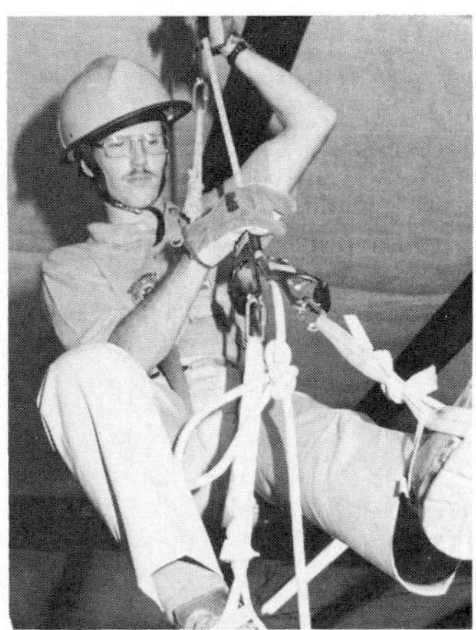

Fig. 10.6(f)

Chicken Loops

One safety feature that should be used when using ascenders that are attached to the foot is called the **chicken loop** *(see Figure 10.7).* In ascending systems where feet are used, ascenders are often attached to the feet via foot stirrups. These stirrups are often simple loops of webbing or rope sling. In addition, chicken loops should be used around the ankles to serve the following purposes:

- It helps prevent the feet from slipping out of the stirrups while the user is ascending.
- Should the person ascending lose an upper body sling and fall over backwards, the chicken loops keep the feet in the stirrups and prevent a fall to the ground.

Chicken loops are constructed of webbing (1 inch or larger) and stitched, or securely tied, into a loop that is larger than the ankle but smaller than the boot. **THE CHICKEN LOOP MUST BE WEIGHT SUPPORTING.**

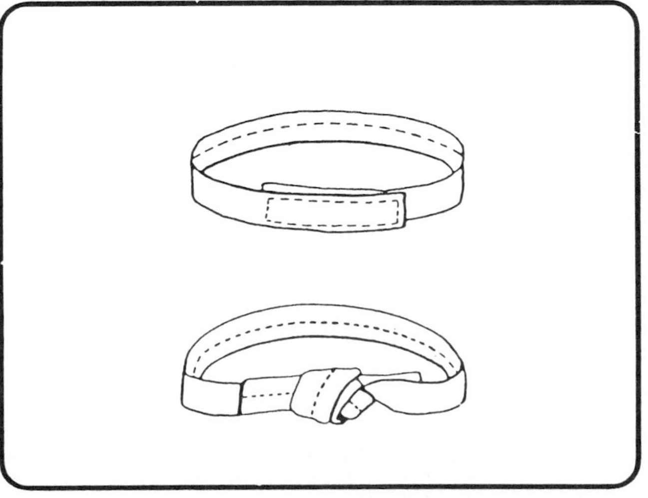

Fig. 10.7 Chicken Loops

Ascending With Example Systems

Fig. 10.8 Ascending Practice System

The following portions of this chapter will explore ascending techniques with some example systems. These systems may be used with Prusik knots (or other comparable friction knots), cams, or handled ascenders. However, in these examples, the systems will be described in use with handled ascenders, since they are more easily manipulated when learning ascending techniques.

WARNING NOTE

Ascending is a *strenuous activity,* to be attempted only by those persons known to be in good physical condition. Those persons with preexisting cardiac or pulmonary conditions, obesity, or other medical conditions that might be exacerbated by exertion, should consult with a physician before attempting ascending.

An Ascending Practice System

Because ascending is a new and unique activity for most people, an **Ascending Practice System** can be very helpful in learning the technique in a safe environment.

This system consists of the following elements *(see Figure 10.8)*:

A. A main line, static kernmantle rope on which the practice takes place. The rope runs over a directional pulley that is securely attached to a beam in the ceiling of a building (or a **very strong** tree limb outside), and down to a securely anchored lowering device, such as a Brake Bar Rack or a Figure 8 descender.

B. In preparation for the practice, only enough rope hangs vertical for the person to get onto the rope.

C. The lowering device (Brake Bar Rack or Figure 8 descender) is locked off until the person begins to ascend. The person operating the braking device allows enough rope through to keep the student high enough off the deck to use his ascenders, but low enough that a fall will not injure him.

CAUTION: THE OPERATOR OF THE BRAKING DEVICE MUST BE THOROUGHLY EXPERIENCED WITH ITS USE AS A LOWERING DEVICE. THE FLOOR UNDER THE PRACTICE ROPE SHOULD BE COVERED WITH A MAT.

D. In addition to a person operating the braking system, an assistant can hold the rope below the climber to help the student as he begins his ascent.

E. When the student completes the ascending cycle, the person operating the braking device lowers the student back to the deck so he can get off the rope. **THE STUDENT MUST ALWAYS STOP ASCENDING WHILE THERE IS ENOUGH ROPE TO LOWER HIM TO THE GROUND.**

A Three Ascender System

Figure 10.9 shows a basic three ascender system using handled ascenders. Because the proportions of each person's body differ, no specific dimensions for the sling attachments can be shown. However, you should tailor the slings for your own use by using the following guidelines:

- The seat attachment (with optional chest harness) should be the top ascender on the rope when you are resting your weight on it. **The ascender must never extend out of your reach up the rope while standing upright in your footloops or sitting in your harness.**
- The ascender for your dominant foot (right foot for right-handed people) should come next on the rope. Its sling should be long enough to allow the ascender to be attached about mid-thigh when standing.

Fig. 10.9
Basic Three Ascender System

- The ascender for the second foot should be the last one on the rope. Its sling should be just enough shorter than the second sling to allow the third ascender to be immediately below the second ascender when both ascenders' slings are tight.

When all the knots have been tied, make certain that they are contoured well and pulled down tightly.

Fig. 10.10 Using the Three Ascender System

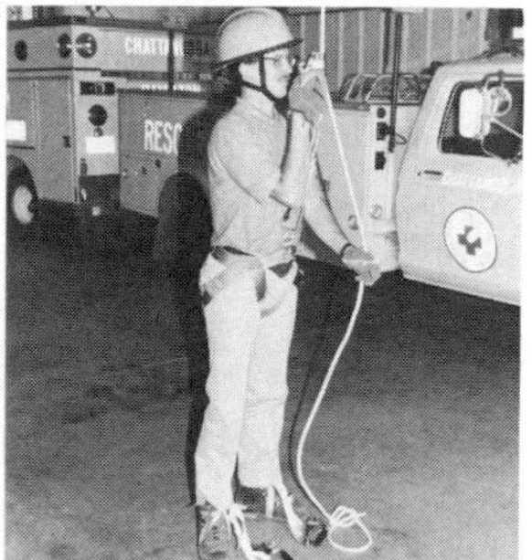

Fig. 10.10(a)

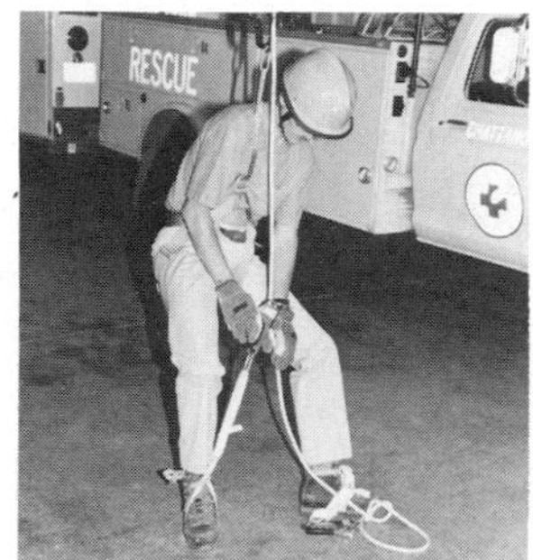

Fig. 10.10(b)

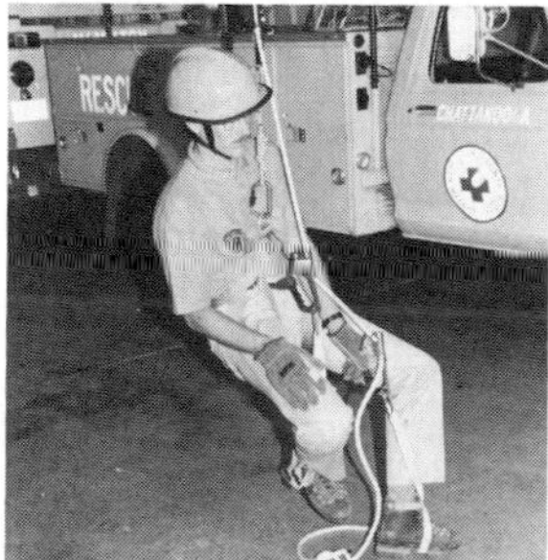

Fig. 10.10(c)

Fig. 10.10(d)

Fig. 10.10(e)

Fig. 10.10(f)

Fig. 10.10(g)

Procedure for Using The Three Ascender System

(see Figure 10.10)

1. For this exercise, use the Ascending Practice System as described above. Before you start, be certain that someone is attending the lowering device and that initially it is locked down tight.
2. With one hand, reach up and pull the main line rope taut. Attach the ascender for the seat harness as high up as you can get it *(10.10a)*.
3. Attach the next ascender down (the one to your dominant foot) under the seat harness ascender *(10.10b)*.
4. Attach the third ascender under the second one *(10.10c)*.
5. To pretest the system, alternately sit down on the seat ascender and on each foot ascender to make certain they are holding *(10.10d)*.
6. If possible, have another person assist you in getting started by having them hold the rope down close to the ground. Otherwise, you may have to hold the rope down yourself. This is necessary in getting started because there may not be enough rope weight initially to cause the rope to slide through the ascenders as they are pushed up. Once you have gotten far enough off the ground, the rope weight will cause the rope to automatically slide through the ascenders as they are pushed up.
7. Shift your weight onto the foot stirrups, and off the top ascender. Push the top ascender up as far as you can *(10.10e)*.

   NOTE: When maneuvering your body up so you can raise an ascender, use your leg strength as much as possible. The more you use your legs and feet, and the less you use your arms and hands, the less fatigued you will become.
8. Sit down so that you are supported by the top ascender. Take the weight off the next ascender in line by lifting the foot attached to it. Raise this foot ascender up as far as it will easily go *(10.10f)*.
9. Take the weight off the remaining foot ascender by lifting the foot attached to it and raising the ascender up as far as it will go.

   NOTE: As you progress up the rope, the person attending the lowering device should slowly let rope out so that you remain a safe distance off the ground. But you should be up high enough so that the rope pulls through the ascenders by its own weight as you raise the ascenders.
10. Continue the cycle by repeating steps 7, 8, and 9 until you start running out of rope or you become fatigued. Have the operator of the lowering device lower you back to the ground *(10.10g)*.
11. Remove all the ascenders from the rope.

Other Ascending Systems

The Three Ascender System, detailed here, is one example of an ascending system. There are dozens of different ascending systems in existence that can be used according to such needs as:

- Distance to travel up the rope.
- Physical strength.
- Stamina.
- Differences in male & female physique.
- Speed.

With a proper amount of research and experimentation, each person can find the type of ascending system just right for him. One good place to begin the search for the proper ascending system is the book, *ON ROPE* by Padgett and Smith (see Bibliography).

Ascending Over an Edge

Usually the most difficult maneuver in ascending occurs when the person ascending has reached the top of a cliff or a building and has to go over the edge.

Remember, as a general rule, as in rappeling, if the strength of the anchors on the main line rope allows it, the higher up the rope is anchored, the easier it is to go over an edge.

The technique for getting over an edge depends on the particular nature of the edge.

A. If the Edge Has a Gradual Rollover:

1. Ascend until the top ascender is about to make contact with the wall.
2. Push yourself away from the wall with one hand and with the feet.
3. Raise the top ascender above the contact point.
4. Be careful that you do not get fingers caught between the hardware or rope and the wall.
5. As you move up and your weight is on the rope above the contact point, the other ascenders should follow more easily.

B. If the Edge Is Undercut:

1. Ascend until the top ascender is about to make contact with the wall.
2. Bring the lower descenders up as far as it is comfortable to do so.
3. Try to work the top ascender over by pushing away from the wall with one hand and with the feet.
4. If this is impossible to do, tie off short into the main line rope.
5. Remove the top ascender from the rope and move it up and over the edge and immediately clip it into the rope.
6. Begin ascending again. It is likely that the two remaining ascenders can now be eased over the edge once your weight is on the main line rope above the edge. If this is not the case, then follow the same procedure as you did with the top ascender. UNTIL YOU ARE IN A SAFE AND SECURE POSITION WITH NO DANGER OF FALLING, NEVER TAKE MORE THAN ONE ASCENDER OFF THE ROPE AT ONE TIME.

Changing Over

"Changing over" means switching from an ascending mode to a rappeling mode, or from a rappeling mode to an ascending mode, while still on the rope. It is a skill that adds to vertical competency, and is particularly useful in emergency situations such as when you rappel to the end of a rope and find it does not reach the bottom.

Equipment Needed for Changing Over

In all cases:

- Two locking carabiners in the seat harness tie-in point—the one that is in use at the time and a spare to be used when changing over.
- Equipment slings. To these are attached the equipment not in use at the time (the rappel device while you are ascending, the ascenders while you are rappeling).

PROCEDURES FOR PERFORMING CHANGEOVERS

Changing over from Ascending to Rappeling
(see Figure 10.11)

1. For this exercise, use the Ascending Practice System described earlier in this chapter. Before you start, be certain that a responsible and experienced person is attending the lowering device and it is initially locked down tight.
2. Using a three ascender system, begin ascending on the rope.
3. When you reach the point where you want to changeover, stop your ascending *(10.11a)*.

4. Remove your rappel device from your equipment sling and clip it into the second locking carabiner that is attached to your seat harness tie-in point and currently not being used in the ascending system. (If the rappel device is a Brake Bar Rack, lock the carabiner *[10.11b]*.)
5. Sit down on your seat harness ascender (the top one) so that your weight is on it.
6. Take your weight off the foot ascenders. Move these ascenders back down the rope so that there is slack in the rope between the foot ascenders and the seat ascender. **DO NOT TOUCH THE CAM SAFETY LEVERS AND DO NOT REMOVE THE ASCENDERS FROM THE ROPE AT THIS TIME.**
7. Lace the rappel device onto the slack rope between the foot ascenders and the seat ascender. Remove all slack in the main line rope between the rappel device and the top ascender *(10.11c)*.
8. Lock the rappel device off securely. Make certain there is no slack between the rappel device and the top ascender. (If you have not done so already, lock the seat harness carabiner that attaches the rappel device to the seat harness *[10.11d]*.)
9. Move a foot ascender back up the main line rope only far enough so that when you put your weight on it, it will remove weight from the seat harness (top) ascender.
10. Shift your weight to the foot ascender and remove weight from the seat ascender.
11. Remove the seat ascender from the rope *(10.11e)*.
12. Sit down so that the rappel device takes your weight. Remove weight from your foot ascender by lifting your foot *(10.11f)*.
13. Remove all remaining ascenders from the rope *(10.11g)*.
14. One by one, remove ascender slings from your body. To secure them, wrap them around the ascender they are attached to, and clip them into your equipment sling.
15. Unlock the rappel device and proceed with the rappel *(10.11h)*.

Fig. 10.11 Changing Over (Ascend to Rappel)

Fig. 10.11(a)

Fig. 10.11(b)

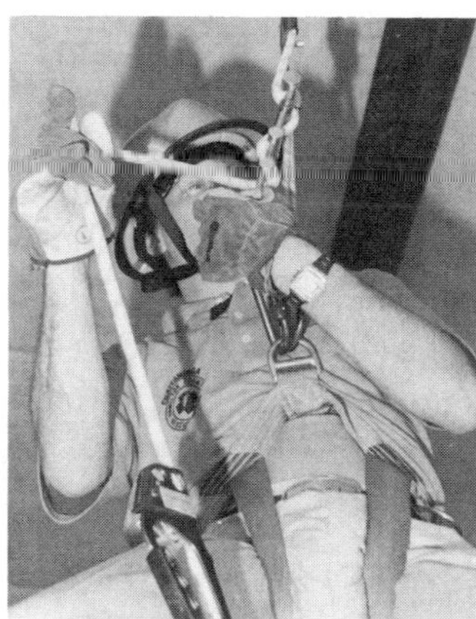

Fig. 10.11(c)

Fig. 10.11(d)

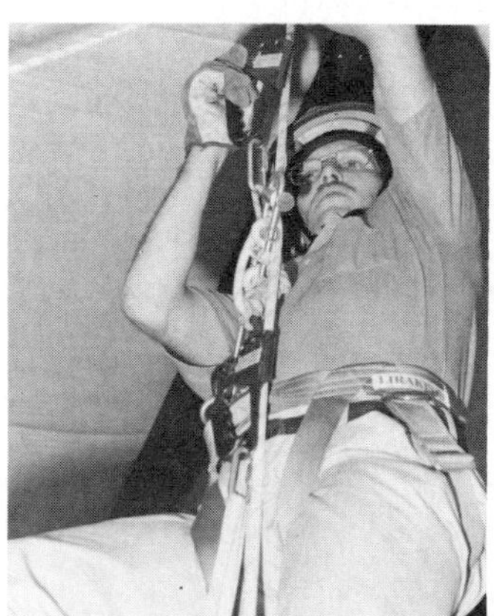

Fig. 10.11(e)

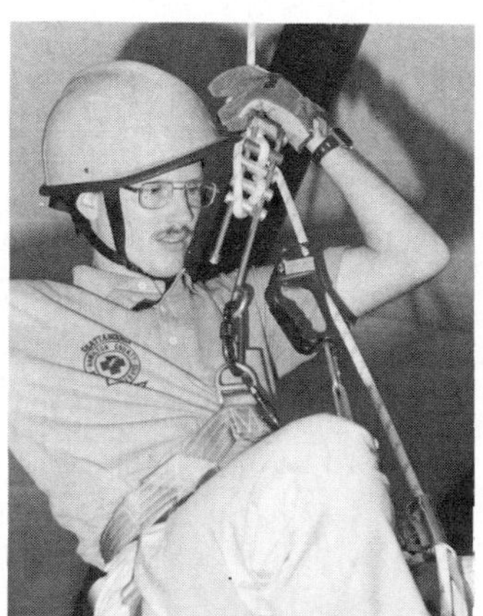

Fig. 10.11(f)

Fig. 10.11(g)

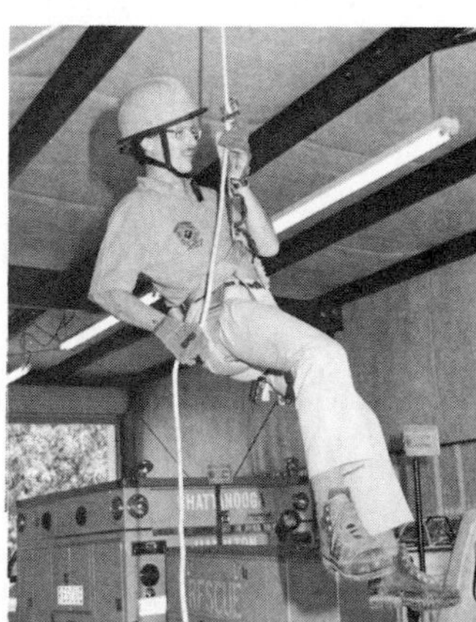

Fig. 10.11(h)

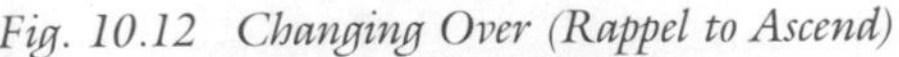

Fig. 10.12 Changing Over (Rappel to Ascend)

Fig. 10.12(a)

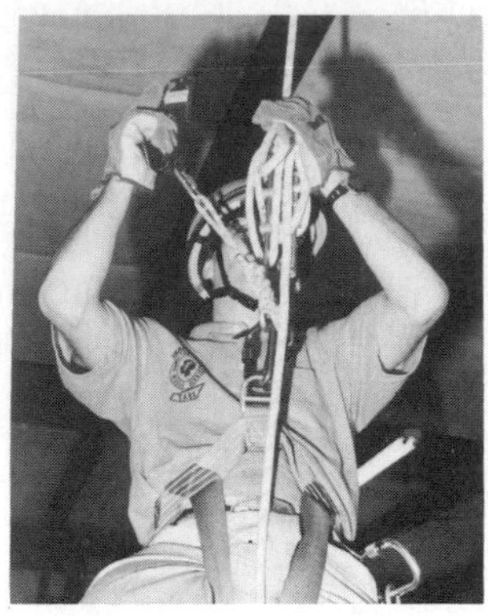

Fig. 10.12(b)

Fig. 10.12(c)

Fig. 10.12(d)

Fig. 10.12(e)

Fig. 10.12(f)

Changing Over from Rappeling to Ascending

(see Figure 10.12)

16. Stop the rappel at the point where you want to begin the changeover. Lock off the rappel device securely, following guidelines in Chapter 9, "Rappeling" *(10.12a)*.
17. Remove the seat harness ascender (top ascender) from the equipment sling. Clip the sling into the spare carabiner in the seat harness front tie in point (the one currently not being used). Lock the carabiner *(10.12b)*.
18. Place the top ascender on the rope as far up as you can push it. It is important that there be no slack in this ascender sling. One way to achieve this is to let slack through the descender until the top ascender sling becomes taut.
19. Place the other ascenders on the rope connected to proper slings via the foot. If any of these interfere with the position of the rappel device, then back them off down the rope a couple of feet below the rappel device *(10.12c)*.
20. When all the ascenders are securely attached, unlock the rappel device and slowly let rope through it. When your weight is off the rappel device, remove the rappel device from the rope *(10.12d, e, f)*.

Extricating a Jammed Rappel Device

The ability to extricate a jammed rappel device from hair, clothing, etc., without the use of a knife, is an essential skill for the high angle technician. The skills and equipment required for this procedure are similar to those used in changing over.

Because of the real possibility of a jammed rappel device, or similar emergency occurring, it is wise to carry the following spare equipment when you are rappeling:

- Two ascenders, any type, or two Prusik slings.
- A spare, large, locking carabiner.

Procedure for Using Ascenders to Extricate a Jammed Rappel Device

(see Figure 10.13)

1. Using the Ascending Practice System, begin a rappel.
2. Assume that the rappel device is jammed *(10.13a)*.
3. Simulate this by locking off the rappel device securely *(10.13b)*.

4. Remove the seat (top) ascender from your equipment sling and clip the end of the sling into a spare locking carabiner. Clip the carabiner into the seat harness front tie in. Lock the carabiner *(10.13c)*.
5. Place the ascender on the rope above the rappel device. Slide it up as far as it will go.
6. Remove a foot ascender from your equipment sling and attach the end of the sling to your foot. Put the ascender on the rope above the rappel device *(10.13d)*.
7. Put your weight on the ascenders and remove the weight from the rappel device.
8. Remove the obstruction (hair, clothing, etc.) from the rappel device. (Simulate this by unlocking the rappel device and causing it to go slack on the rope *[10.13e]*.)
9. Replace the rappel device on the rope and lock it off so that there is no slack in the main line rope between the rappel device and the next ascender up.

   CAUTION: If you are using a Brake Bar Rack, make certain that when you lock it off, the braking hand rope does not get trapped *below* the top bar (see Figure 10.14). **It must remain *above* the top bar. Otherwise, the device will become jammed when the rope tension comes onto it.**
10. Put your weight on the foot ascender and remove your weight from the seat ascender.
11. Remove the seat ascender from the rope *(10.13f)*.
12. Shift your weight off the foot ascender and onto your rappel device *(10.13g)*.
13. Remove the foot ascender from the rope *(10.13h)*.
14. Remove both ascenders and slings from your body, secure them, and clip them into the equipment sling.
15. Unlock the rappel device and continue the rappel *(10.13i)*.

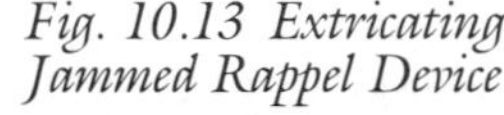
Fig. 10.13 Extricating Jammed Rappel Device

Fig. 10.13(a)

Fig. 10.13(b)

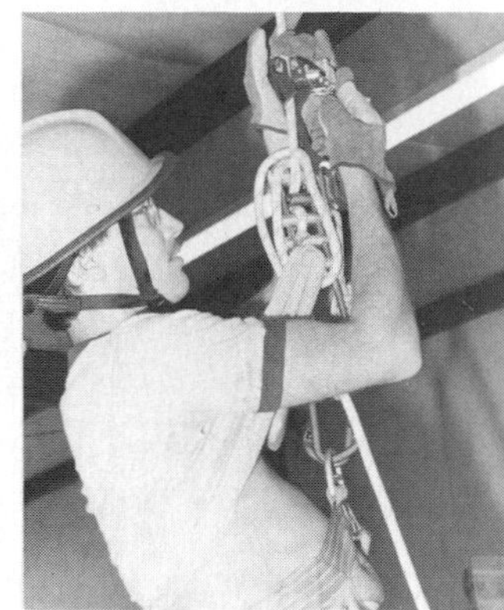
Fig. 10.13(c)

Fig. 10.13(d)

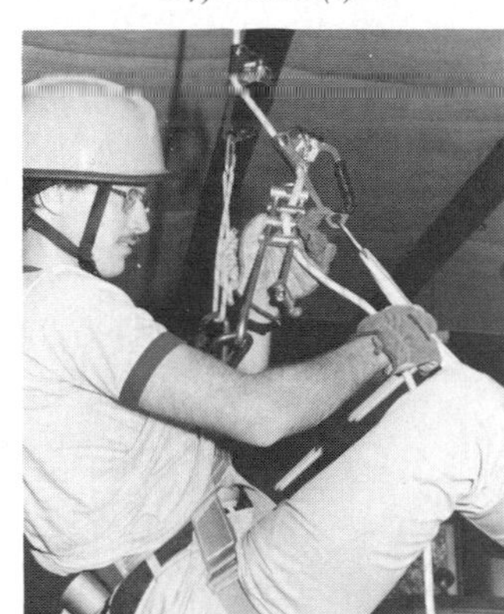
Fig. 10.13(e)

Fig. 10.13(f)

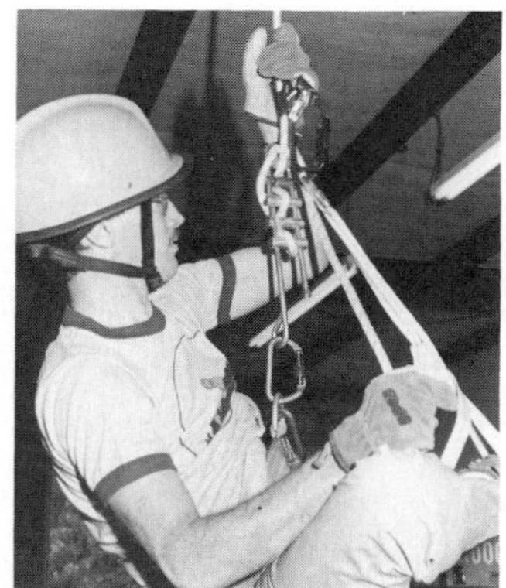
Fig. 10.13(g)

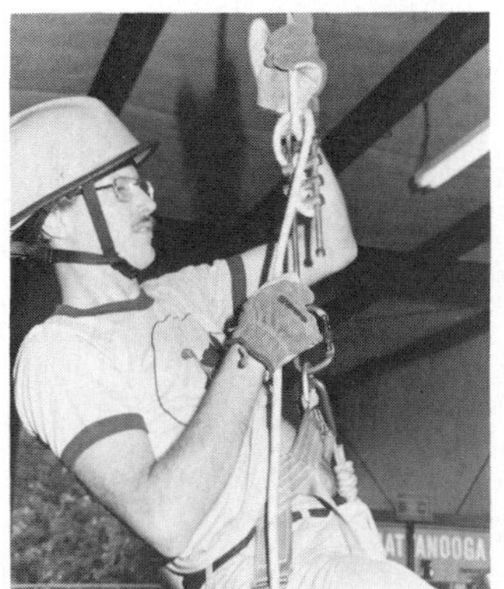
Fig. 10.13(h)

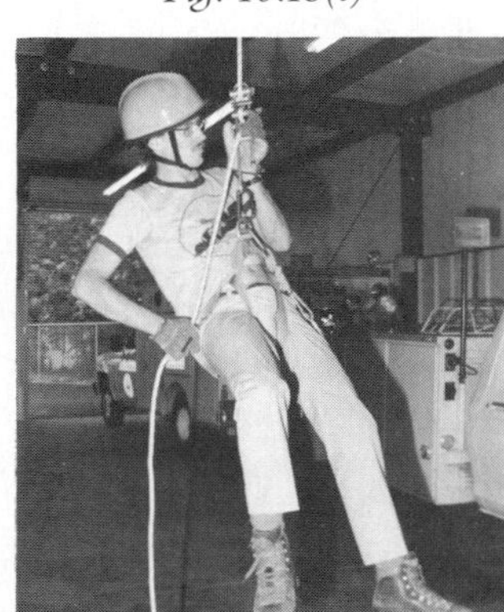
Fig. 10.13(i)

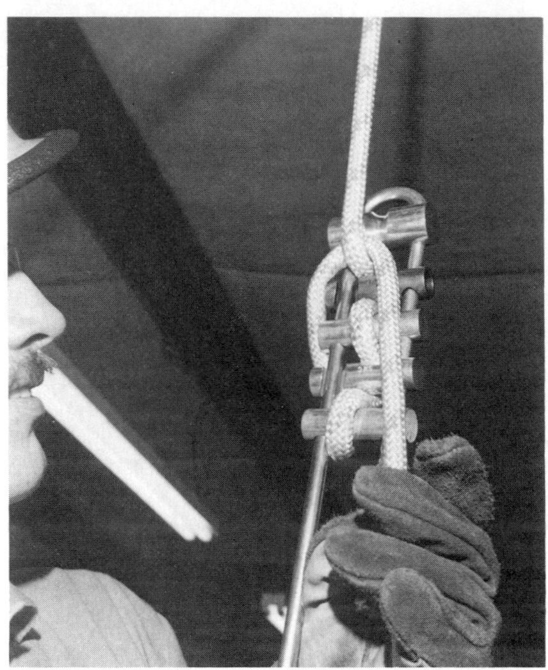

Fig. 10.14 Avoid Trapping Rope Below Top Bar

QUESTIONS *for* REVIEW

1. What are the two basic types of ascenders?
2. What are the differences between cams and handled ascenders?
3. Ascenders are attached to the user's body with _____, connectors made either of _______ or of ________.
4. What is the minimum number of ascenders required for ascending?
5. ________________ were the first type of rope grab devices used in ascending.
6. The Prusik knot will operate more efficiently if the rope it is made from is a ____________ diameter than the main line rope to which it is attached.
7. Name three possible modes of failure when using handled ascenders.
8. When you are ascending with a handled ascender, you must never touch the _________ ______ unless you intend the ascender to come off the rope.
9. List at least three characteristics of a good ascending system.
10. In considering the "three points of attachment" rule, what would be two ways in which a person could be attached?
11. __________ __________ __________ is the safety procedure of tying directly into the main line rope.
12. What is the safety feature used to prevent the feet from slipping out of an ascender sling?
13. To conserve strength during ascending, what parts of the body should be used more than others?
14. Usually the most difficult maneuver in ascending occurs when a person has to _______________________.
15. What kind of procedure would be appropriate if you rappeled to the bottom of a rope and found that the rope did not reach the bottom?

Chapter 11

The Rope Rescuer

Rope Rescue is the providing of aid to those in danger of injury or death in an environment where the use of rope and other related equipment is necessary to perform the rescue safely and successfully. Rope rescue is also sometimes called **technical rescue,** and certain aspects of it are also sometimes called **high angle rescue or vertical rescue.** There are also other related rescue disciplines that may involve some rope rescue skills, but which require additional skills and experience not addressed in this manual. **Cave Rescue,** for example, may involve rope rescue skills, but also involves hazards such as close confinement, darkness chill, and, at times, water hazards. **Swiftwater Rescue** may involve rope-handling techniques, but often requires skills in boat handling and knowledge of the character of running water. Different rope-handling techniques are used in swiftwater operations to avoid dangers not encountered in high angle work. A seat harness that may be adequate for high angle work could, for example, be dangerous when used in swiftwater.

A **Rope Rescue Technician** is one who is trained in the necessary skills for rope rescue, has shown that he is competent in the skills, and continuously trains to maintain these skills. Before progressing to qualification in rescue skills, the person should have evidence of some minimum personal vertical knowledge and skills. He should be able to:

- Demonstrate the proper use and care of rope.
- Demonstrate the proper use and care of other equipment employed in the high angle environment.
- Demonstrate the ability to tie correctly, confidently, and without hesitation those knots necessary for effective and safe work in the vertical environment (see Chapter 6, "Knots," for examples).
- Demonstrate the ability to rig safe and secure anchors.
- Demonstrate the ability to safely and confidently belay another person.
- Demonstrate the ability to rappel safely, confidently, and under control; the ability to tie off a rappel device to operate safely with hands free of the rope and then return to a safe and controlled rappel; and the ability to operate on the rope with the body in any position, including an inverted one.
- Demonstrate the ability to ascend safely; the ability to tie correctly, confidently, and without hesitation a friction knot and how to use it, and the uses and limitations of mechanical ascenders; and the ability to safely ascend a fixed rope using either friction knots or mechanical ascenders.
- Demonstrate the ability while on rope to confidently and safely change over from rappeling to ascending and from ascending to rappeling; and the ability to extricate oneself from a jammed rappel device (or similar problem) without the use of a knife.

In addition to personal vertical skills, the rope rescue technician must train and qualify in other areas before qualifying in the area of rescue. Among these skills are:

- Emergency Medical Skills. Team members should be trained **at least** to a level of Red Cross Advanced First Aid or the DOT First Responder. There should be enough team members trained **at least** to the Emergency Medical Technician level so that there will be two EMTs on each field team.
- Team Skills. Some examples are stretcher handling techniques and lowering and hauling systems.
- Communication Skills. These include not only the ability to communicate electronically, but the standard voice communications required in specialized team operations such as rescuer lowering and hauling systems.
- Other skills depending on the environment. These may include such skills as land navigation, snow and ice travel, or survival, for example.

While it is difficult to exactly define all the characteristics of an effective rescuer there tend to be certain traits that do stand out.

One of the most significant is that the focus of the rescue operation is the **subject,** the person who is being rescued. The rescuers must realize that the rescue subject is the reason for everyone's being involved in the rescue. This is a human being in distress, a fellow person in physical and, perhaps, mental pain. A rescuer may contribute to that distress by regarding the rescue subject more as an object than as a person. The rescuers must continually communicate with the rescue subject in their care, even when the subject is unconscious. Hearing is the last sense to go in the unconscious person, and many persons will recall conversations that took place around them when they were unconscious.

Use Low Risk Methods First

The approach to the subject should be the one that involves the least amount of danger both to the subject and the rescuers.

Some examples of this lower risk approach include the following:

- Always evaluate the situation before approaching it. Depending on the specific environment, hazards to rescuers could range from live wires, to rocks falling, to hostile persons.
- Don't rush. Rushing into a rescue situation is the sign of an inexperienced rescuer. Rushing causes mistakes that endanger both the rescuers and the subject. Move carefully and meticulously; don't bump into other rescuers.
- Choose the route to the subject that is the least dangerous.
- In rope rescue, the simplest way of doing the job is usually the most effective way. The more complicated a rescue system is, the greater chance there is of something going wrong.
- Instead of setting up operations directly above the subject, where rocks or hardware might fall on him (and other rescuers), move slightly off to one side. Keep all nonessential personnel away from the area where they might dislodge rocks and other objects.
- Appoint a team Safety Officer to oversee equipment use, anchoring, rigging, belays, and personal rappeling, ascending and seat harness tie-ins. The Safety Officer should be among the most experienced of the team members and be able to oversee all aspects of the operation.
- All rescuers must wear protective gear such as helmets to prevent head injury from falling objects and to lessen injury, should the rescuer take a fall. All personnel qualified for high angle operations should be wearing their seat harness so they are ready for any immediate needs.
- Before rescue operations begin, immediately set up safety lines. Anyone near the edge must clip into one of them. If there needs to be continued access up and down a steep slope, a securely anchored safety line should be set from top to bottom off to the side from the rescue site.
- Allow only those personnel near the edge who have a reason to be there.

Prepare for Self-Rescue

Whatever the rescue situation, the rescuer should always keep in mind that anything that can go wrong, **probably will.** Thus, the individual rescuer should always be ready for something to go wrong at any time and be prepared to extricate himself. This preparation should include not only mental preparation, but also physical preparation with equipment. Every rope rescuer should carry with him a small assortment of carabiners, a couple of slings, and either Prusik loops or ascenders in order to extricate himself from any difficult situation.

Back Up Other Rescuers

In the same way, each rescuer should be ready at any time to extricate any other rescuer who has gotten into a difficult situation. Whenever any team member is not at work performing a task, he should be directing his attention to the activity at hand. Everyone, no matter how intelligent and experienced, has an occasional lapse. All team members should be ready for any unsafe condition to develop.

Care of Equipment

Just as the sign of a good mechanic or carpenter is evident in the manner he cares for his tools, the sign of a competent rope rescuer is evident in the manner in which he cares for his ropes and equipment. But in the case of vertical gear, it is even more important. For upon these kinds of tools, lives will hang. Among the considerations for avoiding loss and damage to rescue gear are the following:

- Do not lay equipment on the edge of a drop. It can easily be kicked or knocked over the edge and be damaged. Or, possibly, it could cause injury to those working below.
- Do not lay equipment on the ground or on the floor. Hardware is easily lost in leaves or dirt, and grit does hardware no good. When first arriving on the site, a sling should be hung on a tree or other projection. All hardware not in use should be clipped into this sling.
- When working on a vertical face, keep all equipment attached to something such as your seat harness equipment loops.
- Inspect all gear after each operation. The time to discover that gear is defective is during inspection time, not during a rescue.
- Belay all lowering and raising of rescue subjects. Any rescuers who request belays must be provided with them.
- Ropes need special care (see Chapters 5 and 6).

Commitment to Working as a Team

One of the most important attributes of a qualified high angle person is his commitment to working as part of a team. The difficult and intricate skills of high angle work require teamwork. But it is important that the essential skills be spread throughout the team. If there are only a few people who have the important skills, there is the chance that those few might be absent or disabled when the rescue call comes.

Teams should not think that they are prepared for any rescue if they continually practice on the same rescue tower or same cliff that has convenient places for anchors. The team should make a point of varying their practice sessions in terms of rescue problems and sites. As they become more experienced, they should create practice environments that are deliberately inconvenient and difficult.

In Fire Departments

Because the rope rescue team is a new concept in some areas, it is sometimes difficult to determine how it fits in with the rest of the department. In large urban departments, it is generally unreasonable and unnecessary to train and equip every firefighter to be a Rope Rescue Technician.

What seems to be a common solution in large departments is to treat the rope rescue group as a specialty team, such as is done with many hazardous materials response teams. In this way, the team has its own officers, its own training, and, particularly important, its own equipment.

Additional arguments for the creation of a specialized unit include the fact that the members train as a team and respond as a team. With special requirements for membership, there tends to be greater motivation towards skills maintenance and better equipment management.

This specialty team concept may be more difficult to develop in rural areas, where there might be smaller departments and where, often times, the team will have to be comprised of personnel from several different units.

In Ambulance Crews

As with fire departments, the nature of the arrangement for high angle rope rescue skills is often dependant on whether the service is in an urban or rural area. In an urban area, the ambulance crew faced with a complicated rope rescue situation might be able to call for a backup from a specialty team, just as they would for a hazardous material or law enforcement problem.

However, ambulance crews in many areas may, in the course of their calls, be routinely faced with special situations such as slope evacuation (see Chapter 13, "Slope Evacuation"). Because of the frequency of these calls, it may be worthwhile to include training for these situations. In terms of gear, this may mean equipping the ambulance crew simply with a length or two of rope and for each person a couple of locking carabiners and a length of webbing.

Law Enforcement

In urban law enforcement, the skills for rope rescue often can be integrated into those experienced tactical teams who already employ high angle tactics for such activities as barricade or hostage operations. The training and equipment is often similar.

In sheriffs' departments, particularly in the western United States, there is often a responsibility for search and rescue. The organizational circumstances will vary. In some cases, paid employees perform the search and rescue tasks. Other departments use a sheriff's auxiliary for the team. Perhaps most common, a paid member of the department serves as the search and rescue coordinator for the county volunteer teams.

Volunteer Search and Rescue Teams

In many areas of the country, particularly in rural regions, high angle rescue problems have traditionally been handled by volunteer search and rescue (SAR) teams. In the western United States, where this tradition is older and more common, these teams often work under the auspices of the sheriff's department. Often these groups are referred to as "Mountain Rescue Teams," and many are certified by the U.S. Mountain Rescue Association.

As is often the case in Emergency Medical Service and fire, the management of volunteer SAR teams can be different from paid units. The greatest challenge usually comes in two areas: a rapid turnover of personnel and difficulty in motivating members to attend training sessions. The latter problem sometimes can be satisfied with the creation of a certification system. This system requires members to attend a minimum number of training sessions to achieve and maintain their certified status.

Ultimately, motivation may depend on the rate of of rescue activity. Those groups answering the greater number of calls are generally the ones that remain viable through the years.

Small Team Management: The Key to Success or Failure

The failure of rope rescue personnel to complete a task in a timely manner can often be laid to the failure of management of the small team situation. The main elements in successful small team management in rope rescue involve leadership, organization, and direction toward the rescue goals.

Leadership

In a rope rescue situation, there should be only one leader. However, there should be several persons on the team who are capable of being a leader when they are needed. The rescue leader is not a dictator. A single person is fallible, so the leader must be able to solicit opinions from other experienced members and to use their advice. The leader must also be flexible. The rescue situation may change, it may not be what the first call reported it to be, or the first plan may not work. So the leader must be able to develop alternative plans.

Goal and Direction

In most cases, a rope rescue is the solving of a puzzle: how does the team rescue the subject using its rope, hardware, skills and ingenuity? But for the team to arrive at the answer, it must know the question.

The question (what are we about to do?) is presented in the briefing. This briefing always should be given to the personnel who are primarily involved in the operation. They should know what the problem is (subject trapped on ledge), what the overall solution will be (raise the subject with a hauling system), what their unit's task will be (provide an anchor system for the hauling system), and what each person's role is to be (have the necessary equipment ready for the riggers).

To best move towards the goal, each task should be assigned to a unit (one or more persons) which has in charge of it a person responsible for seeing the task done and for communicating with the other leaders. Every one involved in the rescue must have a clear idea of what their task is.

Strategy and Allocation of Resources/Time

As will be constantly emphasized in this manual, the focus of the rescue activity must be the subject of the rescue. Thus, the priority must be to have someone take medical charge of the subject and evaluate his medical condition as soon as possible.

But while this is happening, all other members of the team should be involved in preparing for the rescue. It will be only a very large team in which there is someone who has nothing to do.

An effective way to accomplish this task of preparation is to divide the job into specific tasks. Each task is assigned to a subgroup headed by persons responsible to the group leader. Among subgroup tasks would be things such as rigging, anchoring, preparation of the litter, evacuation, crowd control, communications, safety, equipment control, and rope management.

The leader, or on-scene coordinator, makes certain that all the actions of the various subgroups are meshing to attain the ultimate goal: the rescue.

Unit Size

One problem encountered in management is that leaders are placed in charge of too many people for them to handle, and they become overwhelmed by making certain that each person completes his assigned task. In other words, they do not have a manageable **span of control**. A manageable span of control is generally thought to be from three to five persons. If a task grows in size or complexity so that the span of control no longer is manageable, then the group should then be divided into subgroups, each with its own leader.

Putting It Together: A Command System

To make certain that a management system is indeed **manageable** and that it fits with other organizations involved in the activity, there must be some sort of management system. One that is being used more and more in North America for the management of emergencies is the **Incident Command System** (ICS). ICS is used for such diverse operations as fires, law enforcement incidents, special events, natural disasters, and hazardous materials spills. It has particular application to incidents involving responses from different agencies or jurisdictions, but operates well on incidents as small as auto wrecks.

Communications

In the high angle rope rescue environment, communications are extremely important for the coordination of the operation and for the safety of all involved. But the conditions in the high angle environment often make normal communications very difficult.

Other sections of the manual describe some of the standard voice communications used in rope work. But in many high angle situations, the ability to hear voice is severely restricted by distances, by the wind, and by other noise factors, such as falling water.

While the solution may seem to be electronic communications, conventional radio systems are also subject to disruption by conditions at the rescue site. Typical problems involve the interruption of radio transmission by rock overhangs and intervening ridgelines.

One of the most common problems encountered by rescue team members relates to the management of a radio on a vertical face where both hands tend to be occupied with other activities.

One solution that is being increasingly used is a special radio harness that fits across the chest *(see Figure 11.1). It takes* only a simple, short hand motion to key the mike. Although the radio is close at hand, it is also protected and out of the way of much of the activity involved during rope work.

One approach that has been successfully employed by some teams for short range communications is the voice-actuated headset. Among possible drawbacks are that the units may not fit well with some helmet designs, and they can be put out of commission in close quarters where elements of the system can get snagged or torn off.

A non-electronic alternative is the use of whistle blasts for communications. Obviously these are limited in the degree of information that can be communicated. But they are often audible where nothing else works. The exact form of these whistle communications has to be worked out and used in practice beforehand.

One more alternative type of communications exists. This is an even more primitive one, but one that might be necessary in certain critical situations. It is what might be called **digital** communication, the use of a yes-or-no signal.

This kind of system might be called for when a subject is high on a cliff or hill, or far across a valley. These are situations where normal two-way communication is impossible and where the subject does not have a radio.

This digital system works with the rescuers employing a loud hailer to ask questions that allow a "yes" or "no" answer. The subject replies with a flashlight blink or a raised arm.

Fig. 11.1 Radio Chest Harness

The Human Side of Communications

Often the greatest detriment to good communication is not an electronic or mechanical failure, but the communicators themselves. In high angle situations, it is particularly essential to communicate in a clear, concise and specific manner. Does "right" and "left," for example, mean as you face the cliff or building or as you face away from it?

It is important to reduce the command vocabulary to as few words as possible and use only those words that are clear, concise, and have few syllables. One should also use the same words for specific actions. As noted in the sections on lowering and hauling systems, the only word for cessation of action is "stop!." There should never be substituted another word, such as "whoa," which could easily be mistaken for **"slow," or even worse, for "go."**

Preplanning

Preplanning is the invisible part of rescue. But it is the essential element that can determine whether the rescue operation succeeds in a fairly smooth manner, or whether it is cursed with confusion and delay.

Among the elements of preplanning are determining and establishing:

- Local needs for rescue.
- Who has jurisdictional and operational responsibility for rescue in a local area.
- How a rescue group fits in.
- What initiates call-out and how it proceeds.
- The command structure on scene.
- Communications, including frequencies.
- How the group relates to other organizations.
- Medical control and protocols.
- Basic procedures in approach to the rescue.

Chapter 12

One-Person Rescue Techniques

PREREQUISITES

Before attempting the activities described in this chapter, you must have demonstrated that you can properly:

1) Use and care for rope.
2) Use and care for other equipment employed in the high angle environment.
3) Tie correctly, confidently, and without hesitation the eight knots described in Chapter 6.
4) Apply the principles of anchoring and rig a safe and secure anchor.
5) Apply the principles of belaying and safely and confidently belay another person using either a Munter Hitch or belay plate.
6) Apply the principles of rappeling: rappel safely, confidently, and under control; tie off the rappel device to operate hands free of the rope and then return to a safe and controlled rappel.
7) Apply the principles of ascending: tie correctly, confidently, and without hesitation a Prusik knot and know how to use it; comprehend the uses and limitations of mechanical ascenders; confidently and safely ascend a fixed rope using either friction knots or mechanical ascenders; confidently and safely change over both from rappeling to ascending and from ascending to rappeling; and extricate yourself from a jammed rappel device (or similar problem).

OBJECTIVES–

At the completion of this chapter, you should be able to:

1. Describe the kinds of conditions one-person rescue techniques might be employed.
2. List the skills and equipment that are required for one person rescue techniques.
3. Discuss the rescue considerations and priorities involved in the following one-person rescue situations: a) subject wearing seat harness, b) subject not wearing seat harness, c) unconscious subject, d) hostile/combative subject.
4. Discuss the medical concerns and priorities involved in the various aspects of one-person rescue.
5. Safely and efficiently perform a one-person rescue of a person wearing a seat harness.
6. Safely and efficiently perform a one-person rescue of a person not wearing a seat harness.
7. Tie on a subject a hasty seat harness and either a hasty seat harness with a chest harness or a hasty body harness.

One-Person Rescue Techniques are procedures in the high angle environment in which a single rescuer has direct physical contact with a rescue subject. Other persons may be involved in the rescue to perform such vital tasks as belaying, or in support capacities. Teamwork and good communications remain essential in one-person techniques.

One-person rescue techniques are usually done without the use of a litter. They often involve the attaching of the rescue subject directly to the rescuer, with the rescuer rappeling to control the body weights of both rescuer and rescue subject. In certain advanced techniques (not covered in this chapter), the rescuer may ascend the rope with the subject attached to himself. Or he may lower the subject using a braking device while the rescuer remains in position on the rope.

One-person rescue is generally only performed when the rescue subject is uninjured or only slightly injured. It is extremely difficult for only one person without a stretcher to rescue a seriously injured subject without making the injuries worse.

For this reason, it is absolutely essential that one-person rescuers evaluate the subject in terms of injury before moving him. To be able to competently evaluate a subject and treat for injury, a one-person rescuer should have emergency medical training at least to the level of Red Cross Advanced First Aid or the DOT First Responder, and preferably to the level of Emergency Medical Technician or better.

One-Person Rescue Situations

One-person rescue techniques might be needed where:

- Only one person is needed to perform the rescue.
- There is a shortage of personnel and/or resources.
- The urgency of the situation means there is no time to await additional personnel.

The need for a one-person rescue might occur in any of the following situations:

- A fire fighter's interior exit is blocked during a fireground operation and there is no time to set up a ladder.
- A high-rise window washer has had an equipment malfunction and is stranded on the side of a building out of reach of ladders.
- A construction worker has become stranded on scaffolding.
- A potential suicide has climbed to an exposed area to jump, but has changed his mind.
- A rock climber has fallen, is only slightly injured, but needs assistance in getting off the face.
- A sightseer or picnicker has slipped onto a ledge and cannot go up or down.
- A hiker has blundered into dangerously steep terrain and is unable to move because of the danger of falling.

Teamwork and Communications

The term, "one-person rescue" does not mean that other rescuers will be kept from the operation. It will be a safer and more efficient operation if other skilled and knowledgeable persons are involved in essential tasks such as belaying, lowering or raising equipment, spotting, etc.

Also, if the rescue subject is to be rappeled or lowered to the ground, then essential personnel will be needed at the arrival spot to attend to the medical needs of the subject. Several people may also be needed to perform a litter evacuation to an ambulance or other medical care.

Skills and Equipment Required in One-Person Rescue

In the performance of one-person rescue techniques, there are often several sets of rope work skills being performed simultaneously, and several different kinds of equipment being used at one time. Consequently, an absolute knowledge of the equipment and the ability to perform the skills are necessary. These only come with constant practice of the skills using the equipment.

With one person rescue procedures also come a greater number of ropes in use at one time, along with more connecting lines, slings, and equipment. This means an increased challenge in rope and equipment management. The rescuer must have the experience and knowledge to be aware of all the rope, equipment, and webbing in use at once, to keep it from damage from rope cross, to keep it from tangle, and to be able to manipulate without hesitation the specific line when needed.

The Belay Question

In some one-person rescue situations, a belay may be desirable, but it may not always be feasible. In fact, there may be some situations where a belay could encumber the operations, or even endanger the involved individuals.

For example: If there were a situation where both the rescuer and the rescue subject each had his own rope, what would happen if both persons were also belayed with additional separate ropes? There would now be a situation in which **four** ropes were coming together. There would be the real possibility of rope tangle and possible damage from rope cross. The problem of rope management would increase as rescuers tried to determine which line went to specific harness tie-in points, and how each line would affect other lines as angles changed with the positions of the people involved.

The question of a belay is one that will have to be answered through intelligent decision-making and on the specific situations and people involved. That kind of intelligent decision-making can only come with the experience of practice with the techniques in differing environments and under varying conditions.

MEDICAL CONSIDERATIONS

Unless immediate life-threatening environmental factors prevent you from doing so, the following medical considerations must be made as a minimum **before** moving a rescue subject and must **continue** during the course of the rescue:

■ Airway

Does the subject have a clear airway?

Do you need to maintain the airway during the rescue?

Does he have objects in the mouth—gum, tobacco, dentures—that need to be removed?

Is he vomiting, or does he have the potential for vomiting? How will you clear the vomitus during the course of the rescue so that he does not aspirate it?

■ Breathing

Is the subject breathing on his own?
Will you be able to monitor breathing during the course of the rescue?

Will you need to perform rescue breathing before and during rescue?

■ Circulation

Does the subject have a pulse?

Will you be able to monitor the pulse during the rescue operation?

Is CPR necessary?
Will it be possible during the rescue operation?

AND:

■ Shock

Does the subject show signs of shock?

What must be immediately done to treat for shock?

What rescue procedures will increase/reduce the possibility of shock?

■ Bleeding

Is there life-threatening bleeding? How can you stop it?

■ Spinal Injury—*Many emergencies in the vertical environment relate to spinal injury.*

Is there a possibility of spinal injury?

Should a cervical collar be used before the subject is moved?

Should you secure the subject to a backboard before moving him?

Would it be wiser to move the subject with a litter evacuation rather than a one-person rescue?

Rescue of a Person Wearing a Seat Harness

This situation assumes that the rescue subject is wearing a secure seat harness that will keep him upright during the procedure.

Equipment Required

(1) Main line rope with adequate safety factor for two persons.

(1) Sewn, manufactured seat harness with thigh/leg supports for rescuer.

(1) Rappel device with enough friction to handle the weight of two persons and, preferably, with variable friction.

(2) Large, locking carabiners (in addition to locking carabiner already in rescuer's seat harness tie in point).

(1) Short (approximately 2 ft.) sling with loop in both ends, or an adjustable rescue pick-off strap that will support one person's weight with an adequate safety factor.

One-Person Rescue Practice System

In the practice of one-person rescue techniques, it is essential that the person acting as practice rescue subject be in a position where he is secure from injury from falling. In the initial practice of the pick-off of the practice subject, the practice subject should be stationed only a short distance off the ground so there is a minimum possibility of injury from falling.

As the practice sessions move farther off the ground, the practice subject should either be initially tethered or belayed until he is in a secure position.

Procedure for Performing One-Person Rescue of Subject Wearing Seat Harness *(Figure 12.1).*

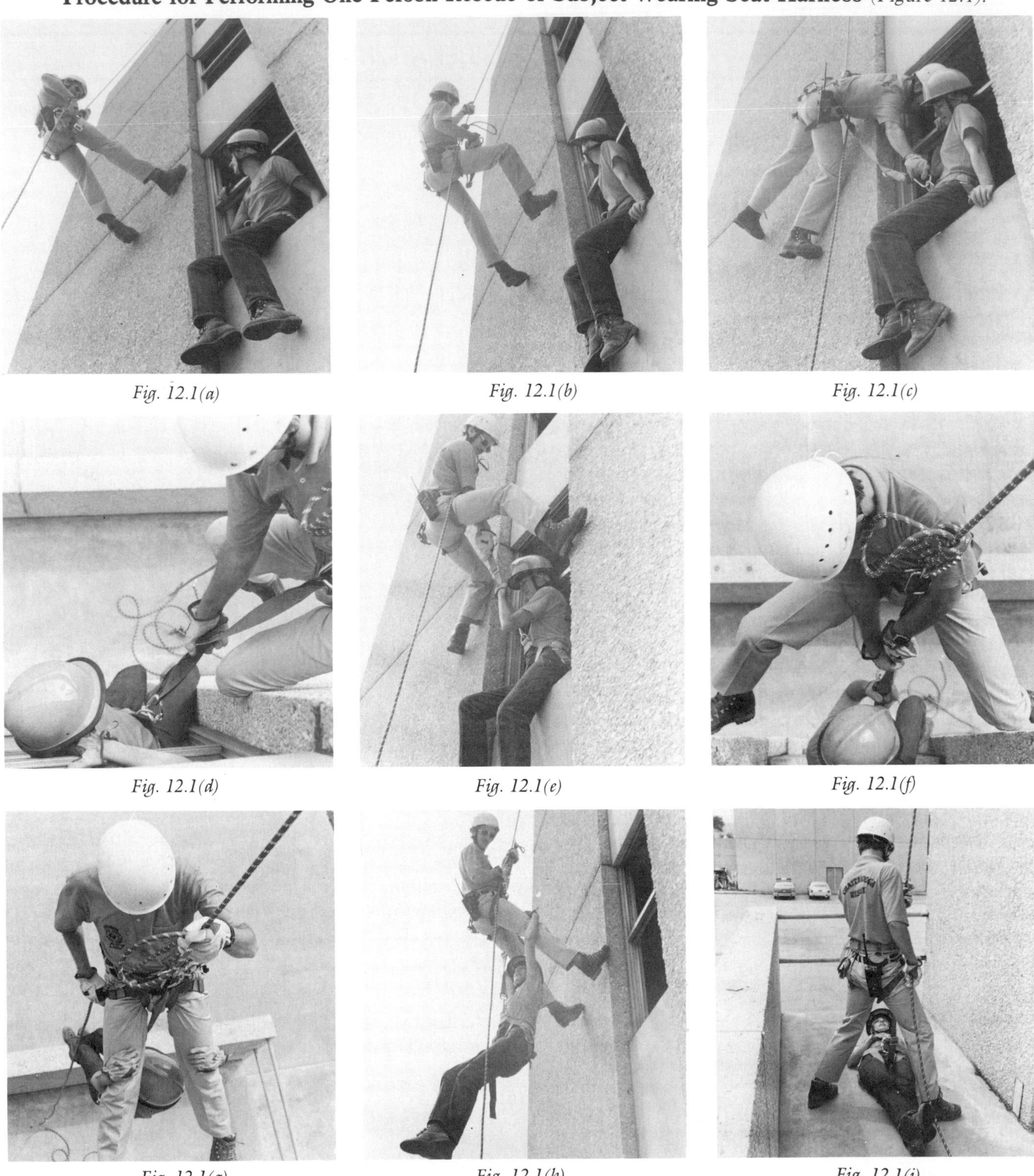

Fig. 12.1(a) *Fig. 12.1(b)* *Fig. 12.1(c)*

Fig. 12.1(d) *Fig. 12.1(e)* *Fig. 12.1(f)*

Fig. 12.1(g) *Fig. 12.1(h)* *Fig. 12.1(i)*

1. Station a practice rescue subject wearing a seat harness in a position of minimum exposure, so that he would not be injured if he fell, for example:
   a) On a low cliff ledge.
   b) In the first floor window of a practice building.
   c) On a structural member, such as a low beam of a bridge or tower.

2. At the top, rig anchor(s) for your main line rope. Your rope will have to be off to the side of the rescue subject with this horizontal distance a compromise between:
   - Being far away enough so that:
     a) your rope will not knock rocks or other debris onto the subject, and
     b) you can be out of reach if the subject attempts to grab you.

WARNING NOTE

During one-person rescue procedures, anchors, rope, hardware, and personnel are subjected to sudden increased loads, shock loading, and loads that may come from directions different from those originally anticipated.

1. Anchors must be rigged for increased and multi-directional loading.

2. Carabiners must be locked, aligned in manner of function, and monitored so that they remain in manner of function.

3. Ropes and slings must have an adequate safety factor.

4. Rescuers must be prepared for sudden increased weight and for providing extra friction on rappel devices.

- Being close enough so that:
  a) you can easily pendulum over to the subject, and
  b) when the two of you pendulum back, it does not excessively shock load your rope and anchors.

AND REMEMBER:

- The anchors may be subjected to shock loading.
- The anchors will be loading from different directions.
- If the pendulum causes a loaded rope to rub unprotected across a sharp edge, the rope may be cut.

3. Don a sewn, manufactured seat harness with leg/thigh supports. Clip a locking carabiner into the seat harness front tie-in point.
4. Attach to the seat harness carabiner a rappel device that has both:
   a) Variable friction, and
   b) Enough control to handle weight of two persons.

   The Brake Bar Rack qualifies on both points. If this is not available, then double wrap a Large Figure 8 with Ears. **USE THE FIGURE EIGHT ONLY IF YOU KNOW FROM EXPERIENCE THAT YOU CAN CONTROL THE COMBINED WEIGHT OF YOURSELF AND THE RESCUE SUBJECT.** Lock the carabiner on the rappel device.
5. Into one end of the rescue sling, clip a large locking carabiner. For the moment, leave this first carabiner unlocked. Into the other end of the sling clip the second large locking carabiner. With the sling and first carabiner attached, clip this second carabiner **DIRECTLY INTO THE RAPPEL DEVICE TIE IN POINT** so that the weight of the rescue subject will be taken **DIRECTLY ON THE RAPPEL DEVICE. DO NOT CLIP THIS RESCUE SLING DIRECTLY INTO YOUR SEAT HARNESS.** *(see Figure 12.2).*

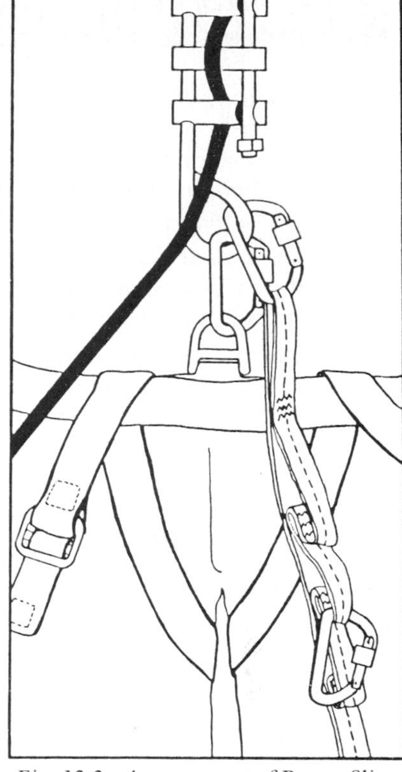

Fig. 12.2 Arrangement of Rescue Sling

WARNING NOTE

When performing one-person rescue techniques, *DO NOT* clip any attachments to support the weight of the rescue subject directly into your own seat harness, primarily for the following reasons:

■ It can result in painful and, perhaps, damaging pressure on the body of the rescuer, particularly in the groin area.

■ It will stress a seat harness in an unnatural manner, perhaps resulting in damage and potential failure.

6. Lock the carabiner that connects the rescue sling to the rappel device tie-in point. Do not yet lock the carabiner that is in the opposite end of the sling.

7. Begin your rappel and the approach to the rescue subject As you reach the proximity of the subject, rappel slowly or stop just out of his reach. You must now do two things simultaneously *(12.1a)*:
   a) Size up the situation, and
   b) Communicate with the subject.

WARNING NOTE

Anytime you go over the edge in a rappel, but *particularly in a rescue,* check for loose personal items or vertical gear. All items and gear must be secured so they will not fall out of pockets, packs, or gear slings.

Other than causing injury to the rescue subject or other rescuers, you may lose an essential piece of equipment just when you need it the most.

Always do a last minute SAFETY CHECK. Make certain that all carabiners are locked or aligned in the correct manner of function. Be certain that knots are tied correctly and anchors are secure. Check for any loose clothing or hair that might be drawn into the descender. Be sure that your helmet is secure.

Size Up:

1. What are the **physical circumstances?**

   Is the subject stable where he is or is he in immediate danger of falling?

   Is there loose rock above the subject that might be dislodged by your rope?

2. What are the **emotional circumstances?**

   Is the subject about to leap onto you?

   Will he follow directions?

   Is he comfortable in the high angle environment?

   Is he experienced with high angle work so that he can assist in the procedure?

3. What are the initial signs of the **medical circumstances?**

   Is he conscious?

   Is he breathing?

   What are the obvious injuries?

   What are his physical complaints?

Communicating with the Subject:

Reassure him.

Tell him who you are.

Tell him what you plan to do.

Ask him if he is injured.

Describe how he can help:

1) Do not move until told to.
2) Do not grab anything unless told to.

8. Stop and tie off your rappel device with yourself about 2 feet above the level of the subject. **ALWAYS ABOVE.** If you are initially too high to reach him, you can always unlock and rappel down a few more inches. But if you are too low, you may not be able to get back up to him easily and quickly *(12.1b)*.
9. Make certain that the subject stays put. If it is feasible, the best position is for him to be sitting. While you are talking with him, begin your initial medical evaluation.
10. Take the large locking carabiner that is at the bottom of the sling attached to your rappel system. Lean over and clip it directly into the subject's seat harness tie-in point. **Do not clip into any connecting features on the harness.** If you are too high, rappel down only the distance needed to make the connection. **Avoid slack in the rescue sling** *(12.1c)*.
11. As soon as you have clipped the rescue sling into the main part of his harness, immediately tighten the carabiner lock. You now have him secured to your system. If using an adjustable pick-off sling, take up all the slack in the sling *(12.1d)*.
12. Conduct your medical evaluation. If you determine that it is safe to move him, prepare to evacuate him.
13. Brace with your legs spread wide and your feet against the wall *(12.1e)*.
14. Tell the subject to do the following:
    a) If he is not sitting and it is possible to do so, tell him to sit.
    b) Have him place his legs together between yours *(12.1f)*.
    c) Have him place his hands on your legs for support *(12.1g)*.
    d) As he holds onto your legs, have him very slowly swing down between your legs until his weight comes onto the sling. While he is doing this, direct him, reassure him, and brace so that you and he do not slip *(12.1h)*.
    e) Tell him to hold onto the sling to steady himself. The final position of the subject should be facing you. His head should be below your buttocks; his full weight onto the sling and the rappel device. The sling should remain between your legs during the remainder of the procedure.
15. When the subject's full weight is on the sling and he has stabilized, unlock your rappel device and begin to rappel very slowly.
16. Keep the subject below you and between your legs. If you are close enough to the cliff/building face, fend away from it with your feet.
17. As you approach the ground, tell the subject that you want him to lie down as his body touches the ground. You will straddle him.
18. When the subject is lying full on the ground, you are straddling him, and you have slack in your rappel device, unlock him from your system by removing him from the short sling *(12.1i)*.
19. Remove your rappel device from the rope.

Rescue of a Subject Not Wearing a Seat Harness

Placing a Manufactured Seat Harness on Subject

There is a good chance that the subject involved in a one-person rescue will not be wearing a seat harness. In such a case, the rescue procedures are essentially the same, except that the rescuer must either place onto the subject a sewn, manufactured harness or tie onto him a hasty harness.

In itself, this placing of the harness for rescue can be a very difficult process. The rescuer will probably be hanging on a rope at a difficult angle (perhaps even upside down), dealing with a frightened and, perhaps, injured subject, under difficult environmental conditions.

Because of these difficulties, the potential rescuer should first practice a one-person rescue on a subject wearing a seat harness, and then try the same thing by placing a seat harness on the subject.

If one is available, a sewn, manufactured harness is preferable to a tied one, with the following considerations:

- The harness should be designed so that it can be placed onto the subject with as little disruption as possible. In particular, it should be done without the subject's becoming unbalanced by having to step into the harness.
- It must be quick and easy to put on. It should not have a multitude of buckles, snaps, and adjustments.

Placing a Tied Seat Harness on a Subject

There are a variety of tied harnesses that might be placed on a subject for a one-person rescue procedure, but the following considerations should be made in deciding which ones to use:

- The harness should be tied with a minimum of physical disruption to the subject.
- The harness should be quick and easy to tie under difficult conditions.
- The harness should be self-adjusting.

The Hasty Seat Harness

The Hasty Seat Harness is one type of seat harness that can be placed on a subject easily with a minimum of disruption. It is created from a length of tubular webbing ranging from 10 feet to 15 feet long, depending on the size of the subject. The webbing is tied into a continuous loop using a water knot before the rescuer beings his rappel.

Tying the Hasty Seat Harness
(see Figure 12.3).

1. Approach the subject from behind. Place the loop across the subject's shoulder so that the sides of the loop hang down along his side and the top of the loop runs across the back of his neck.
2. With both hands, reach around the sides and under the arms of the subject and the vertically hanging sides of the loop. **From this point until you have the subject clipped into your system you must keep your arms in this position around the subject in case he should slip or fall.**
3. Now reach down with either or both hands. Go in between the subject's legs **from the front** and grasp the bottom of the loop. Take it firmly in both hands.
4. Pull the loop back through the subject's legs and up toward the front of his waist.
5. As you pull the loop up through his legs, let the top section of the loop running across his shoulders fall down his back. If necessary, you can help this along with your chin or head.
6. Continue pulling on the lower end of the loop. As you pull the slack out of the loop from behind, the webbing will slide down your arms and past your hands to form the harness.
7. To cinch down the webbing, take a loop in each hand and pull each one to an opposite side, so that the webbing is contoured around the body. Make certain that the webbing remains taut.
8. Bring the two loops back to the center together and clip a locking carabiner across them together.

Tying of Hasty Seat Harness *(Figure 12.3).*

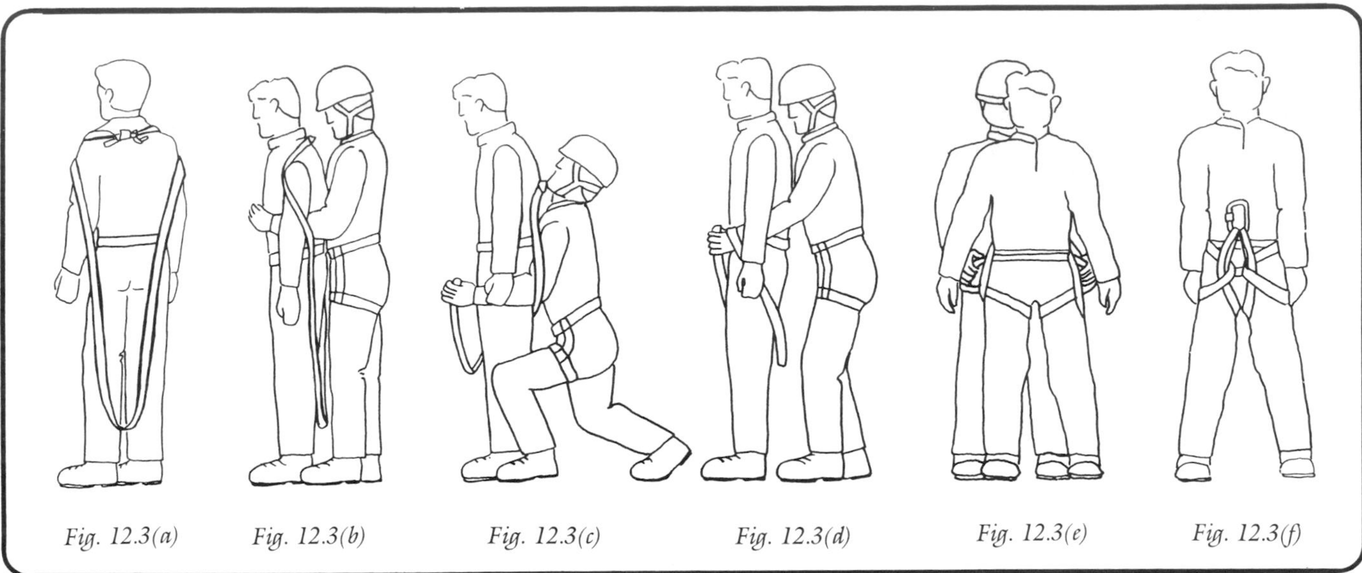
Fig. 12.3(a) *Fig. 12.3(b)* *Fig. 12.3(c)* *Fig. 12.3(d)* *Fig. 12.3(e)* *Fig. 12.3(f)*

Rescue of an Unconscious Subject

MEDICAL CONSIDERATIONS

An unconscious subject should be rescued using a litter unless there are overriding considerations such as an immediate threat to life.

In addition to the primary survey for the ABCs (**Airway, Breathing, Circulation**), there are particular medical considerations associated with an unconscious subject:

- If there is a particular threat of airway blockage, there must be continued attention to keeping it clear.
- If unconsciousness is due to trauma (such as a fall) or the cause is unknown, it must be assumed that the subject has a spinal cord injury, and there must be spine immobilization.

Because the subject may not be able to describe his injuries, there are potential hidden injuries, such as fractures.

When a person becomes unconscious, hearing is the last sense to go. So, always talk with an unconscious subject, even though you suspect he might not be able to hear. This communication may eventually elicit a response and may prevent him from becoming combative.

If a litter cannot be used for an unconscious subject, then special one-person rescue techniques may be employed.

An unconscious subject will not be able to hold himself upright, or fend off from the face of a building or cliff, so there must be special considerations for a rescue harness used for an unconscious person:

- The use of a full-body harness that is sewn and manufactured, or
- The addition of a chest harness to the hasty seat harness, or
- A tied full-body harness.

Requirements for a full body/combination seat/chest harness include the following:

- It must hold the subject upright.
- It must prevent the subject from sliding out.
- It must be simple to put on, without large numbers of buckles or knots.
- It must be usable in adverse conditions—cold, dark, wind, etc.

A Rescue Chest Harness

Figure 12.4 illustrates a rescue chest harness. Unless there is no other alternative, this chest harness must not be used alone. It must be used in combination with a rescue seat harness, such as the hasty harness. It may also be used with a sewn, manufactured seat harness to help hold a subject upright.

Fig. 12.4 Manufactured Chest Harness

Tying The Rescue Chest Harness

Figure 12.5 illustrates the tying of a quick chest harness. This can be used in combination with a seat harness for a one-person rescue procedure.

1. Take a continuous loop of webbing tied with a water knot.
2. Twist the loop into a Figure 8 .
3. Lay the loop across the subject's back with the loop crossing on the back at armpit height.
4. One at a time, put the subject's arms through each loop.
5. Bring each loop to the center of the chest.
6. Clip the loops together with a carabiner. Or, if the harness is too loose, pull one loop through the other and clench it down. Clip a carabiner through the long loop.

Fig. 12.5 Tying Rescue Chest Harness

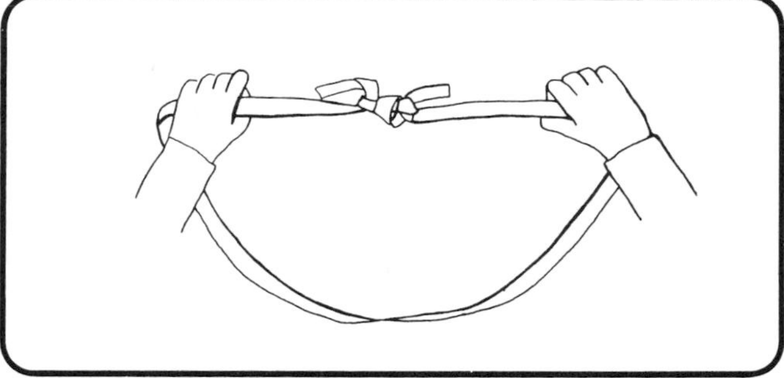

Fig. 12.5(a)

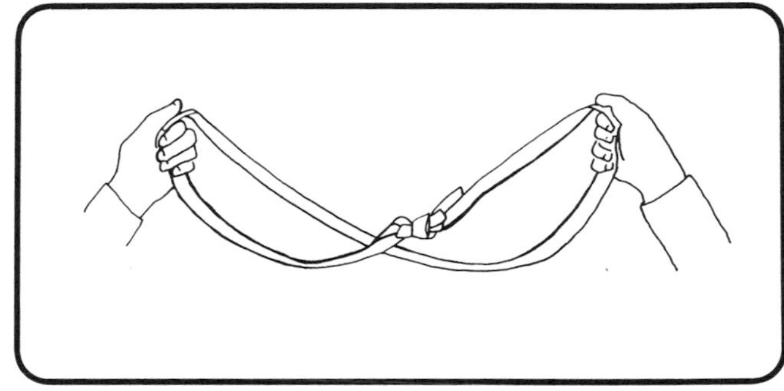

Fig. 12.5(b)

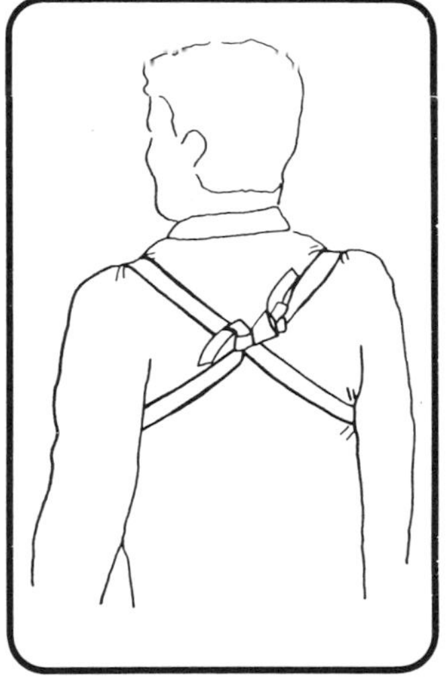

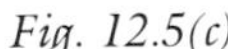

Fig. 12.5(c)

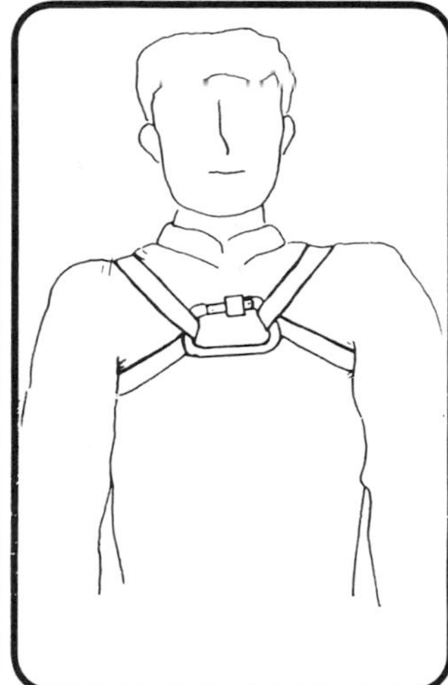

Fig. 12.5(d)

Combining the Rescue Chest Harness Into the System

7. Clip the carabiner from the chest harness into **THE END OF THE RESCUE LOOP** (where you have previously clipped the carabiner for the seat harness. **DO NOT CLIP THE TWO CARABINERS TOGETHER** so you will be able to make adjustments in either the seat harness or the chest harness.

QUESTIONS for REVIEW

1. In deciding whether or not to do a one-person rescue, how does the extent of the subject's injuries affect the decision?
2. Under what conditions might a one-person rescue be needed?
3. Under what conditions in a one-person rescue might a belay not be feasible?
4. Name the major medical conditions to be considered before moving a rescue subject in a one-person rescue.
5. List the minimum equipment needed for a one-person rescue of a person wearing a seat harness.
6. What considerations should be made for the following elements of a rescue system during a one-person rescue:

   Anchors ________________________.

   Carabiners ________________________.

   Ropes and slings ________________________.

   Rescuers ________________________.

7. When performing a one-person rescue, to which of the following should the rescue sling be attached: (a) rescuer's seat harness, or (b) descender tie-in point?
8. As you approach the subject in a one-person rescue, what two things should you do simultaneously?
9. What three considerations are involved in the size up of a subject in a one-person rescue?
10. When you rappel to a subject in a one-person rescue, what should be your physical position in relationship to the position of the subject?
11. Name two considerations for a seat harness to be placed on a subject in a one-person rescue.
12. Name three criteria for a tied seat harness to be placed on a rescue subject during a one-person rescue.
13. What are the major medical considerations for an unconscious subject of a one-person rescue?

Chapter 13

Slope Evacuation

PREREQUISITES

Before attempting the activities described in this chapter, you must have demonstrated that you can properly:

1) Use and care for rope.
2) Use and care for other equipment employed in the high angle environment.
3) Tie correctly, confidently, and without hesitation the eight knots described in Chapter 6.
4) Apply the principles of anchoring and rig a safe and secure anchor.
5) Apply the principles of belaying and safely and confidently belay another person using either a Munter Hitch or belay plate.
6) Apply the principles of rappeling: rappel safely, confidently, and under control; tie off the rappel device to operate hands free of the rope and then return to a safe and controlled rappel.
7) Apply the principles of ascending: tie correctly, confidently, and without hesitation a Prusik knot and know how to use it; comprehend the uses and limitations of mechanical ascenders: confidently and safely ascend a fixed rope using either friction knots or mechanical ascenders; confidently and safely change over both from rappeling to ascending and from ascending to rappeling; and extricate yourself from a jammed rappel device (or similar problem).

OBJECTIVES–

At the completion of this chapter, you should be able to:

1. Describe what constitutes slope evacuation and list some examples of where it could be used.
2. List the elements of a slope evacuation.
3. Select equipment to be used in a slope evacuation system.
4. Describe what is involved in the medical considerations for a subject before and during a slope evacuation.
5. Describe what is involved in the packaging of a subject for slope evacuation.
6. Discuss the functions of the following: litter tenders, litter captain, brakeman, and haul team.
7. Discuss the functions of rope, braking systems, safety cam, haul cam, and pulleys in a slope evacuation.
8. Describe the need for sure communications in slope evacuation.
9. List the elements of a hauling system in slope evacuation.
10. Discuss the principles of a 1:1 hauling system, a counterbalance haul system, and a 2:1 haul system.
11. Act as a litter tender in a slope evacuation.
12. Package a rescue subject for slope evacuation.
13. Rig a litter for slope evacuation.
14. Rig the braking and belay systems and the safety cam for slope evacuation.
15. Repeat from memory voice communications used in slope evacuation.
16. Act as a member of a haul team in slope evacuation.
17. Act as a rope handler in slope evacuation.
18. Act as a brakeman in slope evacuation.
19. Rig a 1:1 or a counterbalance hauling system for slope evacuation.

TERMS– *relating to slope evacuation that a Rope Rescue Technician should know:*

Brakeman—Person who operates the braking device that controls the rate of descent of a litter in slope evacuation.

Counter Balance Haul System—A procedure for hauling that uses a 1:1 ratio and a haul team that moves in a direction opposite to the load.

Haul Team—The group of persons who provide the power to raise the load.

Litter Captain—The person in slope evacuation who manages the litter team and coordinates the litter movement with other members of the rescue team.

Litter Tender—The person who physically manages the litter in slope evacuation.

Multi-Pitch—More than one rope length.

Packaging—The placing of a rescue subject in a litter so that the primary medical considerations are cared for and the subject is physically stabilized in the litter.

Pitch—One rope length.

Rope Handler–The person in a lowering operation who assists the brakeman with rope management.

Safety Cam–A cam type ascender or knot that is placed on the rope in a litter hauling system to prevent the rope (and litter) from unintentionally slipping back down the slope.

Slope Evacuation—The movement of a rescue subject over terrain that is so rugged or angled that it requires the litter to be safetied with a rope and its descent controlled by a braking device or its ascent assisted with a hauling system.

Tree Wrap—A technique of running a rope around a tree trunk to create friction for a braking effect in a litter lowering.

1:1 Hauling System—A procedure for hauling where the force needed to haul is roughly the same as the load being hauled.

While the term **slope evacuation** may not provoke the image of excitement or challenge that comes with **cliff rescue,** it is in many areas the most common type of rope rescue performed by emergency service personnel.

It is also often the type of rescue with which emergency service personnel have problems. This may be because many people are inexperienced with slope evacuation and are unaware of the potential problems involved. Consequently, many rescuers are often poorly equipped and trained for the problems and hazards encountered in slope rescue.

The division of where **slope evacuation** ends and **high angle evacuation** begins is not always easily defined. But the essential differences are in the way that the litter and the rescue personnel are used:

Slope Evacuation:

- Personnel have most of their weight on the ground.
- Weight of litter at least partly rests on rescue personnel.
- There may be four or more litter attendants.
- Litter is attached to rope at one end.

High Angle Evacuation:

- Personnel have their weight supported by litter and rope.
- Weight of litter is supported by rope.
- At the most, there are two litter attendants.
- Litter hangs from vertical rope.

Examples of Slope Evacuation

- Road cuts and fills.
- Loose rocky slopes ("scree" or talus).
- Hills.
- Snow/icy slopes.
- Rugged, broken terrain.

Slope evacuation includes any inclined or rugged area over which a litter must be carried and where it is difficult or dangerous to do so without the assistance of a rope.

Elements of Slope Evacuation

- A slope evacuation will usually consist of the following elements *(see Figure 13.1).*
- The Litter

  Requirements:

  - Strength—

    To withstand stresses of supporting weight of rescue subject and rescuers while being supported by rope. Must be able to withstand blows from hitting rocks, trees, and other hard objects.
  - Tie-in points—

    To attach ropes, rescuers, and patient protection.
  - Rigid or semi-rigid

    To protect patient and aid in handling.

    Examples of Litters for Slope Evacuation:
  - Wire basket stokes.
  - Plastic Basket.
  - Semi-rigid (SKED™, Reeves™).
- A safe and secure anchor system is a critical part of any slope evacuation (see Chapter 7, "Anchoring" for a review of the criteria for safe and secure anchors).
- **Rope**

  To maintain greater control over the operation, most rescuers prefer a rope with a minimum of stretch, such as static kernmantle. The rope is generally attached to the head end of the litter. Up-slope from the litter the rope is attached as in A or B on the next page.

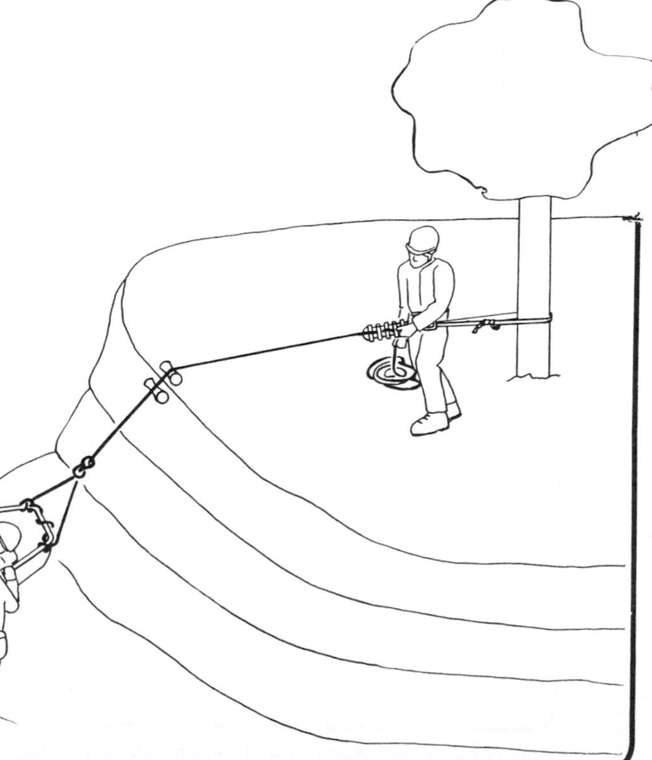

Fig. 13.1 Elements of Slope Evacuation

A—IF LITTER IS BEING LOWERED DOWNSLOPE

- **Braking System**

  This imparts friction to the rope to make the descent of the litter easily controlled by one person known as—

  The Brakeman

  This person controls the braking system and the rate of descent of the litter.

B—IF LITTER HAS TO BE RAISED UPSLOPE, THE ROPE RUNS THROUGH—

- **A Hauling System**

  This enables the litter to be easily raised by the—

  Haul Team

  These people provide the force to safely and efficiently raise the litter up slope.

- **Safety Cam**

  This device prevents the rope (and litter) from inadvertently sliding back downslope.

WHETHER LITTER IS BEING RAISED OR LOWERED:

- **Litter Tenders**

  Because the rope takes much of the litter weight, usually no more than four attendants are used in slope evacuation. More people would get in each other's way. Litter attendants connect themselves to the litter with:

- **Litter Tie-Ins**

  The litter tenders are better able to maintain footing and stability by being tied into the litter. They lean back onto the tie-ins, putting their weight onto the litter. In turn, their weight is taken by the rope, braking device, and anchors.

DEPENDING ON THE MEDICAL CONDITION OF THE RESCUE SUBJECT, THERE MAY ALSO BE:

- **A Medical Attendant—**

  Usually the member of the team with the highest level of medical training and experience.

Litter Rigging for Slope Evacuation

There are two commonly used techniques for attaching the rope to the head of the litter:

- **Tying the Main Line Rope Directly to the Head of the Litter**

If the length of the slope evacuation is only one pitch (one rope length) and the rope will not have to be detached from the litter during the operation, then the main line rope may be tied directly onto the head of the litter.

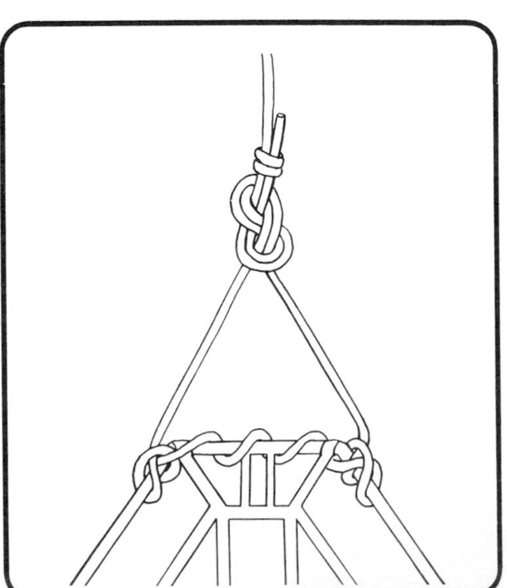

Fig. 13.2 Direct Main Line Attachment/Litter

Figure 13.2 illustrates a typical system for attaching a main line rope to the head of a litter for slope evacuation. This attachment consists of a very large loop created at the end of the rope by a Figure 8 Follow Through knot. This loop around the litter can be created with the following procedure *(see Figure 13.4)*:

1. At the end of the rope that is to be attached to the litter, measure off twice the distance between outspread arms (a total of approximately 10 feet).
2. At this point into the rope, tie a Simple Figure 8 knot.

> **WARNING NOTE**
>
> **When connecting a main line lowering or hauling rope to the end rail of a litter, the force must be spread out evenly along the rail. *NEVER ATTACH A MAIN LINE LOWERING OR HAULING ROPE TO A SINGLE POINT OF THE RAIL OF A LITTER.* Many litters are susceptible to failure if sudden forces pull at a single point of the rail. This is particularly true of litter rails that are butt-welded. A sudden force at such points can cause the weld to break and the litter rail to fail.**

3. Run the rope around the head rail of the litter several times so that it spreads the forces all along the rail.

Fig. 13.3 Tying Main Line to Litter

SUGGESTION:

For a more stable tie-in, start the wrap of the litter rail with a clove hitch and end the wraps with a second clove hitch on the opposite side. This will keep the bridle from slipping around as the direction of load shifts and adds some backup if one portion of the litter rail should fail, and on plastic litters it spreads the load better.

4. Bring the end of the rope back to the Simple Figure 8 knot.
5. Tie a Figure 8 Follow Through knot using the Simple Figure 8 knot. **BE CERTAIN TO LEAVE SEVERAL INCHES OF TAIL PAST THE KNOT.**
6. Center the knot so that both legs of the loops pull evenly onto the litter rail.
7. Make certain that the Figure 8 Follow Through knot is contoured well and pulled down tightly. After doing this, make certain there remains several inches of tail or the knot is backed up.

ALTERNATIVE APPROACH

If local policy mandates that a Bowline knot be used to create a loop in the end of a rope, then the Bowline knot may be used in place of the Figure 8 Follow Through knot, but with the following considerations:

- Make certain that the Bowline knot is tied correctly *(see Figure 6.8, page 57).*
- Back up the Bowline knot with a "keeper knot" such as the Barrel knot.
- Monitor the Bowline knot so that it does not "capsize" when being pulled over an obstruction such as a rock, a tree, a building edge, etc.

■ Tying a Closed Loop Directly to the Head of the Litter

During some slope evacuations, such as multi-pitch (more than one rope length) operations, rope will have to be removed and reattached to the litter. In such cases, it is more practical to leave a closed loop of rope tied at the end of the litter. To attach the lowering/hauling line to the loop, tie a Figure 8 Overhand knot in the end of the main line rope and clip it to the loop with a large, locking carabiner *(see Figure 13.4).*

To create a closed loop in the end of the litter:

1. Take a length of rope about 6 feet long.
2. Run it around the head rail of the litter to spread the force around the rail.
3. Tie the two ends of the rope together with a Figure 8 Bend knot (or Grapevine knot).

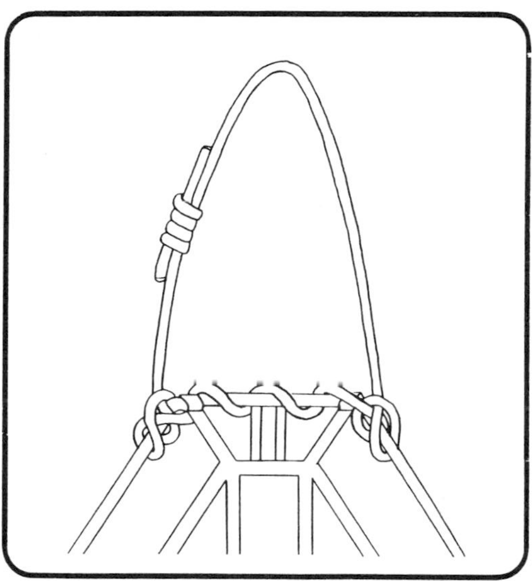

Fig. 13.4 Tying Loop Onto Litter

CAUTION: The angle made in this loop when it is attached to the main line rope must not be more than 90 degrees. (See Chapter 7, "Anchoring.") If the angle is more than 90 degrees, make a larger loop.

4. Adjust the knot so that it is off to the side and not in the center where the main line rope will attach.

ALTERNATIVE APPROACH

In the place of rope, tubular webbing may also be used to create the loop. However, you must take care in tying such a loop with webbing, since knots tied in it have a tendency to work their way out. Tie the two ends of the webbing together using a Water knot. **Once the Water knot is tied, be certain that there are several inches of webbing protruding past the knot**. Because of the tendency of webbing to work out of a knot, this knot should be backed up and monitored.

Packaging the Subject for Slope Evacuation

MEDICAL CONSIDERATIONS

Unless immediate life-threatening environmental factors prevent you from doing so, the following medical considerations must be made as a minimum **before** moving a rescue subject and must **continue** during the course of the rescue:

- **Airway**

  Does the subject have a clear airway?

  Do you need to maintain the airway during the rescue?

  Does he have objects in the mouth—gum, tobacco, dentures—that need to be removed?

  Is he vomiting, or does he have the potential for vomiting? How will you clear the vomitus during the course of the rescue so that he does not aspirate it?

- **Breathing**

  Is the subject breathing on his own?
  Will you be able to monitor breathing during the course of the rescue?

  Will you need to perform rescue breathing before and during rescue?

- **Circulation**

  Does the subject have a pulse?

  Will you be able to monitor the pulse during the rescue operation?

  Is CPR necessary? Will it be possible during the rescue operation?

AND:

- **Shock**

  Does the subject show signs of shock?

  What must be immediately done to treat for shock?

  What rescue procedures will increase/reduce the possibility of shock?

- **Bleeding**

  Is there life-threatening bleeding? How can you stop it?

- **Spinal Injury—*Many emergencies in the vertical environment relate to spinal injury.***

  Is there a possibility of spinal injury?

  Should a cervical collar be used before you move the subject?

  Should you secure the subject to a backboard before placing him in a litter for transport?

Packaging The Subject in the Litter

The major considerations for packaging a subject for slope evacuation include:

- The medical condition of the subject.
- The subject not be further harmed by being carried in the litter.
- The subject be protected from environmental factors (cold, wetness, debris).
- The subject be physically stabilized so he does not shift whatever the angle or nature of litter movement.
- The subject remain as comfortable as possible since it may be a very long evacuation.

Protecting The Subject in The Litter

- Underside:

  This is a particular concern in the wire basket stokes. This litter is uncomfortable to be in and offers little protection on the bottom from protruding objects, such as twigs, branches, stones, etc. To protect the bottom and make it more comfortable, line it with material such as Ensolite™ and/or blankets.

  Be certain to pad hollow spaces along the body, such as behind the knees and the small of the back.

- Topside:

  Topside protection should be for wind, cold, and rain. Additional blankets over the top will help keep the subject warm. But wind and precipitation mean the need for a waterproof/windproof outer layer.

- Face and eyes:

  A rescue subject in a litter cannot shield his face from branches, or his eyes from falling debris or rain. Face/eye protection may include a litter shield, a face shield, or, at the least, goggles and helmet.

Physically Stabilizing the Patient in the Litter

During a slope evacuation, the litter will be lifted, tilted and carried at an angle. The subject must be packaged so that he does not slip lengthwise in the litter, slide from side to side, or come out of the litter.

Lengthwise Stabilization

While most litters come equipped with some form of foot plates, many of these are unreliable in preventing the patient from sliding down when the litter is inclined.

One alternative is the feet tie-in *(Figure 13.5)*, which can be constructed of 1-inch tubular webbing. They should be tied off securely on the side rails to prevent the subject from sliding down in the litter.

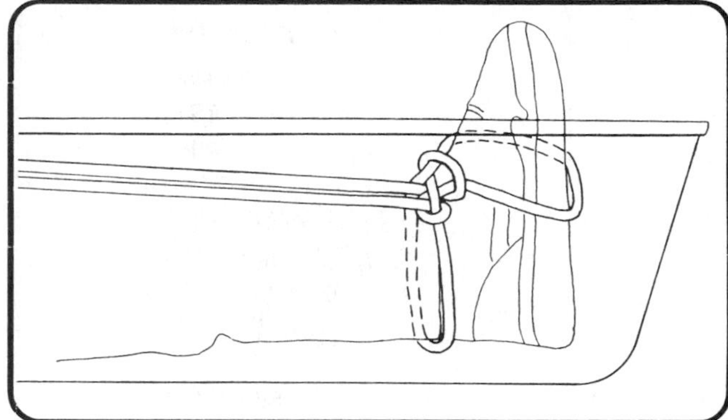

Fig. 13.5 Foot Tie-in to Litter

However, if the subject is suffering from a fracture to a lower extremity, then a foot tie-in cannot be used on that extremity. In this situation, a thigh tie-in *(Figure 13.6)* is a possibility. **MAKE CERTAIN THAT THIS DOES NOT OBSTRUCT BLOOD CIRCULATION THROUGH THE MAJOR VESSELS OF THE LEG TO THE FOOT.** To make certain a thigh tie-in does not obstruct circulation, it will be necessary to regularly palpate pulse points in the foot.

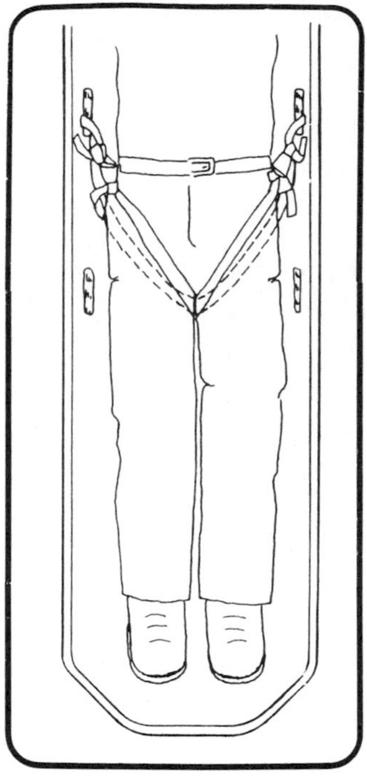

Fig. 13.6
Thigh Tie-in to Litter

Securing the Subject in the Litter

Litters often come with tie-in straps, but they may be inadequate for any of the following reasons:

- The buckles may not fasten securely or be strong enough.
- The straps may be old and rotten.
- The straps may be missing.

Figures 13.7(a)through 13.7(c) illustrate an alternative technique for securing the subject in the litter:

1. Begin with at least 30 feet of 1-inch tubular webbing. (If webbing is not available, then a rope is an alternative.) Find the center.
2. At the center, tie the line with a Girth Hitch onto the center of the foot railing of the litter.
3. Begin to lace the webbing or rope through opposite points along the side of the railing. If possible, do not run the line over the main rail. This can expose the line to greater abrasion. With the typical wire basket model, the line may be run around the larger uprights that connect the rail to the body of the litter.

   On the plastic style stokes litter, a pre-rigged rope is provided around the inside of the litter for tie-in points.

   (NOTE: As this litter comes from the factory, this tie-in rope is created in a nearly continuous loop. For added safety, remove this rope and replace it with one that is securely tied-in every few feet as it runs around the litter.)
4. Cinch the line back on itself as shown in Figure 13.8.
5. At the head of the litter, bring the material together and tie it securely with a Water knot.

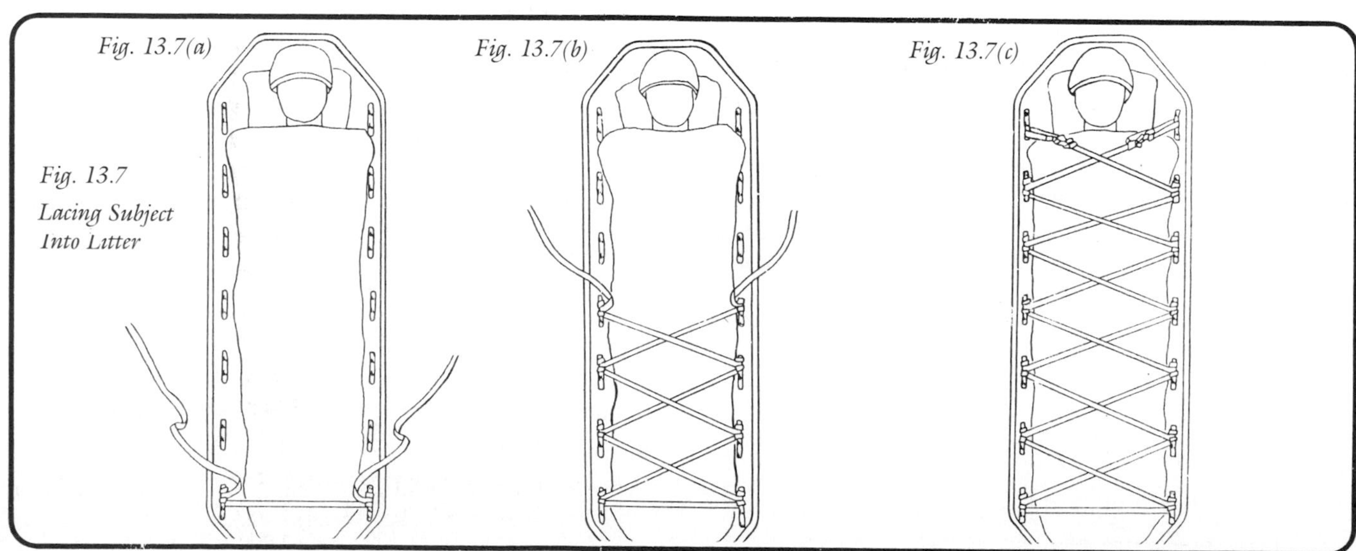

Fig. 13.7(a) *Fig. 13.7(b)* *Fig. 13.7(c)*

Fig. 13.7
Lacing Subject Into Litter

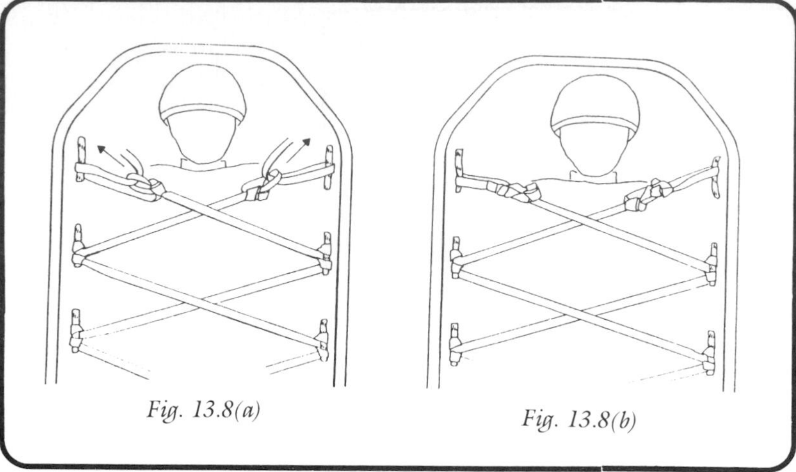
Fig. 13.8(a)

Fig. 13.8(b)

Fig. 13.8 Avoiding Webbing on Neck in Litter

CAUTION: DO NOT lash webbing horizontally across the upper chest or neck. Should the subject slide down in the litter, he could be strangled by a line across this area. Instead, return the ends of the webbing back toward the foot area of the litter, as shown in Figure 13.8.

Immobilizing the Head

You must immobilize suspected cervical injuries by following accepted medical protocols. **If you are CERTAIN** there are no cervical injuries, then you may immobilize the head area of the subject through other means. If the subject is not wearing a helmet, then the head area can be packed with blankets, packs, clothing or other soft material. If the subject is wearing a helmet, you can immobilize it with duct tape. (Webbing and rope tend to slip off the slick surface of a helmet.)

Additional Packaging

If the litter has a leg divider, then you must sufficiently pad the top of the divider to protect the subject's groin.

To prevent side-to-side movement of the trunk, pad the spaces along the sides of the subject with soft material such as sweaters, clothing, blankets, etc.

For the comfort of the subject, pad under the hollows of the body such as behind the knees, the small of the back, and (unless there is cervical immobilization) behind the neck. Also pad bony parts such as the occiput back of the head).

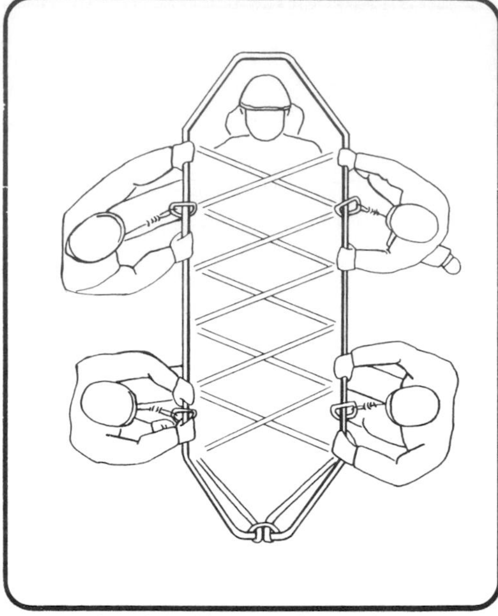
Fig. 13.9 Litter Tender Positions

The Litter Team for Slope Evacuation

Before attempting a slope evacuation, a litter team member must realize that slope litter management is very different from managing a litter on level terrain.

On level, even terrain, the normal number of persons carrying a litter is six (plus any additional personnel attending to the medical needs of the subject). The full weight of the litter is on them. They are totally supporting their own weight as they maneuver their way across the terrain.

On slopes there may be difficult and treacherous footing, making it difficult to safely handle the litter. So the approach to litter management on slopes must be different.

Characteristics of Litter Management on Slopes

- Much of litter weight is taken by rope.
- Litter movement is controlled by the rope system.
- Attendants have much of their weight taken by the litter/rope system.

Optional Additional Personnel in Slope Evacuation

- Medical Attendant
  Supervises medical care of subject and, where it relates to medical considerations, the movement of the subject. He is usually positioned at the head of the litter. Where the slope is at a greater angle, he may be safetied into the litter or main line rope with an adjustable sling.
- Scouts
  Move ahead of the litter to clear a path or warn the team of loose rock, branches, briars, snakes, etc.

Litter Tender Positions in Slope Evacuation

Figure 13.9 illustrates the positions for the four litter attendants during slope evacuation. Characteristically they are:

- Bodies turned towards the litter or slightly uphill.
- Depending on the terrain, both hands gripping litter rails, or one hand gripping the litter rail and the other being used for balance.
- Leaning back against tie-ins, their weight taken by the litter/rope system.
- Bodies perpendicular to the slope.
- Allowing rope system to determine litter rate of descent or ascent.
- Equal spacing so they do not bump one another.

Litter Tender Strategy for Slope Evacuation

On the slope above, the brakeman, in communication with the litter team, determines the rate of descent of the litter. The litter team members lean back into their tie-ins and allow the litter/rope system to take their weight. If a litter attendant slips, he continues to hold onto the litter rail and pulls his body taut on his tie-in. The rope system and the other team members can usually keep the litter stable and prevent him from falling. The litter team, in concert with the rope, act as a sort of self-equalizing table and provide stable transportation for the subject.

If more than one litter attendant loses footing or the terrain becomes particularly treacherous, the team captain can call a temporary "stop." This will give the team the chance to regain its stability.

Litter Tender Tie-Ins

The litter attendant tie-ins are critical for maintaining litter attendant stability and position at the litter. The litter attendant clips one end of the tie-in directly into his seat harness and the other end directly into the litter rail. The simplest form of the tie-in can be either:

- A daisy chain with multiple clip-in points,
- A continuous loop of webbing or small diameter rope, or
- A section of small diameter rope with a loop in each end.

Because each person's body proportions, such as trunk size and arm length, are different, the tie-in length will vary. But generally the tie-in will be no more than 2 feet long.

The optimum length for the tie-in should be one that will hold the litter attendant in position to have both hands on the litter rail, but maintain his freedom of movement.

An option to the fixed length tie-in is one of an adjustable length. The adjustable length tie-in enables the attendant to change the space to adapt to varying circumstances, such as changing terrain. It also is an advantage if different team members have to use it.

The tie-in is clipped directly onto the litter rail with a large carabiner. The carabiner should be placed on the rail so that it will support the attendant and not slide down the rail when the head end of the litter is raised. This can be done on basket litters by clipping the carabiner over the main rail on the head end side of one of the larger uprights that connect the rail to the body of the litter. On plastic litters, there is not as much choice of where to clip in. But one possibility is the spot where the plastic meets the rail. This junction will prevent the carabiner from sliding down toward the foot end when the litter is tilted.

WARNING NOTE

Older model plastic litters may not withstand the stresses of rope rescue:

- **The plastic material from which their body was constructed is brittle and sometimes fails.**
- **The plastic body of the litter is not as well-fastened to the rail as later models.**

Also, sunlight will deteriorate the plastic material in litters stored outside, such as on the tops of vehicles. These litters should not be used in rope rescue.

Not all locking carabiners have gates that will clear the rails of the stokes type litter. The larger rails on wire basket models and the rails on plastic litters will accept only certain models of a few carabiner brands. Consult manufacturer/distributor specifications before purchasing carabiners for this purpose.

ALTERNATIVE APPROACH

An alternative to attaching the tie-in to the litter rail with a carabiner is to attach it by girth hitching to the rail. The disadvantage in this technique is that it cannot be removed or attached as quickly as one attached with a carabiner.

Brake/Anchor Systems for Slope Lowering

The braking system is essential for a lowering operation during slope evacuation. It in turn is dependent on safe and secure anchors. (See Chapter 7, "Anchoring," for a review of the criteria for safe and secure anchors.)

The location and number of brake/anchor systems depends on whether the length of the lowering is single pitch (one rope length) or multi-pitch (more than one rope length).

One Pitch Slope Lowerings

A one pitch slope evacuation usually means that only one set of brake/anchors is needed. They are usually placed slightly above where the subject is loaded into the litter.

One pitch lowerings are usually easier and less complicated than multi-pitch lowerings because in one pitch lowerings:

a) Only one set of brakes/anchors is required.

b) There is no changeover to different sets of brake/anchors.

c) Less people are involved.

But even with a single pitch lowering there is a need for sufficient number of personnel to divide the labor and make the operation faster and more efficient. For example, while the rigging team is setting the brakes and anchors, the litter and medical teams can be assessing the subject for medical condition and packing him into the litter.

Multi-Pitch Slope Lowerings

Multi-pitch slope lowerings will be more complicated due to:

a) Multiple anchor/brake systems.

b) Changeovers from one anchor/brake system to another.

c) More personnel involved.

d) Coordination of movement of personnel.

In a multi-pitch lowering, anchor/braking systems will have to be rigged at strategic points down the slope. Among the factors that determine the placement of the anchor/brake systems along the slope are:

- The length of the main line lowering rope, minus—
  - a) the length needed on the lower end of the rope to tie into the litter, and
  - b) approximately 20 feet from the top end of the rope. This must always be kept as spare length before the rope runs out.
- The availability of anchor points.

The rigging of anchors for a braking system can be very time-consuming. In a multi-pitch evacuation, the litter team should not be forced to stop the operation each time the rope runs out while waiting for new anchors to be rigged. More preferable is one of two alternatives:

- If there is enough equipment, and enough skilled personnel for anchor rigging, then the anchor/brake systems can be rigged ahead of the litter team, or
- Leapfrogging.

Leapfrogging is the practice of using two sets of rigging teams and brake teams to alternate the rigging of anchors and the operation of the brakes. While the first team is operating the first set of brakes, the second team is rigging the second set of anchors. After the litter rope is detached from the first set of brakes, the first team derigs the first anchors and moves them down to the position for the third anchor/brake system.

This can be a very effective system and can speed the evacuation of a rescue subject. But for it to be effective, the rigging teams must be skilled at anchoring, knowledgeable of the equipment, and self-reliant. If they do not posses these qualities, then the entire operation may be interrupted as the litter is stopped and rescuers wait for the riggers to finish establishing the next brake/anchor sets.

Braking Systems for Slope Evacuation

Brake Bar Rack

The Brake Bar Rack is useful for slope evacuation in the following circumstances:

- Steeper Slopes—
  The Brake Bar Rack has spare friction.
- Terrain that varies between steep and gentle—
  The Brake Bar Rack can easily vary friction.

Figure 8 Descender

The Large Figure Eight with Ears can be used in more gentle terrain. Depending on the nature of the terrain, it can be double-wrapped.

Tree Wrap

The so-called **Tree Wrap** can be used where there is a shortage of equipment for braking systems, but only where there are strong, large diameter trees. The primary disadvantage of the tree wrap is that it tends to increase the difficulty of rope management.

Fig. 13.10 Using the Tree Wrap

Fig. 13.10(a)

Fig. 13.10(b)

Fig. 13.10(c)

Fig. 13.10(d)

Fig. 13.10(e)

Fig. 13.10(f)

WARNING NOTE

Before using a tree wrap brake system for slope evacuation, you must consider the following:

- **The tree wrap system must only be used by those who have thoroughly practiced it under realistic conditions.**
- **It can only be used on *live*, large diameter trees that have strong root systems.**
- **It can cause permanent damage to trees.**
- **It can contaminate ropes with tree sap.**

Using the Tree Wrap Braking System

Figure 13.10 illustrates a brakeman using the tree wrap braking system.

1. Attach the main line rope to the litter either directly with a large loop at the end of the rope, or clipped into a continuous loop that is attached to the litter head rail. The litter should be slightly down-slope of the tree wrap location so that the litter does not interfere with the braking operation.
2. a) If the rope is coiled:
   stack the rope uphill of the tree with the uphill end of the rope on the bottom of the stack and the litter end of the rope feeding off the top.

   b) If the rope is bagged:
   deploy the rope bag up-slope of the tree with the uphill end of the rope in the bottom of the bag and the litter end of the rope feeding out of the top of the bag *(Fig. 13.10a).*
3. Stand slightly to the side of the tree, with your back to it and the right side of your body pointed down-slope *(Fig. 13.10b).*
4. The rope should be on your side of the tree, running from the stack or bag to the litter, and between you and the tree.
5. With the rope behind you, pick it up and hold it behind you loosely in both hands.
6. Slowly begin to move around the tree backwards, still holding the rope as before: loosely in both hands, just below the buttocks in the rear *(Fig. 13.10c).*
7. As you go around the tree, the path of the rope from uphill will be off the stack (or out of the bag), around the downhill side of the tree, through your hands, across your buttocks, back around the lower part of the tree, and then to the litter *(Fig. 13.10d).*
8. As you need slack, pull it from the rope stack or bag. The rope handler can assist.

9. Up-slope of the tree, you will have to step over the rope coming out of the stack (out of the bag). Keep the rope evenly contoured around the tree, and prevent it from crossing itself on the tree *(Fig. 13.10e).*
10. The amount of friction needed depends on the:
    - Steepness of the slope,
    - Weight of the litter and the team, and
    - Circumference of the tree.

    A large tree, for example, will rarely require you to wrap it as much as 360 degrees.
11. Once you have enough friction, have the litter team pick up the litter, and take slack out of the rope between the litter and the tree. They then lean back into the rope. If there is sufficient friction, they can begin moving down slope with you controlling the speed of their movement *(Fig. 13.10f).*
12. To increase friction, go farther around the tree to wrap the rope more, and by gripping the rope with your hands.
13. Reduce friction by moving from around the tree, to wrap the rope less, and by gripping the rope less with your hands.

Rope Management in Slope Evacuation

Good rope management is particularly critical to ensure smooth operation of the brake system in slope evacuation. Without good rope management, kinks in the rope and tangles coming from the stack or bag could jam the brakes to slow the operation or bring it to a complete halt.

To help provide good rope management, a rope handler should assist the brakeman, particularly in a tree wrap lowering. The rope handler assists the brakeman by feeding him the rope and removing kinks and tangles before they reach the brakes.

The Belay Question

The question of whether to belay in a slope evacuation must be answered on the scene by trained and experienced personnel. They consider the following questions:

- How steep is the slope?
- Is the footing particularly loose and treacherous?
- What would be the consequences of a fall by the litter team?
- Are the main line brake anchors questionable?
- Are there plenty of anchors?
- Is there thick underbrush/large boulders?

  (The angle made by the main line rope and the belay line in slope evacuation is similar to an advancing wedge. If there is thick underbrush or large boulders, the advancing wedge could tangle.

Communications

Good communication in slope evacuation is particularly critical, especially between the litter team and the brakeman. It is essential that communication be simple, to the point, and clear.

For these reasons, voice communications should be limited to a few people. Among these would be the litter captain and the brakeman. In those cases where there are problems in communication, such as distance, wind or waterfall, a relay person may be necessary. **Everyone else should keep quiet.**

To avoid confusion, voice communication must be limited only to those few standardized commands necessary for litter movement. If a belay is being used, then the standard belay commands are included.

Voice Communications for Slope Lowering

"On Belay."	*(Litter captain to belayer.)*
"Belay On."	*(Belayer to litter captain.)*
"Down Slow." ***or*** **"Down Fast."**	*(Litter captain to brakeman.)*
"Stop!"	*(Generally the litter captain to the brakeman, but may be given by anyone who sees danger or potential problem developing.)*
"Stop? Stop? Why Stop?"	*This is given by litter captain to brakeman. It is given when, for an unexplained reason and without command from the litter captain, the rope has stopped moving. It could be that the brakeman is still letting out rope, but the rope is jammed somewhere. Obviously the potential for a very serious problem.)*
"Two – Oh."	*(Given by the brakeman to the litter captain. It means there is only about 20 feet of rope left. The litter team should set the litter down at a convenient spot so a new brake/anchor set can be established.)*
("Off Belay.")	*(Litter captain to belayer. The litter has been set down in a secure spot. It and the litter team are in no danger of falling.)*
("Belay Off.")	*(Belayer to litter captain.)*

A Typical Slope Lowering

The following sequence outlines the basic elements of a slope lowering operation.

Preparation

Before movement of the litter begins, all major elements of the slope evacuation system are in place and are prepared:

The subject has been medically assessed, treated, and his condition is being monitored.

- The rope has been properly attached to the litter.
- Secure anchors have been set and brakes attached to them.
- The rope is wrapped in the brakes and locked off.
- The brakeman has the rope in hand and is ready to run brakes.
- The rope handler is ready to feed the rope to the brakeman.
- Belay line is set in belay device.
- Belayer is ready to belay.
- Litter attendants have tie-ins attached to themselves and to the litter and are ready to lift the litter.

1. The litter captain directs the litter team to lift the litter. He says, **"Pretest."** The team removes slack from the rope by holding the litter downhill against the rope and leaning into their own tie-ins. The brakeman holds the braking system fixed. This is to pretest the system and make certain everything is in order and the litter team has their tie-ins correctly set.
2. If everything is in order, the litter captain gives the voice communication, **"Down Slow."** The brakeman begins to let the rope through the braking system. The rope handler feeds rope to the brakeman. The litter team leans into the system and moves down hill. (The litter captain may also issue a **"Down Fast"** if he wants to move down slope faster.)

   If anything begins to go wrong—a kink slips past the rope handler and jams the brakes, a litter attendant begins to loose a boot, whatever, anyone can call, **"Stop!"**
3. The rope handler warns the brakeman that the rope is running out. The rope handler shouts, **"Two – Oh!"**
4. The litter captain looks for a good place to set the litter down. When the litter is at the place he has chosen, he communicates, **"Stop!"** He then directs the team to set the litter down. When the litter is secure and the team in a stable position, he communicates, **"Off Rope."**

IF MULTI-PITCH EVACUATION:

5. When the litter gets into a spot convenient to the second set of brakes, the team stops there and sets the litter down. The second brakeman and rope handler are already in position.
6. *Alternative A:*

   The main line rope is now trailing from the litter uphill to the first set of brakes, but has been removed from the first set of brakes. If the rope is unlikely to tangle, the brakeman simply attaches the same rope to the second set of brakes. The rope handler will pull the rope down as needed and feed it to the brakeman.

   Alternative B:

   If the main line rope now trailing uphill is likely to snag, the brakeman detaches it from the litter. He then attaches a second rope, which is stacked or bagged near the second set of brakes. The team at the first set of brakes will derig the brakes/anchors, recover the first rope as they come downhill, and then rig the third set of brakes/anchors (leapfrogging).

   (If belays are being used, then the second belayer gets ready to belay.)
7. The teams repeat the cycle.
8. As soon as the litter team sets the litter down and becomes secure at the end of the first pitch, the first brakeman and rope handler (and first belayer) quickly derig the first brake anchor set. They then quickly leapfrog down-slope past the second brake/anchor set to rig the third brake/anchor (and belay) set. They are ready when the litter team reaches their position and detaches the rope from the second brake/anchor set. This cycle continues until the litter safely reaches an objective such as a road and waiting ambulance or a clearing with a helicopter.

Hauling

Not all litter evacuations will be going **down**-slope. Many of them will have to go **up**-slope. **Hauling** techniques use many of the same principles as lowering. However, hauling systems may involve slightly more complex rope work and may require the use of additional personnel.

Mechanical vs. Human Power

There are two general approaches to hauling systems: mechanical power or human power. In the area of mechanical power, there are several excellent powered winches suitable for rescue work. With practice at using them and understanding of their potential and limitations, these winches can help perform a slope hauling rescue safely and efficiently. But, as with other types of hauling/lowering systems, there is always the potential for human or mechanical failure, so winches should always be belayed or safetied in some other way.

One other potential problem with powered winches is too much power at the wrong time. If there is a jammed litter or a hand caught between a rock and a litter, the powered winch may not slow down until damage is already done. This chapter will concentrate on simple systems that use human power efficiently to haul litters during slope evacuations.

The All- Important Safety Cam

Whatever the type of hauling system, it is essential that there be a safety system to prevent the litter from falling back down slope in case of mechanical or human failure in the system. One of the commonly accepted safety system devices is the **safety cam**, which usually employs a cam-type ascender.

> **WARNING NOTE**
>
> **Never use handled ascenders for hauling systems in which more than one person's body weight may be involved. The use of this type of ascender can result in failure in two potential ways:**
>
> - **The frame or other portion of the ascender may fail.**
> - **The sharp teeth on the cam can tear the rope sheath.**

Using Safety Cams

Safety cams should always be connected to a safe and secure anchor system that is, if possible, **separate** from the anchor system supporting the haul system.

On some models of cams there is an arrow with the caption, "Up." This is the indicator for direction of use when ascending a rope. **But in hauling systems, this arrow should always point along the rope towards the load** (the litter, the rescue subject). Some older model cams do not have this arrow inscribed on the shell. But if you look at the shell in profile, you will see that the shell vaguely resembles an arrowhead *(see Figure 5.23)*. Again, the arrow should point in the direction of the load.

Positioning of the Cam

Though the safety cam should be on a separate anchor from the hauling system, it should be close to, and parallel to, the main line rope. This will help prevent shock loading and reduce the slack that interferes with haul system efficiency.

Setting of the Safety Cam

When rigging a safety cam, you have two primary concerns:

- That the cam clamp on the rope should a safety be needed.
- That the cam not ride up the rope as the rope moves. This could result in dangerous shock loading.

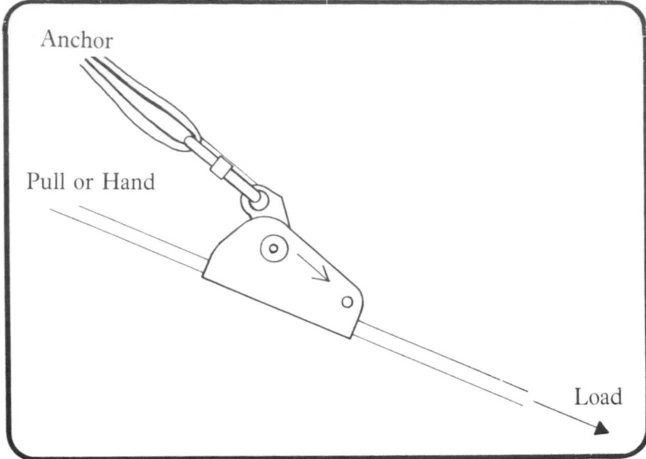

Fig. 13.11 Setting Safety Cam

Fig. 13.12 Safety Cam With Tender

How the cam is specifically rigged depends in part on the specific type of cam used and in part on the specific circumstances of the rigging *(see Figures 13.11 and 13.12)*.

Free-Running (Not Spring-Loaded) Cam

Figure 13.12 shows one method of ensuring that the cam stays in place and clamps on the rope when needed. This technique uses the services of a person known as the **cam attendant**. As the main line hauling rope moves up, the Cam Attendant makes certain that the cam does not travel up the rope. He does this by holding the back side of the shell with the palm and with fingers extended out of the way of the cam. He holds the shell of the cam in this manner so that fingers or gloves do not get caught and the cam suddenly shock loads. A finger caught by the cam could get injured, and a glove caught in the cam could prevent it from clamping the rope.

While the use of a cam attendant can be effective, it does require extra manpower. And there is always the potential for human failure or inattention.

A second technique of using a free-running cam as a safety is shown in Figure 13.13(a). A bungee (elastic) cord is clipped into the empty hole that is usually found in the "point" of the arrowhead. The other end of the bungee cord is anchored securely to a convenient spot towards the load. A great deal of tension is not required on the bungee cord. But there should be enough to:

a) Keep the cam from riding up the rope.

b) Keep the cam closed on the rope.

NOTE: Tie the bungee cord **only to the shell,** and not to the cam itself. Otherwise, the system will not function as needed.

A third alternative is shown in Figure 13.13 (b). This method uses a short piece of cord. It is attached to a weight and the weight is hung so that it will pull the shell of the cam toward the load.

Fig. 13.13 Setting Safety Cam for Automatic

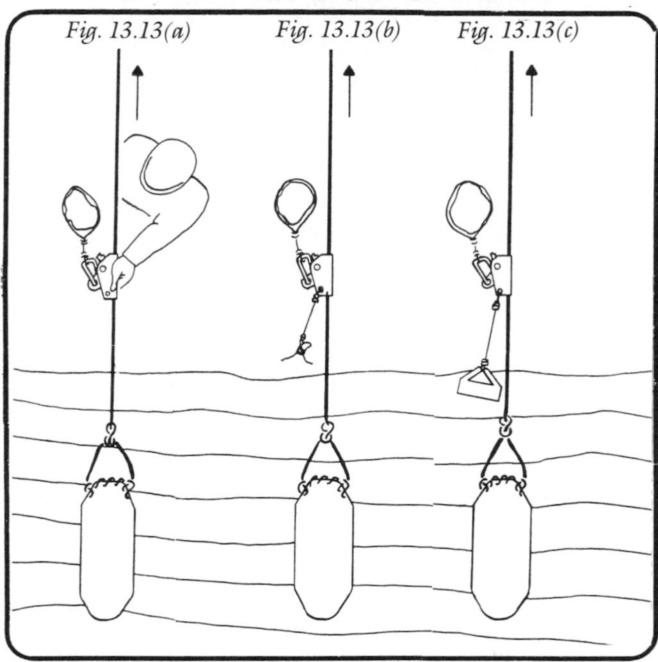

Spring-Loaded Cam

If a spring-loaded cam is properly rigged, then the cam will close onto the rope when needed. But the problem remains of how to keep the shell from riding up the rope as the rope moves. Again, the solution would be similar to that of the free running cam: a cam attendant, a bungee cord, or a weighted cord.

ALTERNATIVE APPROACH

Under certain conditions a properly used Prusik may be employed in the place of a cam for hauling. However, the following precautions should be taken:

- A three-wrap Prusik should be used;
- A tandem Prusik should be used;
- The Prusik should be constructed of material that holds a knot well, has an adequate safety factor, and is resistant to heat damage; and
- The Prusik should be tied, placed on the rope, and operated by persons experienced in the use of Prusiks as cams.

For further information, see page 119 and information on Prusik knots in Chapter 10, "Basic Ascending Techniques."

A 1:1 Mechanical Advantage Hauling System

One of the simplest hauling systems to be used in slope evacuation is the 1:1 hauling system. In essence, the 1:1 simply means that the force needed to haul the load (litter, subject, and attendants) is about the same as the weight of the load.

Figure13.14 illustrates the elements of a basic1:1 hauling system. They include:

- **The litter system,** essentially the same as in lowering—

  The subject packaged in the litter.

  The litter team attached with tie-ins.

- **The rope system,** which is—

  Attached to the head of the litter, and in turn—
  Runs through a safety cam, which is attached to different anchor from—

- **Pulley** *(Optional),* which can be used to change the direction of the rope to a more convenient angle for:

- **Haul team,** whose members are attached to the main line haul rope by way of Figure 8 on a Bight knots in the rope, or ascenders on the rope, which are attached to their seat harnesses.

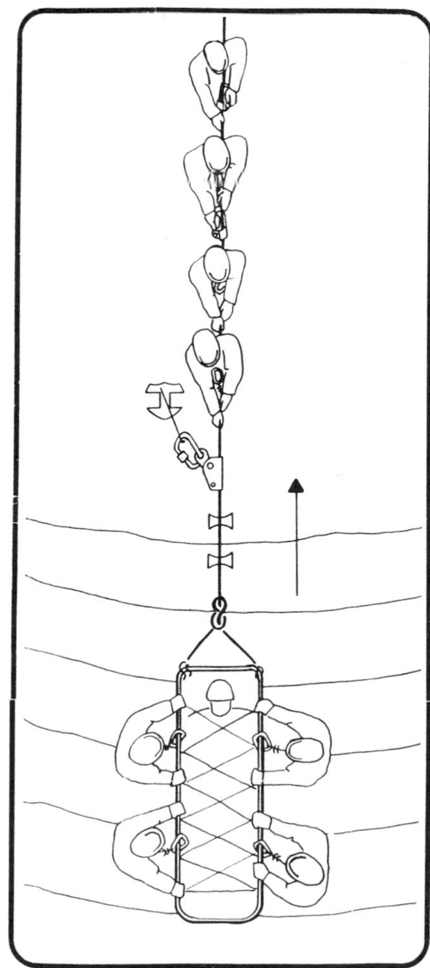

Fig. 13.14 Elements of 1:1 Hauling System

Pulleys Used in 1:1 Hauling Systems

Pulleys used in 1:1 haul systems do not add any mechanical advantage. (In fact, some advantage may be lost through the pulley's inherent friction.) Pulleys can, however, make things more convenient for a haul team by changing the direction of the rope pull to:

- Enable a team to pull along the contour of a hill instead of straight up.
- Allow the haul to be in a more convenient place, such as a clearing or along a road *(see Figure 13.15)*.

(See Chapter 5, "Basic High Angle Hardware," for more information on pulleys.)

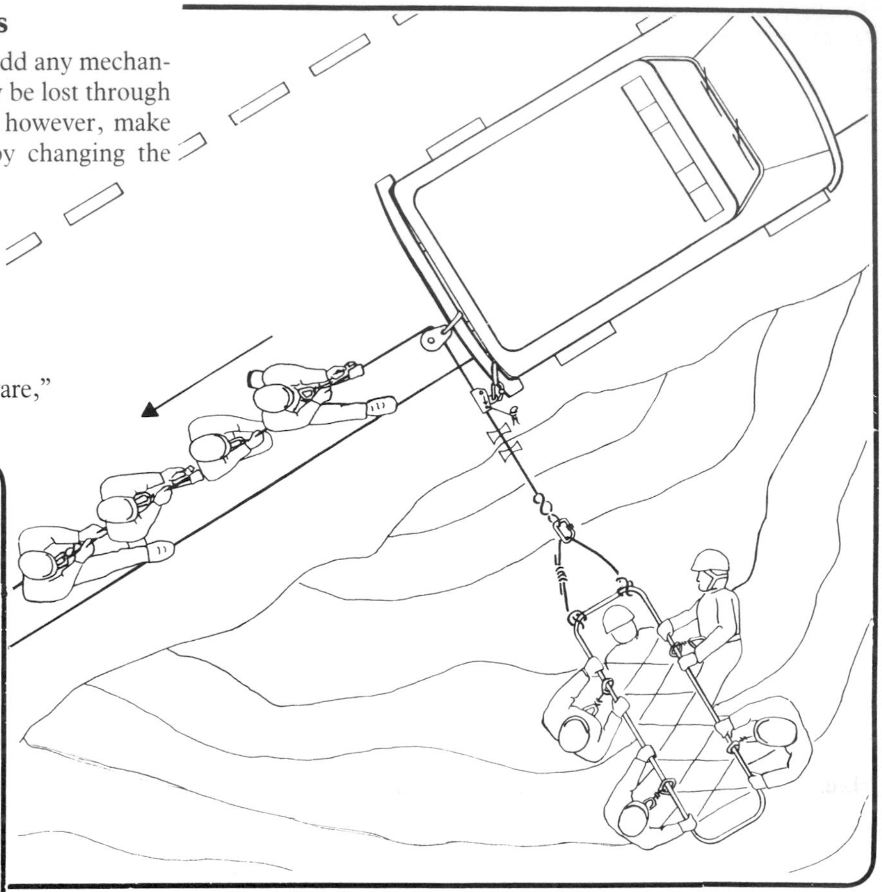

Fig. 13.15 Making Haul More Convenient.

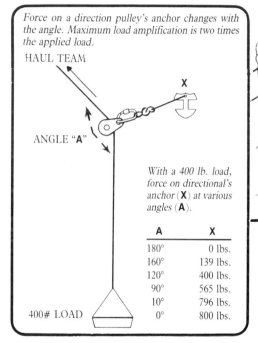

Force on a direction pulley's anchor changes with the angle. Maximum load amplification is two times the applied load.

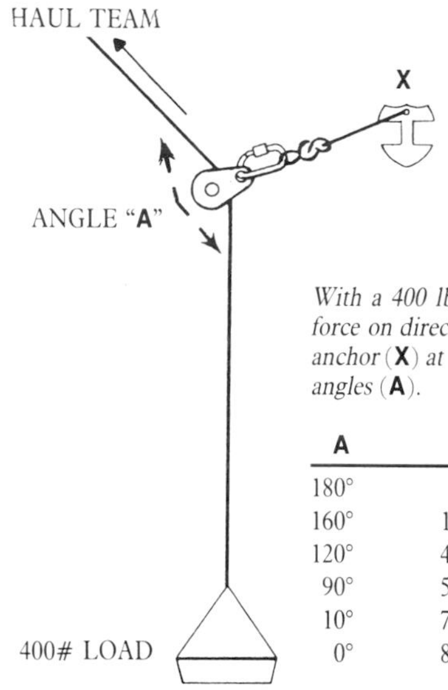

HAUL TEAM

X

ANGLE "**A**"

400# LOAD

*With a 400 lb. load, force on directional's anchor (**X**) at various angles (**A**).*

A	X
180°	0 lbs.
160°	139 lbs.
120°	400 lbs.
90°	565 lbs.
10°	796 lbs.
0°	800 lbs.

Fig. 13.16 Forces on Anchors of Hauling System

WARNING NOTE

Anchors established for pulley directionals must be stronger than if they were simply supporting a weight equal to the load being hauled. The greater the angle in the rope created at the change in direction, the greater the force on the pulley and the anchor system *(see Figure 13.16).*

The Haul Team

Haul team members should not be selected on the basis of brute force ability, but according to their:

- Intelligence.
- Ability to follow commands.
- Responsiveness.

WARNING NOTE

Hauling systems create enormous forces on the rope rescue system. Unnoticed problems can quickly result in catastrophic system failure. Personnel must constantly monitor for potential problems including, but not limited to:

- **Knots on moving rope that jam in cracks.**
- **Broken gear.**
- **Systems reaching their limit.**
- **Pinned arms and legs.**

There must be good communication between team captains and the haul team.

The haul team must be aware that what seems a normal speed for them seems *very fast* for those being hauled (rescue subject, litter attendants). For all these reasons, the haul team must:

- **Pull slowly unless told otherwise.**
- **At all times, be prepared to stop instantly.**

Communications for Hauling

The voice communications for hauling movement are initiated by the litter captain. Except for one special exception, **no one** else initiates communications for hauling. **Everyone else stays quiet.**

The Haul Commands

To avoid potential disastrous confusion, haul commands are limited to a few standardized ones.

WARNING NOTE

Never substitute voice haul commands, unless there is an overriding reason and all team members have previously been informed of the change.

Use only terms that are crisp and distinct and have no chance of being mistaken for other words.

For example, do not replace the word "stop" with "whoa," which can easily be mistaken for "slow," or even worse, "go."

"Haul." *(Begin Hauling. Needs to be spoken only once. The team will continue to haul until given another command.)*

or

"Haul Slow." *(A variation when there is an emphasis on slow movement, such as when the litter is about to reach the top.)*

"Set." *(The haul team immediately stops hauling and gently eases back on the load. This is to set the safety cam. It may be for reason of safety or, more commonly, to get another bite on the rope.)*

"Slack." *(The safety cam is set, so the haul team slacks on the rope. This enables any part of the system to be reset, and the haul team to get another bite on the rope.)*

(If a belay is employed, then the normal belay communications are also used.)

THERE IS ONE COMMAND THAT *ANYONE* MAY GIVE WHEN THEY SEE SOMETHING GOING WRONG:

"Stop!" (All movement stops immediately. The haul team holds tension until told otherwise.)

The Counter Balance Haul System

In any hauling system, the major force that rescuers are fighting is gravity. The Counter Balance Haul System is a method of using **gravity** to fight **gravity**.

Strictly speaking, a counterbalance hauling system is also a 1:1 haul system. Without considering loss due to friction and other inefficiencies, the force used to haul is the same as the weight of the load it is hauling. But the difference is this: instead of the haul team going off the side (as in the example on page 164) the haul team takes advantage of gravity by going downhill. The load (the litter system) still goes uphill.

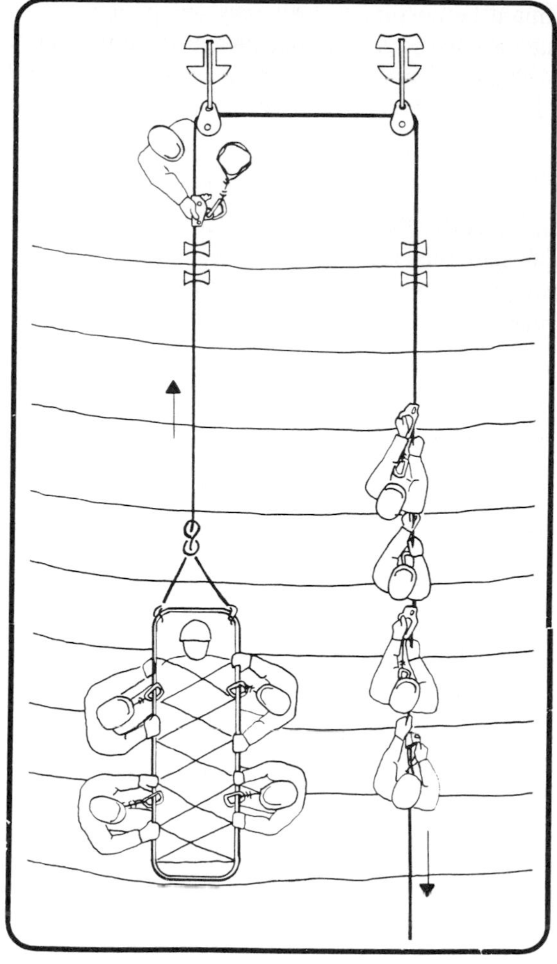

Fig. 13.17 Elements of Counter Balance Haul

Figure 13.17 illustrates the elements of a Counter Balance Haul System. It includes the following:

- **Litter System**

  The subject medically assessed, treated, packaged in the litter, and his condition being monitored.

  The litter team attached with their tie-ins.

- **The Rope System.**

  The bottom end of the rope is attached to the head of the litter.

  As it gets to the top, the rope first runs through the safety cam, which is on its own anchor system.

  Just above the safety cam, the rope goes through the:

- **Pulley/Anchor System**

  Note that in this example there are two pulleys:

  The first one that initially changes the rope direction 90 degrees, and then—

  The second pulley that changes the rope direction another 90 degrees, so that the rope goes back down the slope.

It might be possible to have only one pulley. But in this situation, two pulleys mean a greater margin of safety and less chance of failure. Remember there are now two loads pulling down slope on the anchors:

1) The weight of the load (litter system), and

2) The force of the haul team pulling slightly.

Consequently, there are two pulleys and two anchor systems to share the load. But having these two anchor systems and two pulleys set a distance apart also means two additional advantages:

- It means less chance of the two rope strands tangling.
- It means less chance for the haul team and the litter team to interfere with one another's movements.

A 2:1 Hauling System for Slope Evacuation

Figure 13.18 illustrates a 2:1 hauling system for slope evacuation.

In theory, the force needed by the haul team to move the litter is only 1/2 the force needed in the 1:1 hauling system previously examined. But with this 2:1 hauling system, the haul team will have to travel twice as far.

The 2:1 hauling system is created by anchoring a rope at the top, running it back down to the litter through a pulley that is attached to the litter yoke, and then running the rope back to the top near the anchor. The haul team might be able to pull uphill on the rope. But Figure 13.19 shows them going off to the side, to make it a little easier on themselves. This sideways pull is made possible by running the rope over a **directional** pulley anchored near the anchor point of the top end of the rope.

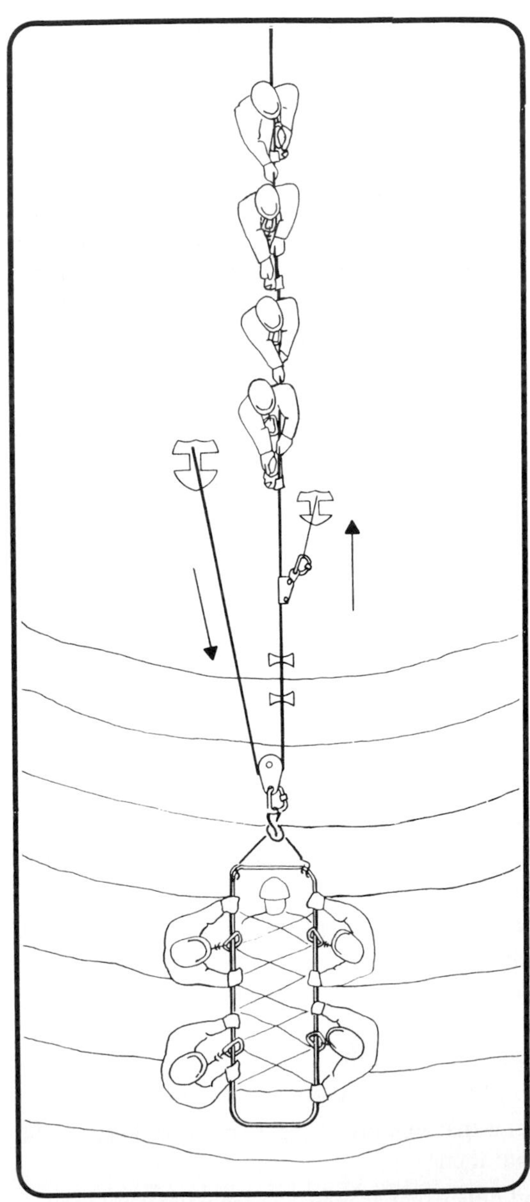

Fig. 13.18 Elements 2:1 Hauling System

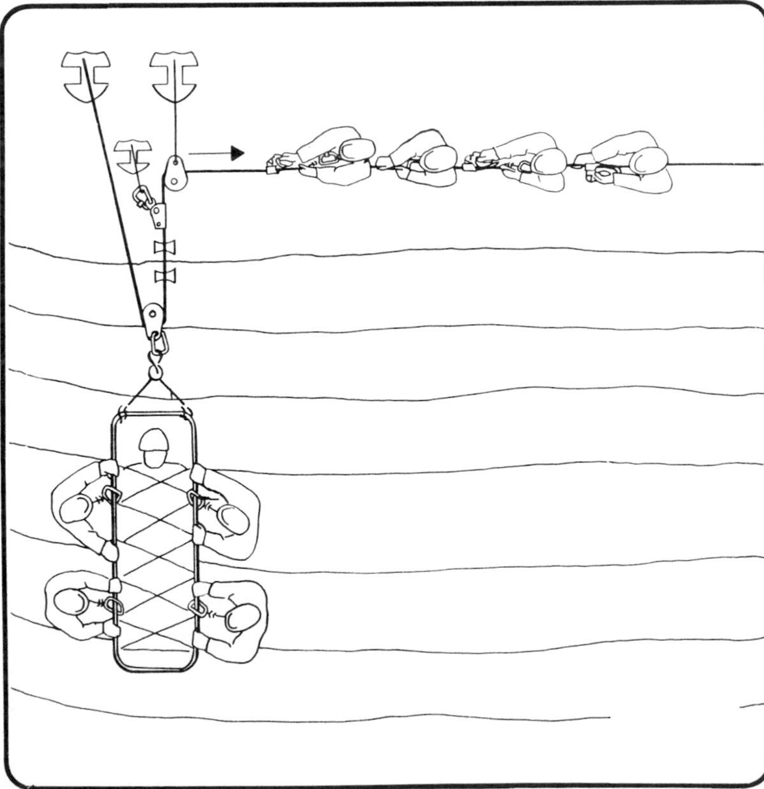

Fig. 13.19 Side Pull on 2:1 Haul System

WARNING NOTE

While a 2:1 haul system can make hauling easier, it does increase rope management problems:

There are now two strands of rope moving along the same path.

There is a pulley moving, which could easily become jammed on underbrush, trees, rock, etc. Rescue personnel must be able to reach this pulley, wherever it jams, to free it.

(See Chapter 15, "Hauling Systems," for a possible solution to this problem.)

Other Hauling Systems

Other types of multiple force haul systems may be used in slope evacuation as:

- The slope becomes steeper.
- Less personnel are available for the haul team.

Fig. 13.20 Personal Safety Lines

Drawbacks to Multiforce Haul Systems

Remember that while multiforce haul systems can make it easier for haul teams to move a litter, all of them have some drawbacks, including:

- More complicated to rig.
- Taking longer to rig.
- Creating greater forces on anchors and other parts of the system.
- Easier to foul and, perhaps, to fail.
- The haul team has to travel a longer distance than with a 1:1 system. Chapter 15, "Hauling Systems," examines multiforce haul systems in greater detail.

Safe Movement of Personnel in Slope Evacuation

One of the problems in slope evacuation is the safe movement of personnel up and down the slope for rigging, medical evaluation, litter rigging, etc. These people may have problems with footing, thus endangering themselves. But they may also knock rocks or other debris down onto the rescue subject and other rescuers.

To avoid this, one of the first actions on the scene should be the immediate establishment of personnel safety lines *(see Figure 13.20)*. These lines should be well-anchored and established off to the side. In this way, personnel can travel up and down them without endangering those below.

Once the safety lines are established, personnel can move up and down, either by hand or with descenders and ascenders.

QUESTIONS for REVIEW

1. List four ways in which slope evacuation differs from high angle evacuation.
2. List three circumstances where slope evacuation might be used in a rescue.
3. Why should you never use a single point of attachment for a main line rope to a stokes litter rail?
4. Under what conditions of slope evacuation would you attach the main line rope directly to the litter rail? Under what conditions would you tie a closed loop into the litter rail?
5. What are four major considerations for the rescue subject in packaging him in a litter for slope evacuation?
6. What kind of knot could be used to make a litter attendant tie-in adjustable?
7. What is the difference between "single pitch" and "multi-pitch" lowerings?
8. Name two factors that will determine the location of brake/anchor systems in a multi-pitch lowering.
9. What are the duties of a rope handler in a slope evacuation?
10. What are three considerations in deciding whether or not to have a belay in a slope evacuation?
11. Repeat from memory and in sequence the voice communications between the litter captain and the brakeman, and the litter captain and a belayer during a slope evacuation.
12. Before movement of the litter begins in a slope evacuation, all major elements of the operation must be in place and prepared. List nine things you would check before allowing movement of the litter.
13. Why should handled ascenders not be used for hauling systems in which more than one person's body weight is involved?
14. When cams are used in hauling systems for slope evacuation, which way should the arrow point?
15. What are two techniques for making sure that safety cams set when needed and do not ride up on the rope?
16. Name three potential problems that could develop in a slope hauling system and which all members of the team should be on the lookout for.
17. What is the voice communication that should be given by any member of the rescue team who sees something going wrong during a slope evacuation?
18. One of the problems in slope evacuation is the movement of personnel up and down the slope for rigging, medical evaluation, etc. This movement can result in possible injury to these people and to the subject and other rescuers because of loose rock and other debris. What is a possible solution to this problem?

Chapter 14

High Angle Lowering

PREREQUISITES

Before attempting the activities described in this chapter, you must have demonstrated that you can properly:

1) Use and care for rope.
2) Use and care for other equipment employed in the high angle environment.
3) Tie correctly, confidently, and without hesitation the eight knots described in Chapter 6.
4) Apply the principles of anchoring and rig a safe and secure anchor.
5) Apply the principles of belaying and safely and confidently belay another person using either a Munter Hitch or belay plate.
6) Apply the principles of rappeling: rappel safely, confidently, and under control; tie off the rappel device to operate hands free of the rope and then return to a safe and controlled rappel.
7) Apply the principles of ascending: tie correctly, confidently, and without hesitation a Prusik knot and know how to use it; comprehend the uses and limitations of mechanical ascenders, confidently and safely ascend a fixed rope using either friction knots or mechanical ascenders; confidently and safely change over both from rappeling to ascending and from ascending to rappeling; and extricate yourself from a jammed rappel device (or similar problem).
8) Apply the principles of slope evacuation: correctly set the rigging in any of the elements of slope evacuation; and safely and confidently assume the role of litter tender, haul team member, rope handler, brakeman, or belayer.

OBJECTIVES–

At the completion of this chapter, you should be able to:

1. Define high angle lowering and describe some examples of it.
2. List the elements of a high angle lowering system.
3. Select equipment for a high angle lowering system.
4. Discuss the medical considerations for a rescue subject before and during a high angle lowering.
5. Describe the functions of the following positions: litter tender, brakeman, rope handler, belayer, and edge tender.
6. Describe the functions of rope, braking systems, belay, edge protection, spiders, and litter tender tie-ins in high angle lowering.
7. Discuss the need for reliable communications in high angle lowering.
8. Describe the function of knot passing in a high angle lowering.
9. List the differences between single strand and double strand lowering systems, along with the advantages and disadvantages of each.
10. Repeat from memory the voice communications employed in high angle lowering.
11. Rig a litter and act as litter tender in a high angle lowering.
12. Rig the braking system and act as brakeman in a high angle lowering.
13. Rig the belay system and act as belayer in a high angle lowering.
14. Perform as rope handler in a high angle lowering.
15. Perform as an edge tender in a high angle lowering.
16. Perform a knot pass in a high angle lowering.

TERMS—*relating to High Angle Lowering that a Rope Rescue Technician should know:*

Auxiliary Tender—The person who rappels alongside the litter as it is being lowered in order to assist in the rescue. Duties may include medical assessment and/or primary treatment of the rescue subject, assistance in getting the litter over the edge, assistance in handling the litter on the vertical face, and in loading the subject into the litter.

Brakeman—The person who operates the braking device for controlling the rate of descent of the load in high angle lowering.

Double Strand Lowering—The use of two ropes attached to the litter in a lowering rigged so that they may be operated independently to change the angle of the litter. Double strand lowerings usually involve two litter tenders.

Edge Tender—The person connected to a safety attachment who works at the edge of a drop in a high angle lowering. His duties include assistance in getting the litter over the edge, reducing edge abrasion to the rope and, when necessary, relaying communications between the litter tender and the brakeman.

Load—The object/person being lowered by rope in a high angle lowering. Some examples include a rescue subject, a rescuer and subject in a litter with attached litter tender.

Pig Tail—A short piece of rope with which the litter tender attaches to the litter system.

Single Strand Lowering—The use of one main lowering rope with a belay in litter lowering.

Spider—The system of attaching a lowering rope to a litter. A spider usually has four or more legs that connect to various points of a litter to equalize loading.

The High Angle Lowering System

High angle lowering is also sometimes called **vertical lowering** or **technical lowering**. All of these terms refer essentially to the same principle: the controlled lowering of a rescue subject using a rope. If the subject's injuries are severe enough, the lowering is done with the subject packaged in a litter. High angle litter lowering is usually done with a rescuer (litter tender) attached to the litter, and with the weight of the subject, litter, and rescuer supported by the rope.

As the rescuers lower the litter, it may run down against the side of a high angle wall, such as a cliff or side of building, or it may be "free" (not touching the wall). This will depend on the nature of the wall where the lowering is performed. If the top of the drop is overhung, the litter will hang out away from the wall. This might be a more difficult operation for the rescuers since they would not have the advantage of pushing against the wall to maneuver themselves and the litter during the rescue.

In some cases where the subject is uninjured or only slightly injured, it may be possible to lower him without the litter, by attaching him to a rescuer.

Some of the environments in which a high angle lowering might be employed include:

- Cliffs.
- Buildings.
- Other structures such as cranes, towers, stacks, or silos.
- Vertical caves.

The vertical lowering may take place either on the outside of the structure or on the inside, such as the interior of a silo or tank.

The Lowering System

Figure 14.1 illustrates the basic elements of a vertical lowering system.

1. The load
   This often will be the subject packaged in a litter. Because of medical considerations and concern for the subject's comfort, the litter is usually lowered in a horizontal position (although certain confined space environments may require it to be lowered in a vertical position with the head up). The load could also be only the subject, if he is uninjured or only slightly injured. Or the load could be the subject attached to a rescuer.
2. The litter tender(s)
   This may be one person, or the situation may require two litter tenders. Each litter tender is attached to the litter by at least two connections: a primary tie-in and a safety.
3. Attachments of the load to the main line rope(s)
   Litter attachments to the rope are known as **spiders**. They are lines that attach at several points to the litter and come together where they are attached to the rope. Spiders

Fig. 14.1 Basic Elements of Lowering System

usually have at least four **legs** that attach to the litter, though in some systems there may be as many as six.

4. The main line, lowering rope(s)

   These lines must have a safety factor adequate for the load they are to lower. In some lowering systems, there will be only one main line rope with a belay. In other systems, there will be two main line ropes.

5. Belay system

   This is attached to the load and acts as a safety should there be a failure in the main line lowering system.

6. Brake device

   These are friction devices that are the same as, or similar to, rappel devices. They provide friction on rope running through them to control the descent of the load.

7. Brakeman

   This person controls the speed of the descent of the load by controlling the rope through the brake devices.

8. Rope handler

   This person assists the brakeman by feeding him the rope and making certain there are no kinks to jam the brake device.

9. Belayer

   This person controls the belay rope through the belay device and catches the load with the belay rope should the main lowering system fail.

   Additional personnel may include:

10. Edge Tenders

    They assist the litter tender(s) in getting the litter over the edge of the drop, prevent rope abrasion on the edge, and, if needed, relay voice communications between the litter tender and the brakeman.

Braking Systems for Lowering

The principles of using a braking device for high angle lowering is similar to rappeling. Whether you use the device for rappeling or lowering, you impart friction to a rope running through the device. And, indeed, some of the devices that are used for lowering are the same as used in rappeling. Some examples are Brake Bar Racks and Figure 8 descenders. But in lowering there may be more than one person's body weight on the load, so THERE MUST BE A GREATER SAFETY FACTOR FOR THE DEVICE when it is used for lowering.

The primary difference between lowering and rappeling is that in most cases during a lowering, the friction device remains stationary while the rope moves through it. In rappeling, the friction device moves, while the rope remains stationary.

Belaying for Lowering Systems

In lowering systems, there should be a belay that is apart from, and on a separate anchor from, the main lowering system. The lowering system and the belay system should be far enough apart so that these two elements do not interfere with one another, such as having their ropes tangle. But they should be close enough together to prevent a dangerous pendulum should the main line fail and the belay be forced to catch the load.

The load caught by a belay in a lowering operation may not fall as great a distance as in climbing. But the lowering operation may involve greater weights. The belay anchor must be able to hold the weight of the rescue load plus any force from shock loading. The belayer must use belay devices that enable the belayer to stop the fall of the rescue load.

WARNING NOTE

In a rescue lowering, a belayer must never use a hip belay or any other belay technique that employs rope friction around the body.

Also, the belayer must never place his body as a link in the belay system when rescue lowering (see Chapter 8, "Belaying").

Communication in Rescue Lowering

As in slope evacuation, the primary exchange of communication signals takes place between the litter tender and the brakeman. If conditions permit, this should be direct communication between the two persons. If there cannot be direct communication because of such factors as distance, high wind, traffic noise, or waterfall, then a relay person might be required. If edge tenders are being used, then one of these persons can be assigned as a communications belay.

Radios

Over longer distances, radios may be a necessity. They also may be required for medical control or for relaying the medical condition of the subject.

The use of the traditional hip belt radio holsters creates some problems in high angle rescue:

- They interfere with seat harnesses and use of equipment.
- They require extensive use of hand and arm motions to use the radio.
- They can result in the radio's being dropped.

One solution to these difficulties is the use of a radio chest harness *(see Figure 14.2)*. These harnesses, which are commercially available, have the following advantages:

- They are in proximity to the face, so they do not have to be taken out of the holster for use.
- The chest area is usually free of other harnesses and equipment.
- A simple hand motion can key the mike.

Fig. 14.2 Chest Radio Harness

Voice Communication for Rescue Lowering

Direct and reliable communication in vertical lowering is critical for the safety of the rescue subject and the rescuers and for the success of the operation. To avoid dangerous confusion, voice communications must be limited only to a few standardized commands necessary for litter movement and for safety. The standard belay commands are included.

"On Belay." *(Litter tender to belayer.)*

"Belay On." *(Belayer to litter tender.)*

"Down Slow." **or** **"Down Fast."** *(Litter tender to brakeman.)*

"Stop!" *(Generally the litter tender to the brakeman, but may be given by anyone who sees danger or potential problems developing.)*

"Stop? Stop? Why Stop?" *(Litter tender to brakeman. This is given when, for an unexplained reason and without command from the litter tender, the rope has stopped moving. It could be that the brakeman is still letting out rope, but the rope is jammed somewhere. This obviously has the potential for creating a very serious problem and requires an explanation from the brakeman.)*

("Off Belay.") *(Litter tender to belayer. The litter, rescue subject, and litter tender(s) are on the ground, or in a secure position, and in no danger of falling.)*

("Belay Off.") *(Belayer to litter tender.)*

In addition, the following voice communications may also be used when needed:

"Slack." *(Litter tender to brakeman or belayer. The rope is too taut, give us some slack.)*

"Tension." *(Litter tender to brakeman or belayer. Take up some rope and make it more taut to help us out here.)*

When a belay is being used, then the tender must specify which line he is talking about:

"Slack on Belay Line."
or
"Slack on Main Line."

SUGGESTION:

Where there is more than one rope at work, such as in a lowering with a belay, it is helpful if you use different colored ropes. For example, if the belay rope is blue and the lowering rope red, line management would be much easier than if they were both the same color.

When the litter has reached the ground, there might also be:

"Off Rope." *(Litter tender to brakeman—I have unclipped the rope from the litter and no longer need the line.)*

Fig. 14.3 Figure 8 Lowering on Slope

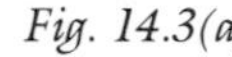

Fig. 14.3(a)

Fig. 14.3(b)

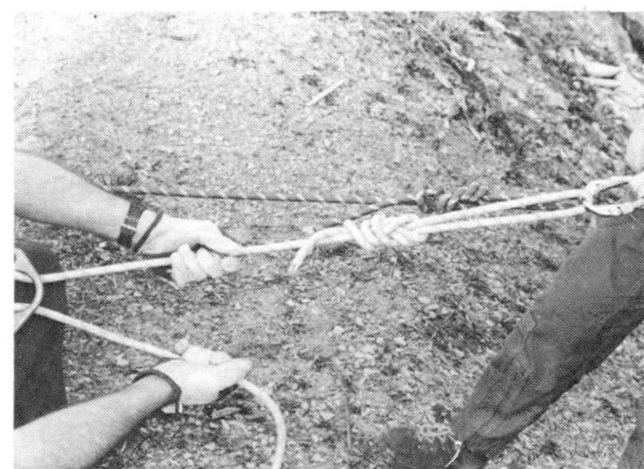

Fig. 14.3(c)

Fig. 14.3(d)

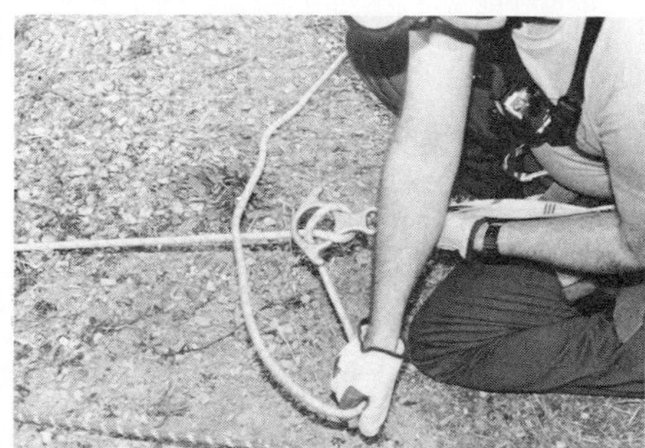

Fig. 14.3(e)

THE PRINCIPLES OF RESCUE LOWERING

(see Figure 14.3)

A. **Lowering One Person Using Figure 8 with Ears** (Steep Slope)
This lowering sequence uses a practice rescuer as a load.

A1. At the top of a short, approximately 45-degree slope, establish a secure anchor, with an anchor sling attached to it. In the end of the anchor sling, attach a locking carabiner. (If a slope is not available, then use a flight of stairs.)

A2. At a separate secure anchor, have a belayer establish a belay station with a sling attached to it. In the end of the sling should be a large locking carabiner with a belay device attached (see Chapter 8, "Belaying," for additional guidelines on belaying).

A3. Have a practice rescuer don a sewn, manufactured seat harness. Into the seat harness tie-in points, have the practice rescuer clip two locking carabiners. One is for the main lowering line, the other for a belay. Have the practice rescuer clip the appropriate carabiner into the belay rope and lock the carabiner.

A4. With the Large Figure 8 descender close to where it will be anchored, the brakeman laces the main line rope onto it as shown in Figure 14.4. The small ring should be clipped into the anchor rope with a locking carabiner and the carabiner gate locked. The large ring should be pointed toward the practice rescuer and laced onto the rope going to him.

A5. The practice rescuer should be facing the brakeman and in a secure position where he is in no danger of falling. The practice rescuer takes the main line rope with a Figure 8 on a Bight knot in it and clips it into a seat harness carabiner. He secures it by locking the carabiner.

A6. The practice rescuer takes the Figure 8 knot in the end of the belay line, clips into the second seat harness carabiner and secures the carabiner. The practice rescuer initiates the belay sequence (Practice rescuer: **"On Belay."** Belayer: **"Belay On."**)

A7. The brakeman pulls the slack out of the section of the main line rope that is between the Figure 8 descender and the practice rescuer.

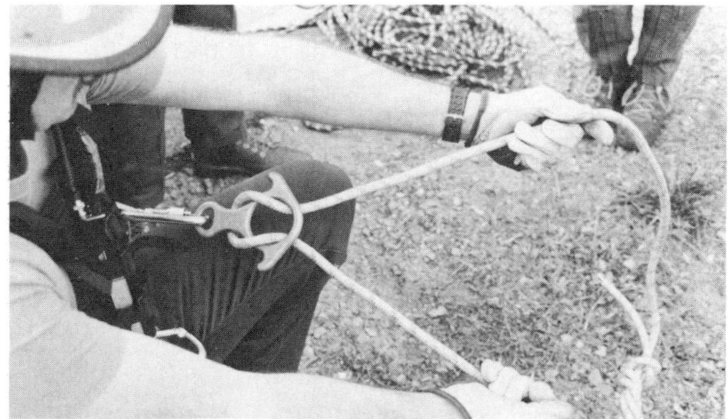

Fig. 14.4 Figure 8 Laced for Lowering

A8. For the remainder of this procedure, the brakeman should be wearing gloves. The brakeman's stance should be as shown in Figure 14.5. The following should all be in a straight line: anchor point, anchor sling, anchor carabiner, Figure 8 descender, the rope between the Figure 8 descender and practice rescuer, and the practice rescuer's seat harness tie-in.

A9. The brakeman's dominant hand (right hand on right-handed people) should be on the slack rope that is feeding into the braking device. This is the **brake hand**. The brake hand must never leave the rope during the lowering operation except when the lowering device is securely locked off. The other hand (the left hand on right-handed people) should be on the rope leading out of the Figure 8 descender toward the practice rescuer. This is the **guide hand**.

Fig. 14.5 Brakeman Stance for Lowering

A10. When the practice rescuer is ready to begin descending, he says to the brakeman, **"Down Slow."** The brakeman begins slowly allowing the rope through the Figure 8 descender as the practice rescuer walks backward downhill. The brakeman should keep the brake hand approximately 18 inches from the Figure 8 descender and slowly allow the rope to slip through his gloved hand.

The guide hand stays lightly at its place on the rope helping to guide the rope out of the Figure 8 descender on its way toward the practice rescuer. Where there is temporarily too much friction for the practice rescuer to pull the rope through the descender, the brakeman's guide hand can help pull rope through the device.

A11. The belayer must control the belay rope so that there is enough slack in the belay line and it does not interfere with the movement of the practice rescuer. But there must not be too much slack in the belay line, so should the main lowering line fail, the belay line can immediately take the load without severe shock loading.

Locking Off

A12. The practice rescuer calls a **"Stop!"** The brakeman holds the rope taut, stopping the descent of the practice rescuer. The belayer maintains an **"On Belay"** status.

A13. Tying off the Figure 8 descender as a braking device is similar to tying it off as a rappel device. DURING THIS PROCEDURE, MAINTAIN A FIRM GRASP ON THE BRAKE SIDE OF THE ROPE AND DO NOT ALLOW THE ROPE TO SLIP THROUGH THE DESCENDER. With the brake hand, swing the brake side of the rope forward toward the load (practice rescuer) until the brake side of the rope is parallel with the rope that goes out of the descender toward the practice rescuer. Still holding the brake side of the rope taut, swing it farther around in an arc until it is trapped between the large ring of the Figure 8 descender and the rope going to the load.

A14. To further secure the rope, pull the brake side of the rope firmly down toward the anchor, across the surface of the Figure 8, and around behind one ear. Pull it firmly between the line going to the load and the large ring, and above the line first locked off. Make certain that the rope lays firmly around the device and there is no slack. Bring the brake side of the rope down and around the Figure 8 again as before, then behind one ear, but do not place it between the line going to the anchor and the large ring. Instead, form a large bight of rope from the brake side of the rope. Bring the bight up parallel with the rope going to the load. Tie an overhand knot with the bight onto the rope going to the anchor. Be certain that the overhand knot is contoured well, is set firmly against the top of the Figure 8, and there is no slack in the knot *(see Figure 14.6)*.

Fig. 14.6 Locking Off Figure 8 Brakes

Unlocking

A15. Unlocking the brake device is the reverse of tying it off. Untie the overhand knot. Take the bight out of the rope and unwrap it from the Figure 8 descender **while making certain that the brake side of the rope remains trapped between the large Figure 8 ring and the rope going to the load. Now, while maintaining**

firm control of the brake side of the rope with the brake hand, untrap the rope and bring it back to its normal position for lowering.

A16. The practice rescuer gives the voice communication, **"Down Slow"** (or **"Down Fast"**). The brakeman lowers at the appropriate speed, while the belayer maintains control of the belay rope.

Getting off Rope

A17. When the practice rescuer reaches the ground or other intended secure position, he calls, **"Stop."** The brakeman stops rope from passing through the descender. (If the lowering rope is too tight for the practice rescuer to disconnect from it, he calls, **"Slack."** The brakeman allows some slack to come into the rope.) The belayer maintains the load on belay until the practice rescuer relieves him by completing the belay cycle. (Practice rescuer: **"Off Belay."** Belayer: **"Belay Off."**)

If the practice rescuer has finished with the rope and is ready for it to be used for other purposes, he disconnects it from himself and signals to the brakeman, **"Off Rope."** The brakeman can then pull the rope back to the top, remove it from the anchor, or do whatever is needed to proceed to the next step.

Fig. 14.7 Figure 8 Lowering on Vertical

Fig. 14.7(a)

Fig. 14.7(b)

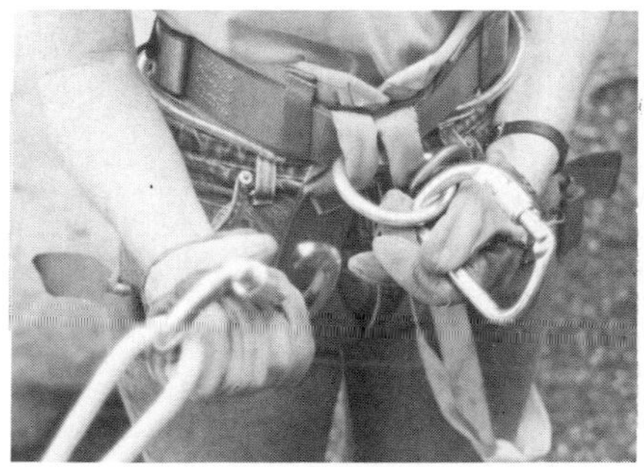

Fig. 14.7(c)

Fig. 14.7(d)

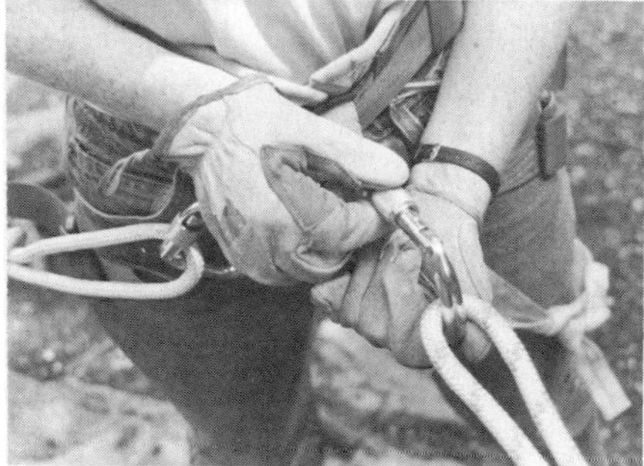

Fig. 14.7(e)

Fig. 14.7(f)

Fig. 14.7(g)

Fig. 14.7(h)

Fig. 14.7(i)

Fig. 14.7(j)

B. **Using a Figure 8 with Ears to Lower a Person down a Vertical Face** *(see Figure 14.7).*

B1. Choose a short vertical face (approximately 20 feet) where the top breaks over gradually into a steep face. On the first try, do not choose a face with a sharp edge or one that is undercut.

B2. Establish a secure anchor point safely back from the edge. If possible, have the anchor point high off the ground. This will assist the practice rescuer in going over the edge. Attach an anchor sling to the anchor point. This anchor sling should be short enough so that the practice rescuer can rig into the main line without being too close to the edge. Have a loop in the end of the anchor sling, and clip a locking carabiner to that loop.

B3. Establish a separate anchor point for a belay. Securely connect an anchor sling into the anchor point. The anchor point and sling should be established so that when the belayer takes position, he has a good field of view of the top and face, but is not in danger of falling over the edge. The belayer should be tied into a safety line that is not part of the belay line. (In the ideal situation, he should be on a separate anchor from the belay. If this is not possible, then he may be on the same anchor point, but his body must not be linked to the belay system.) (See Chapter 8, "Belaying" *[14.7a].)*

B4. On the edge, place an edge roller or a rope pad to protect the main line lowering rope. The edge roller or rope pad should be anchored so that it remains in place and protects the rope as the rope passes over it *(14.7b).*

B5. Now follow steps A3 through A9 above for attaching the practice rescuer into the belay and lowering system *(14.7c, d & e).*

CAUTION: IF THE PRACTICE RESCUER IS AT ALL HEAVY (ABOVE 160 POUNDS) THE BRAKEMAN SHOULD MAKE HIS LOWERING WITH THE FIGURE 8 DESCENDER *DOUBLE WRAPPED.*

B6. When the practice rescuer is ready, he should give the **"Down Slow"** voice communication to the brakeman. The brakeman should now slowly allow the rope to pass through the Figure 8 descender. The practice rescuer leans back against the rope and begins to move backwards towards the drop *(14.7f).*

Getting Over The Edge

B7. Getting over the edge will likely be the most difficult part of the operation for the practice rescuer. To make this operation go smoothly, he should remain in close communication with the brakeman. He should not be in a hurry, but approach the edge deliberately and slowly. At the edge, he may want to give a **"Stop"** to the brakeman and examine the nature of the situation. The practice rescuer should approach this situation as he would a rappel: feet a shoulder's width apart, body perpendicular to the slope, facing the brakeman, but looking over his shoulder at what is about to come.

The brakeman should be prepared for the greater weight that will very quickly come onto the rope and the braking device as the practice rescuer goes over the edge. The belayer should be particularly alert at this point. He should make certain that he does not create rope tension that might pull the practice rescuer off balance.

B.8 Follow steps A11 through A17 above *(14.7g, h, i & j).*

C. **Using a Brkae Bar Rack to Lower a Practice Rescuer on a Slope** *(see Figure 14.8).*

C1. Follow steps A1 through A6 above except do not use a Figure 8 descender. Instead, attach the eye of a Brake Bar Rack to the main anchor with a locking carabiner. The rack should be aligned so that the bars are in a horizontal plane, with the bend of the rack pointing toward the drop. The top bar should have its slot that clips onto the frame facing down. Lock the carabiner.

C2. Unclip all the bars except for the top bar (the one next to the bend of the rack).

C3. Lay the main line rope **on top of** the top bar. **DO NOT RUN THE ROPE BETWEEN THE TOP BAR AND THE RACK**. Pull slack out of the rope between the practice rescuer and the top bar. This should be done gently so that the practice rescuer is not pulled off balance. With one hand, hold the rack. With the other, pull the rope over the top bar and around the bottom back towards the practice rescuer. At the same time slack is pulled out of the rope between the rack and the practice rescuer, slack should also come out of the anchor sling between the eye of the rack and the anchor point.

C4. While continuing to hold the rope taut with one hand, use the other hand to engage the second bar. Slide it up to jam the rope between it and the top bar. As you lace the rope onto the rack, be certain that the rope runs on the side of the bar that is opposite the notch that clips it to the frame of the rack. This keeps the bars pressed against the rack frame.

C5. While one hand is holding the second bar in place against the rope, use the other hand to bring the rope around the second bar, as was done around the top bar, so that the rope holds the second bar in place.

C6. Continue this procedure until all the bars needed to lower the person down the slope have been laced onto the rack.

C7. To lower with the Brake Bar Rack, the brakeman's dominant hand (the right hand on a right-handed person) should be on the portion of the rope that is feeding into the Brake Bar Rack. ***This is the brake hand AND SHOULD NEVER BE TAKEN OFF THE ROPE***

UNTIL THE PRACTICE RESCUER IS OFF ROPE OR THE BRAKE BAR RACK IS SECURELY TIED OFF. The other hand should be on the bars of the rack, cradling them. This is the ***guide hand*** and it can be used to help change the amount of friction by manipulating the bars.

Fig. 14.8 Brake Bar Rack Lowering on Slope

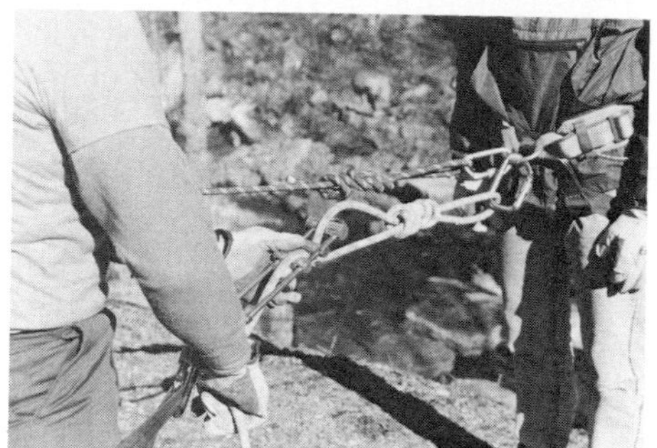

Fig. 14.8(a)

Fig. 14.8(b)

C8. After the rope is laced up in the Brake Bar Rack, take the rope in the brake hand and pull it forward hard in the direction of the load. This should do two things:

a) Take any slack out of the anchor sling.

b) Jam the bars together toward the top of the rack.

This is known as the **quick stop position**. It prevents rope from going through the rack. Use this position whenever the lowering has to be stopped quickly.

Now, hold the rack in the quick stop position until the practice rescuer is ready to start being lowered.

Lowering The Practice Rescuer

C9. When the practice rescuer is ready, he initiates the belay cycle. (Practice rescuer: **"On Belay."** Belayer: **"Belay on."**)

C10. When the practice rescuer is ready to descend, he says to the brakeman, **"Down Slow."** The brakeman then begins to allow rope through the Brake Bar Rack. The practice rescuer should lean against the rope to help it move through the brake system. If the rope is not moving, then the brakeman can encourage it by reducing friction.

Changing Friction When Using the Brake Bar Rack for Lowering

C11. As in rappeling with the Brake Bar Rack, it is best to begin a lowering with more bars engaged than you expect to need. If there is a need to reduce friction so that the practice rescuer can go faster, first try spreading the bars apart along the length of the rack with the

Fig. 14.8(c)

Fig. 14.8(d)

Fig. 14.8(e)

Fig. 14.8(f)

Fig. 14.8(g)

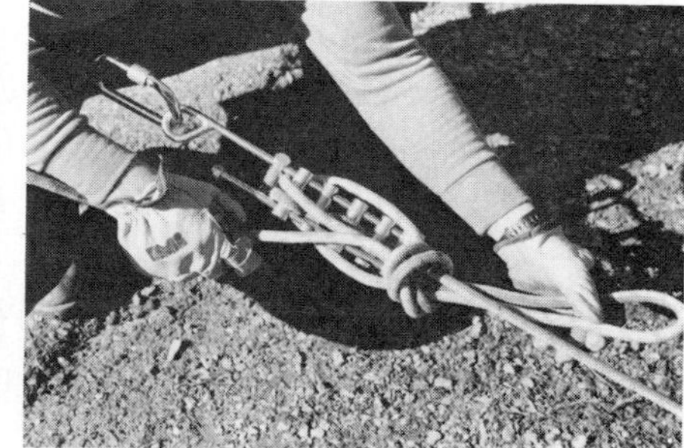

Fig. 14.8(h)

guide hand. The farther apart the bars are from one another, the lower the friction.

If you still have too much friction, then disengage the bottom bar. Do this in the following manner: take the end of the rope in the brake hand and swing in an arc from the quick stop position, first back towards the anchor and then under the rack and towards the load. This releases the bottom bar from under the rope, but still has the rope pressing the fifth bar against the other bars. Now take the guide hand, squeeze the two legs of the rack together, disengage the last bar and move it out of the way. Using the guide hand, spread the remaining bars apart to lessen friction.

If there is still too much friction, repeat the procedure with the next bar up, the fifth bar. Swing the rope in an arc in a direction opposite from before so that it uncovers the fifth bar, but holds the remaining four bars in quick stop. Unclip the fifth bar and slide it back on the rack towards the eye and out of the way.

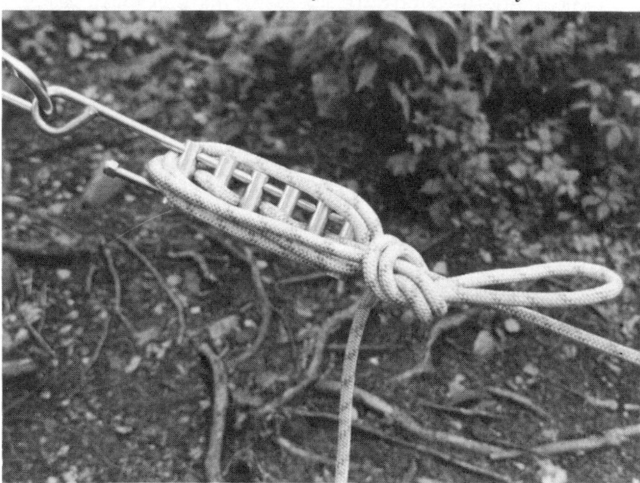

Fig. 14.9 Locking Off Rack

Locking Off

C12. The practice rescuer calls **"Stop!"** The brakeman stops the rope from going through the Brake Bar Rack. He brings the rack into the **quick stop** position as described in C8 above.

C13. If the practice rescuer is to be stopped for an extended length of time, then the brakeman can tie off the Brake Bar Rack. With the brake hand holding the rope in the "quick stop" position, he pulls the rope across the top bar and between the curve of the rack and the strand of the rope going to the anchor *(see Figure 14.9)*.

C14. Next he brings the rope back towards the eye of the rack, holding the rope firmly so that it keeps all the bars locked together. He then brings the rope through the two legs of the rack and across the bottom bar.

C15. Then he again brings the rope towards the practice rescuer, in the same path as before. He should pull it firmly so that all the rope strands are taut and the bars locked together.

C16. The Brake Bar Rack now should be locked in the "stop" position. To secure the rope, the brakeman forms a large bight with the strand of rope in his brake hand. He can use his guide hand to assist in forming this bight.

C17. Treating the bight as one strand of rope, he uses it to tie and overhand knot across the strand of rope that is going to the practice rescuer. He should cinch the Overhand knot down firmly against the top bar of the rack. **There must be no slack in the strands of rope running over the rack, and no space between the bars**. The Brake Bar Rack is now locked off.

Unlocking the Brake Bar Rack

C18. The practice rescuer says, **"Down Slow"** (or **"Down Fast"**). When unlocking the Brake Bar Rack, the brakeman must **ALWAYS KEEP A FIRM GRIP ON THE BRAKE SIDE OF THE ROPE AND ALLOW NO SLACK IN THE BRAKE SIDE OF THE ROPE**. To untie the overhand knot, the brakeman pulls slowly on the braking end of the rope, holding his guide hand at the center of the bight so that it comes out slowly.

C19. With his brake hand firmly on the rope, the brakeman pulls the brake side of the rope in a 180-degree arc until it is pointing in the direction of the anchor.

C20. Now, still grasping the rope firmly with his brake hand, the brakeman pulls the rope from between the two legs of the rack, from around the top bar and then back to the normal braking position. **THE ROPE MUST BE KEPT TAUT BY THE BRAKE HAND DURING THESE STEPS**. With the guide hand, he spreads the bars apart and continues lowering the practice rescuer.

Getting Off Rope

C21. When the practice rescuer reaches the ground or other intended secure position, he calls, **"Stop!"** The brakeman stops the rope from passing through the Brake Bar Rack. The belayer maintains the load on belay until the practice rescuer relieves him by completing the belay cycle. (Practice rescuer: **"Off Belay."** Belayer: **"Belay Off."**)

(If the rope is too taut for the practice rescuer to disconnect the main line from his seat harness carabiner, he calls, **"Slack."** The brakeman allows some rope through the Brake Bar Rack to allow slack in the main line.)

(If the practice rescuer has finished with the rope and is ready for it to be used for other purposes, then he disconnects from it and signals to the brakeman, **"Off Rope."** The brakeman can then pull the end of the rope back to his position, remove it from the anchor or do with it whatever else is needed.)

Fig. 14.10 Lowering on Vertical With Rack

Fig. 14.10(a)

Fig. 14.10(b)

Fig. 14.10(c)

Fig. 14.10(d)

Fig. 14.10(e)

Fig. 14.10(f)

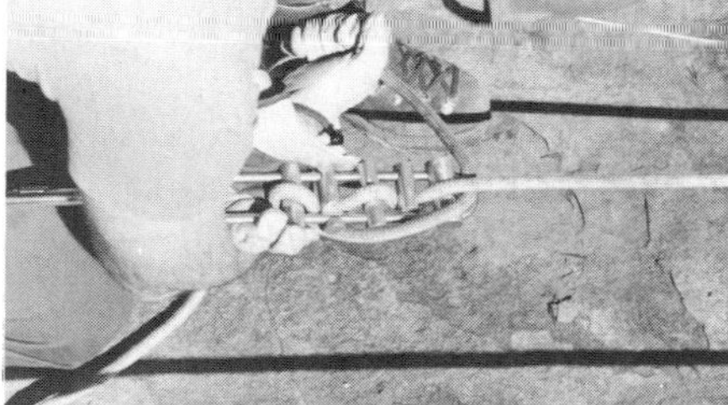

Fig. 14.10(g)

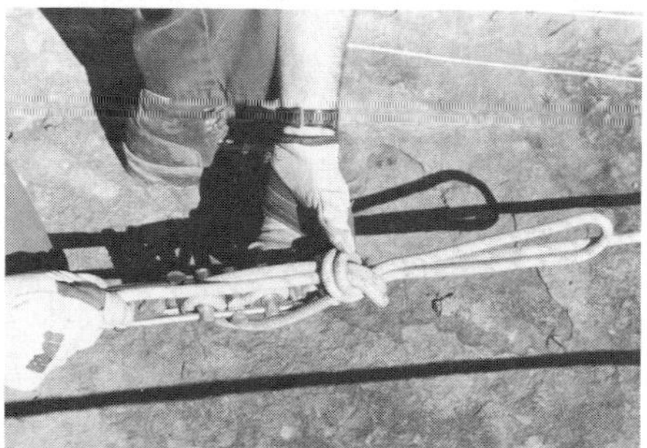

Fig. 14.10(h)

D. **Using a Brake Bar Rack to Lower a Practice Rescuer down a Vertical Drop** *(see Figure 14.10).*

D1. Choose a short vertical drop (approximately 20 feet) where the top breaks over gradually into a steep face. On the first try, do not choose a face with a sharp edge or one that is undercut.

D2. Establish a secure anchor point safely back from the edge. If possible, have the anchor point high up off the ground. This will make it easier for the practice rescuer in going over the edge, and for the brakeman to control him at that point. The anchor sling should be positioned so that the practice rescuer can rig into the main line without being too close to the edge. However, the brake system should be close enough to the edge for the brakeman to hear voice communications from the practice rescuer. At the end of the anchor sling clip a locking carabiner.

D3. Establish a separate anchor point for a belay. Connect a belay sling securely to the anchor point. The anchor point and sling should be rigged so that the belayer has a good field of view of the top and face, but is not in danger of falling over the edge. The belayer should be attached to a safety line that is not part of the belay link to the practice rescuer. In the ideal situation, the belayer should be on a separate anchor from the belay. If this is not possible, then he may be on the same anchor point, but his body not a link in the belay system. (See Chapter 8, "Belaying.")

D4. On the edge of the drop, place an edge roller or rope pad to protect the main line lowering rope. The edge protection should be anchored so that it remains in place and protects the rope and it runs over the edge under load.

D5. Now follow steps C1 through C9 above for attaching the practice rescuer into the belay and braking systems and for lacing the Brake Bar Rack onto the main lowering line.

D6. When the practice rescuer is ready, he should give the voice communication, **"Down Slow,"** to the brakeman. As the practice rescuer leans back against the rope and begins to move slowly backwards towards the drop, the brakeman should allow the rope to pass through the Brake Bar Rack. In order for the practice rescuer to move, it may be necessary for the brakeman to spread the bars of the Rack apart and, possibly, even remove one or two of the bars. But the brakeman must remember: **AS THE PRACTICE RESCUER STARTS OVER THE EDGE, AND HIS FULL WEIGHT COMES ONTO THE ROPE AND THE BRAKING SYSTEM, MORE FRICTION WILL BE NEEDED IN THE BRAKE BAR RACK**.

D7. See notes in step B7 above for **Guidelines on Getting over the Edge**.

D8. On the short vertical drop, continue with steps C11 through C20 above.

Litter Lowering Systems

Litter lowering systems are among the most spectacular of the rope systems. But they require superior skills in vertical techniques, in rope management, and in teamwork, as well as a complete knowledge of equipment. For these reasons, litter lowering systems must be thoroughly planned, worked out, and practiced before they are needed in a real rescue. Otherwise, a greater tragedy than has happened already could result.

Safety Factors in Litter Lowering Systems

Litter lowering systems have higher loadings than those systems that involve the load of only one person. The increased loadings mean greater stresses on the entire vertical system, including ropes, carabiners, knots, anchors, braking systems, and belay system.

These loadings include the combined weight of litter, hardware, and other rescue gear; the rescue subject; and one or two litter tenders. (See Chapter 3, "Rope and Related Equipment" on how to calculate safety factors.)

Position of Litter for Lowering

When possible, a litter is lowered in a horizontal position. This is usually the more comfortable and reassuring position for the rescue subject, it is less likely to complicate most medical conditions and it makes it easier to tend to the medical needs of the subject.

There will be, however, conditions where the litter may have to be lowered in a vertical position. These conditions usually relate to confined space.

Single Strand vs. Double Strand Lowering

There are a multitude of different techniques for lowering litters, but they are generally divided into two types:

- **Single Strand Lowering**
  One main line for the litter plus a belay line.
  One litter tender.
- **Double Strand Lowering**
  Two main lines for the litter. Possibly a belay line.

There are advantages and disadvantages to both systems.

Single Strand Lowering With Belay

- *Advantages*
  - Simpler rope work and brake management.
- *Disadvantages*
  - May not have adequate safety factor for weight of two litter tenders.
  - More difficult to tilt litter from horizontal to vertical position.

Double Strand Lowering

- *Advantages*
  - Can be used where two litter tenders are needed, such as:
    a) complicated medical management of subject, or
    b) vertical face is too difficult for one tender to manage litter (such as overhangs & gullies).
- *Disadvantages*
  - Greater stress on brake systems and anchors.
  - More difficult rope management.

Rigging Litter for Single Strand Lowering

Figure 14.11 illustrates the rigging of a litter for a single strand lowering.

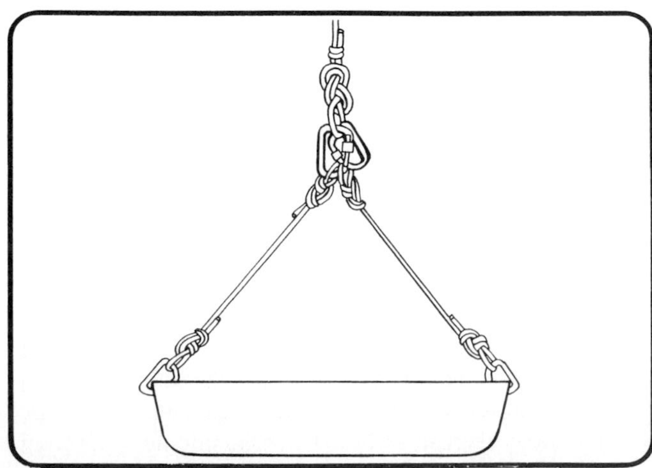

Fig. 14.11 Litter Rigging View From End Showing 2 of 4 Fixed Legs

The Spider

The spider joins the main line lowering rope to the litter. It consists of a group of lines that are first attached at separate points to the litter rail and then attached together at the main line lowering rope. ***For a single strand lowering, there should be a minimum of four legs for a spider.***

NOTE: When employing the SKED™ litter system in a high angle lowering, use the litter's horizontal lift slings as a spider and follow the directions of the manufacturer.

Spider Material

Spiders may be constructed either from webbing or rope. Most spiders are made of rope, since it is easier to handle.

The lower end of the spider legs attach to the litter rails with large locking carabiners. The spider should not be tied directly into the litter rail, since rope or webbing could abrade through when rubbed over rock. The carabiners also give greater flexibility for attaching or detaching the spider to the litter rail.

NOTE: Not all large locking carabiners will fit easily over litter rails. Before purchasing carabiners for this use, measure the diameter of the rail and consult the specifications of the carabiner manufacturer or distributor.

The carabiner gates should be set inward towards the center of the litter. This helps prevent the lock nut on the locking carabiner from being rubbed open on the face of the cliff or wall.

The spider should be adjusted so that the subject rides in the litter slightly head up (unless there are medical reasons to have the head down). If the litter rides head down, it will add to the anxiety and disorientation of the subject.

Creating a Spider from Rope

The simplest way to create a leg of a spider is from an approximately 7-foot length of rope. Tie a Figure 8 on a Bight knot in each end. Create four of these and make certain that after the knots are tied, the spider legs are all exactly the same length.

Attaching Spider to Main Line Lowering Rope

a) At the end of the main line lowering rope, tie a Figure 8 on a Bight knot.

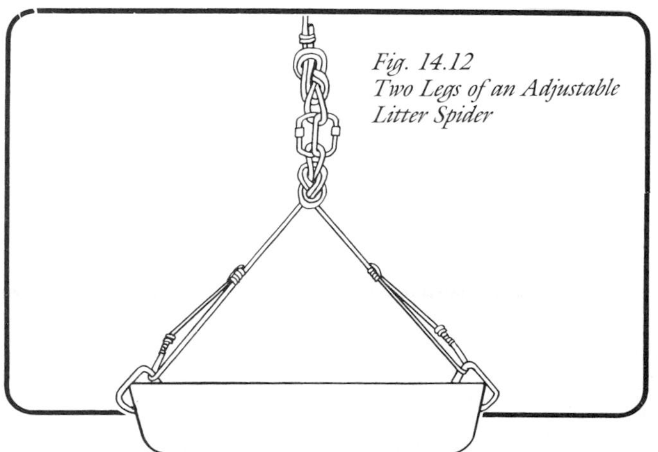

Fig. 14.12
Two Legs of an Adjustable Litter Spider

b) Clip a large locking carabiner into this knot.

c) Now bring all four Figure 8 on a Bight knots in the ends of the spider legs together. Clip the same carabiner across them and lock the gate.

d) Now take a second large locking carabiner and clip in through the same way as the first carabiner.

e) Reverse and oppose the gates of the two carabiners. (This safety precaution is added because the carabiners will be rubbing against the face of the wall and there is the chance that one lock nut will become unscrewed.)

ALTERNATIVE APPROACH

Figure 14.12 illustrates an adjustable litter spider. The adjustable spider can be used in situations where the litter needs to be tilted on its axis. One example of such a situation would be where the cliff is not completely vertical, but lies at a steep angle. To compensate for this angle, but keep the litter and rescue subject on a horizontal, this litter spider can be adjusted.

Materials Needed:

(2) Lengths of static kernmantle ropes, each 12-feet long.

(4) Lengths of Prusik material, each 4-feet long.

(4) Locking carabiners with gate openings large enough to fit over litter rail.

To Create Adjustable Spiders:

1. At the mid-point of each of the static kernmantle ropes, tie a Figure 8 on a Bight knot. Take the Figure 8 on a Bight knots and hold them together with the rope strands hanging down. There should be a total of two Figure 8 on a Bight knots with four rope strands of equal length hanging down from them.
2. Take the center of each piece of Prusik material and tie a Prusik knot onto each strand of static kernmantle rope about halfway between the Figure 8 on a Bight knot and the end.
3. At each of the four ends of the static kernmantle ropes, attach the two ends of a length of Prusik material. Do this with a Figure 8 Bend (or Grapevine) knot.
4. When this is done, there should be four spider legs. Each one should have a loop in its bottom end that is adjustable by sliding the Prusik knot up or down on the static kernmantle material.
5. In each loop at the bottom made by the Prusik material, clip a large locking carabiner.
6. Now hold the completed spider over the litter.
7. Clip each carabiner into a point on the litter rail that gives equalized loading when the litter is loaded.
8. Clip the two Figure 8 on a Bight knots at the top of the spider into a main line lowering rope with two locking carabiners with their gates reversed and opposed.

Rescue Subject Tie-In

During a high angle rescue, the subject in the litter should be wearing a harness. A safety sling runs from the subject's harness to the carabiners at the top of the spider. This safety sling is designed to catch the subject if there is a failure in the litter. Always leave slack in the safety sling so the subject will not be pulled upwards if the litter tilts.

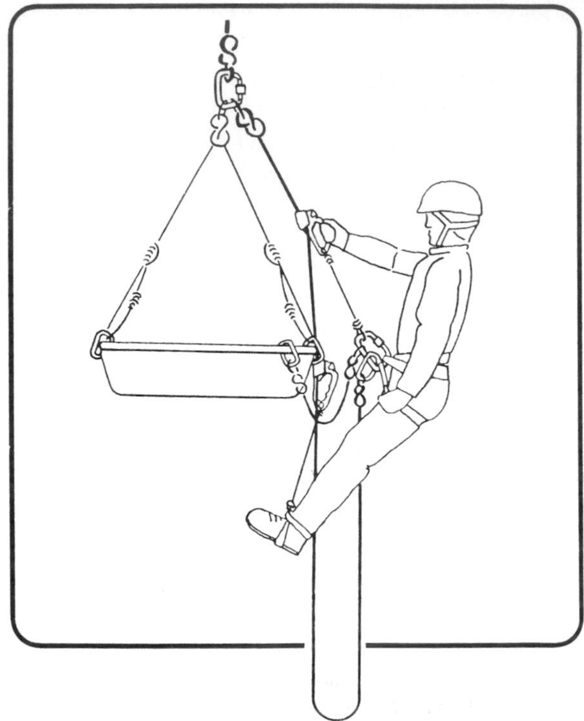

Fig. 14.13 Litter Tender Tie-in

Litter Tender Tie-Ins

A litter tender tie-in serves the following primary purposes:

a) To support the weight of the tender so that he can have his hands free for such tasks as managing the litter and attending to the rescue subject.

b) To provide safety from falling.

c) To allow freedom of movement for the tender. This will allow him to move around the litter to clear possible tangles, to clear obstructions under the litter, and to reach all portions of the subject.

Figure 14.13 illustrates a litter tender tie-in system using two ascenders.

The main attachment to the litter system is a "pig tail." It is made of an approximately 12-foot long piece of rope. It is attached with a Figure 8 on a Bight knot to the carabiners at the end of the main line lowering rope (the same attachment as used for the spiders).

The litter tender is attached to the pig tail with two ascenders. One ascender is attached with a sling to the tender's seat harness, while the other ascender is attached with a sling to the tender's foot. To prevent the ascenders from accidentally slipping off the end of the pig tail, the lower end is brought back up and clipped into the tender's seat harness.

This rigging of the pig tail gives the litter tender the freedom of movement needed in litter lowering. He can move with his ascenders above the level of the litter to clear possible tangles in the spider rigging. Or he can move down below the level of the litter to remove obstructions, such as loose rock.

ALTERNATIVE APPROACH

An alternative litter tender tie-in is a daisy chain. The end of the daisy chain is clipped into the carabiner at the top of the spider, and the litter tender clips into the appropriate "pocket" on the daisy chain to give the proper position on the litter.

The daisy chain tie-in offers the following advantages:

- It is simple to use.
- It can be used by personnel who have no experience with ascenders.

But offers the following disadvantages:

- Once the litter tender is hanging in the daisy chain and the litter is over the side, the tender cannot easily and safely change his position on the daisy chain.
- The litter tender lacks the mobility to move below the litter to clear obstructions or perform other tasks.

Procedure for Lowering a Litter *(Single Line with Belay)*

On Using a Live Practice Subject for Litter Lowering

The use of a live practice subject for litter lowering practice can be very dangerous. There is of course the danger of practice rescuers dropping the practice subject, but in natural areas there is often an extreme danger due to falling rocks. To ensure safety and simplify the operation, first time litter lowerings should use a dummy or weights in place of a live person.

Many rescue teams, once they have become experienced, do use live practice subjects. They feel that it is important that all persons directly involved in litter lowerings eventually gain experience as the "victim" strapped into a litter that is being lowered. This will be the only way they can fully comprehend the feelings of the rescue subject.

However, live practice subjects should **only be used after the rescue team has had the amount of quality practice needed to ensure the safety of the person in the litter**.

The Main Line Lowering System

As noted earlier, both the main line anchor and the belay anchor must have a safety factor appropriate to the load that will be on the system.

In addition, the system must be rigged at the top for safety and convenience. On the main line anchor system, the lowering device (such as the Brake Bar Rack) must be attached to the anchor system so that:

a) The brakeman is close enough to the edge to hear voice communication from the litter tender, but

b) There is enough room at the top between the brakes and the edge so that the litter can be rigged safely and the tender tied in safely.

Belay Systems for Litter Lowering

Figure 14.14 illustrates a belay system for litter lowering. It consists of belay attachments at two points:

a) The belay line is first attached to the top of the spider. This is done by moving back from the end of belay rope approximately 12 feet and tying a Figure 8 on a Bight knot. Clip a large locking carabiner into this knot and then clip the carabiner into the carabiners that are at the top of the spider.

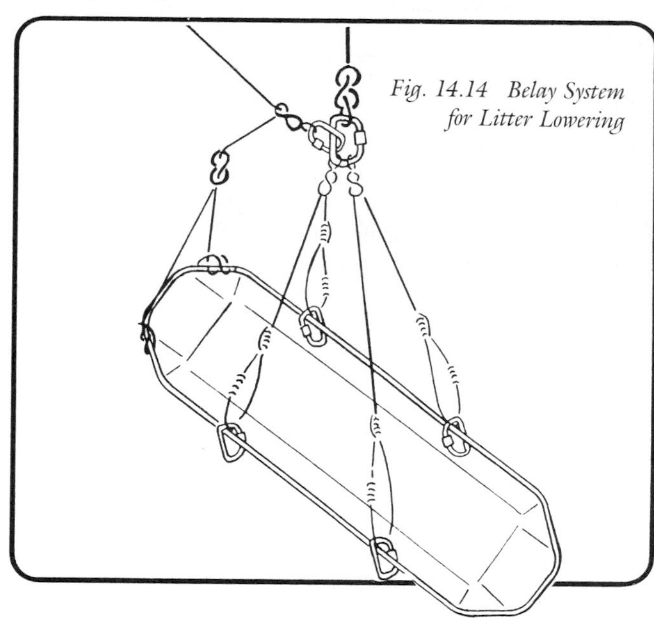

Fig. 14.14 Belay System for Litter Lowering

Should anything along the main line system (such as the anchor for the brakes) fail, the belay could catch at this point. It would still maintain the litter in its normal horizontal position.

b) Take the end of the belay line and attach it to the head rail of the litter in the same way as used for the main line attachment in slope evacuation (see page 176).

This is backup for possible failure at the top of the spider. Should this type of failure occur, then the first belay attachment (the Figure 8 on a Bight knot at the spider) would not catch. The second line of safety would be the belay line attachment at the head of the litter. Should this attachment catch, then the litter would go into a vertical position, with the head up.

NOTE: All those involved with litter lowering must keep in mind what would result from the belay's catching the head of the litter and holding the litter in a vertical position:

a) Never tie the belay line directly onto the subject in the litter. If the belay line caught the subject directly, it would be pulling on his harness while he supported the remainder of the load (litter and tender[s]).

b) The end of the belay rope must always go to the head of the litter. This is to ensure that, should the litter go vertical, the rescue subject remains head up.

c) The litter tender safety sling that is attached to the litter rail must be attached towards the head end of the litter.

Otherwise, if this safety sling is attached at the foot end and the litter goes vertical because of the failure of the spider, the following would occur: The tender's pig tail attachment would come loose. The litter would go vertical. The tender safety would catch at the foot end of the litter. The litter tender would be dangling helplessly below the foot of the litter.

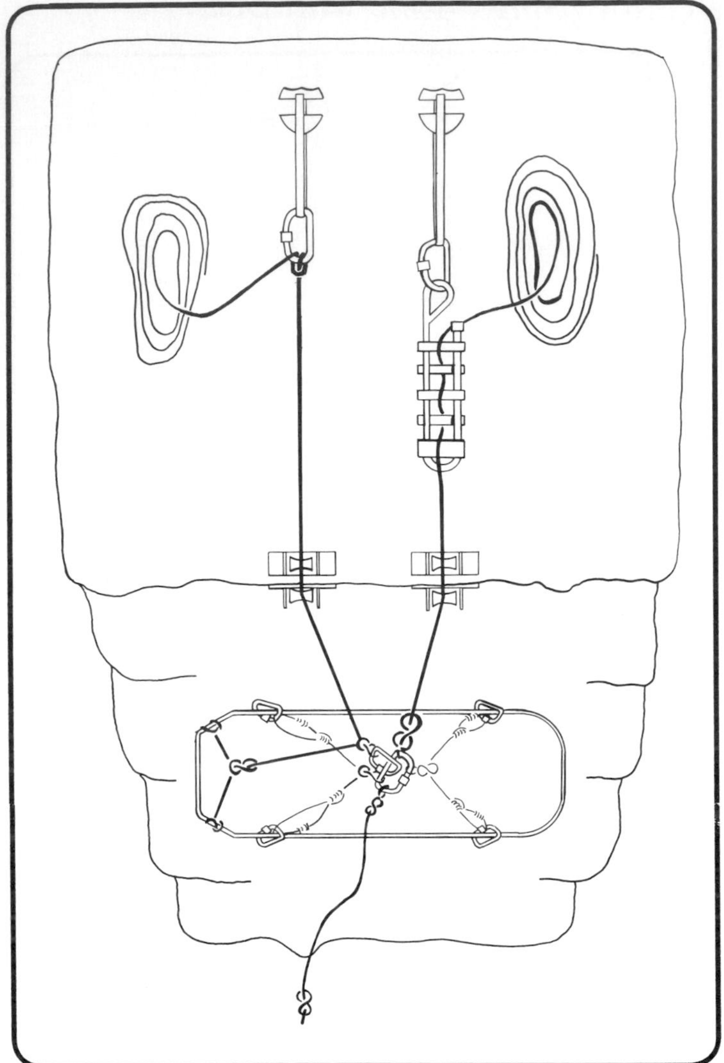

Fig. 14.15 Litter Rigging for Lowering

Things That Almost Always Go Wrong In First-Time Litter Lowerings

Litter lowerings involve complicated rope work. It seems that every first attempt at lowering a litter runs into some typical problems. These problems tend to be the same ones for every group of rescuers. The following is a list of the most common problems:

a) **Line cross**. For example: belay or lowering line running across a fixed safety for edge tenders or anchor line for edge protection.

b) **Litter tender attachments are wrong side of litter**. That is, the tender's rigging is attached to the side of the litter by the wall, when it should be **away** from the wall.

c) **Edge protection is not put in proper place or is knocked out of place when the litter is put across the edge**. The protection is no longer protecting the rope from damage by abrasion.

d) **The spider and tender riggings are twisted around one another**.

e) **Edge tenders are not in position to assist operation**. They are not in position to assist the litter tender in getting the litter over the edge or in restoring edge protection to its proper place.

Rigging the Litter for Lowering *(see Figure 14.15)*

1. At the end of the main line anchor sling, attach a locking carabiner. Into this carabiner clip a Brake Bar Rack. The rack should have its eye towards the anchor, with the bend in the rack, and the top bar towards the edge of the drop.
2. At the end of a belay anchor sling, attach a locking carabiner and a belay device. Run a belay rope through the device with its end where the litter will be rigged.
3. Lace the main line lowering rope through the brakes (the Brake Bar Rack) with the end of the rope where the litter is to be rigged.
4. Rig the litter for a single strand lowering as described above. Attach a four-legged spider to the litter. Have ready a pig tail and a separate safety sling for the litter tender.
5. In the end of the main lowering line, tie a Figure 8 on a Bight knot. Be certain there are several inches of tail once the knot has been tied and tightened down. Attach all legs of the litter spider to the Figure 8 on a Bight knot with two large locking carabiners. Attach the litter tender pig tail with a Figure 8 on a Bight knot to the two carabiners in the end of the main line rope. Lock the two carabiners and align them reversed and opposed. Attach a separate safety sling for the litter tender to the head end of the litter.
6. Connect the belay line to the litter as described above, including attachments both to the top of the spider and to the head rail.

ALTERNATIVE APPROACH

Some rescue teams prefer to lower a litter using two brakes instead of a brake and a belay device. (In Figure 14.15, imagine the belay device on the left replaced by a brake system that resembles the one in the right of the drawing.) In this way, should one anchor fail, the second brake system converts to a belay that can more easily be held by the brakeman, and the device can better be used for a controlled lowering than for a belay device. The two brake system used in this manner is still considered a "Single Strand" lowering, since both lowering ropes come together at a point at the top of the litter spider, instead of to two separated spiders connected to ends of the litter (a "Double Strand" lowering described later in this chapter).

The advantage of a double brake lowering is that it equalizes the load between two anchors instead of one. Also, should one anchor fail, there may be less severe loading, since both ropes are under load instead of one having some slack, as would be the case with a belay.

One disadvantage of a double brake lowering is that both lowering ropes come together under load. It is conceivable that in a massive rock fall, both lines could be cut.

In constructing a double brake lowering system, the two brake stations should be set close together so there is a small angle between the rope. In this way, there would be less of a pendulum should one side fail. Also, both brakemen need to be close together so that there is close communication on keeping equal loading on the two braking devices.

---

7. Load the litter with a dummy or weight equivalent to a large person. Tie the dummy or weights in so that they will not spill out should the litter capsize.
8. Have the litter tender attach himself to the pig tail with a seat harness ascender and a foot ascender as described above. Have the litter tender attach to the end of the pig tail by tying a Figure 8 on a Bight knot in the end and clipping it into his seat harness front tie in point with a locking carabiner.
9. The litter tender should also attach himself with a safety sling to a point on the litter rail near the head end.
10. The litter tender initiates the belay cycle. (Litter tender: **"On belay."** Belayer: **"Belay on."**)
11. Before the litter goes over the edge, the litter tender makes a final check of the rigging.
12. After checking to make certain that all the other lowering team members are ready and alert, the litter tender says to the brakeman, **"Down Slow."** The brakeman begins letting rope through the Brake Bar Rack and the rope handler begins feeding rope to the brakeman. The belayer controls the belay so that there is some visible slack in the rope. The belay rope should not interfere with the litter lowering, but there should be enough tension so that should the main line system fail, the belay will catch the litter system with little shock loading.

    NOTE: At the top, before going over the edge, there will probably be little weight on the brake system to pull the rope through. So the litter tender may have to lean back, pulling the litter with him, while the brakeman lessens the friction. But remember: **ONCE THE TENDER AND LITTER GO OVER THE EDGE, THERE WILL BE GREATER WEIGHT ON THE SYSTEM AND GREATER FRICTION WILL BE NEEDED.**
13. **Getting over the Edge**

    Getting over the edge is often the most difficult step in litter lowering. As with rappeling the first time, the best approach is the slow, deliberate one. Whenever the litter tender begins to feel unbalanced, he calls a **"Stop!"** to regain his equilibrium. The brakeman and belayer must remain very alert to his needs.

    The general strategy is for the litter tender, attached to the litter, to back slowly over the edge, pulling the litter with him. He leans back on his connections to the litter system, and in turn, the lowering system.

    To make the operation as smooth as possible, the tender should avoid any shock loading of the system. To facilitate this, the brakeman should lower **very slowly** as the litter tender moves his way back over the edge.

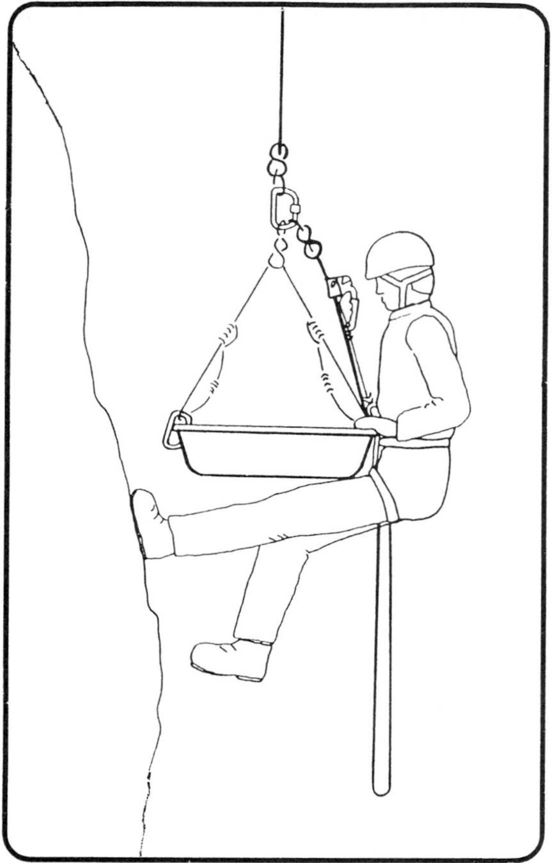

Fig. 14.16 Posture for Litter Tender

As he moves back, the litter tender should try to keep all slack out of the main lowering line, the spider legs, and his tie-ins. Leaning back hard against his tie-ins will help in part.

If the top of the drop is flat, the tender may have to lift up the litter by the rail nearest to himself, with the litter tilted, so that the spider legs are evenly taut.

Should slack begin appearing in the system faster than he can cope with it, the tender should call a **"Stop!** so that he can remove the slack.

Once the litter is completely over the edge and the litter and tender are hanging by their attachments, then all the slack is likely to come out of the system.

Position of the Litter Tender Hanging from Litter

14. The primary duties of the litter tender are:
    - Attend to the medical needs of the subject in the litter.
    - Ensure a good ride for the subject.
    - Communicate with and reassure the subject.
    - Prevent the litter from hanging up.
    - Shield the subject from environmental factors.

    For the litter tender to perform these duties, the best posture is a natural one for sitting in a seat harness, with the litter a few inches above his lap *(see Figure 14.16)*. The litter must not rest on the tender's lap or legs. This confines his movements and if his feet are against the wall, the litter could settle on his legs and pin them.

Both hands should be grasping the litter rail closest to him to help maneuver the litter. If the tender needs to tilt the litter (such as to clear the subject's airway) he can reach across with one hand, grasp the opposite rail, and pull it over towards him to tilt the litter.

If the litter is against the wall, then the tender should have both feet against the wall. As he keeps his feet against the wall, he can use the leverage to pull the litter away from the wall by grasping the near rail. This will help keep the litter from bumping against the wall and from snagging.

15. At mid-point on the wall, the litter tender calls a **"Stop!"** The brakeman brings the brake side of the rope forward towards the load, forcing the bars together and creating a "quick stop" on the Brake Bar Rack. If the stop is going to be a lengthy one, the brakeman has the option of tying off the Brake Bar Rack. But the brakeman should be ready to respond if the litter tender is not initially in position and needs some additional, very short lowerings to reposition the litter.
16. When the litter tender is ready to lower again, he calls a **"Down Slow"** (or **"Down Fast"**) and the lowering procedure continues.
17. Once the litter is on the ground, the litter tender calls a **"Stop!"** If the rope is too taut for him to disconnect the litter system from the main line, he calls, **"Slack."** The brakeman allows slack into the main rope.

When the litter tender is in a secure position with no danger of falling, then he concludes the belay cycle. (Litter tender: **"Off Belay."** Belayer: **"Belay Off."**)

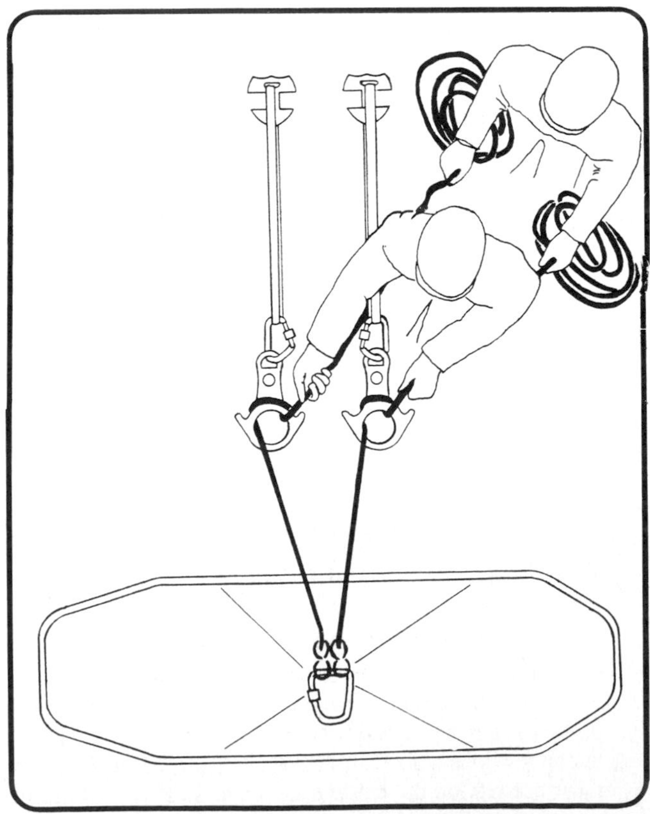

Fig. 14.17 Figure 8 Lowering for Litter

When the litter tender (or others on the ground) have unclipped the litter from the rope and no longer need the line, then the tender gives the voice signal, **"Off Rope."** The brakeman (or others at the top) can then pull the rope back up, or, if appropriate, remove it from the anchors.

ALTERNATIVE APPROACH

Lowering a Litter with Figure 8 Descenders

One alternative to the Brake Bar Rack for lowering a litter is the use of **two** large Figure 8 descenders. Though this involves the use of **two** ropes, the pair of lines are considered together as one main line going to the litter. Combined with the use of one litter tender, this approach is in the category of a single line lowering.

The Figure 8 litter lowering system has the same litter rigging both for spider and for litter tender tie-ins. The difference is in the main line lowering system. This employs two large Figure 8 descenders, each with a rope running through it.

There are some disadvantages to the Figure 8 lowering:

- Friction cannot be varied once the lowering has begun.
- The Figure 8 lowering system is more difficult to lock off.

Figure 14.17 illustrates a litter lowering system using two large Figure 8s and consisting of the following elements:

- Two large Figure 8 descenders, each with its own anchor. The Figure 8s should be only a short distance apart, and equal distance from the edge.
- Two ropes run through the brakes and come to a point at the top of the litter spider. There they are tied with Figure 8 on a Bight knots and clipped into the same two large locking carabiners. The carabiners are then reversed and opposed.
- One brakeman operates the brakes so that the lines run at the same speed.
- Because Figure 8 descenders tend to twist rope, a rope handler is necessary to help avoid tangles getting into the brakes.

MEDICAL CONSIDERATIONS FOR PATIENTS IN HIGH ANGLE LOWERING

The whole purpose of the high angle rescue is the **patient** who will be in the litter. The proper medical care and packaging of the patient will play a major role in determining the outcome: whether the patient is harmed further by the rescue and if he survives the injuries.

Airway

Airway, the "A" in the A, B, Cs of primary patient assessment, is, along with "B" of **breathing,** the most crucial of the concerns. If the patient has an airway blocked by the back of the tongue blocking the airway, by blood, tissue, or vomit; or if he stops breathing, he will expire in a few minutes. The litter attendant should be constantly concerned with the state of the patient's airway and breathing, and be ready to grab the opposite litter rail, tilt the litter toward him and clear the airway. If the patient is unconscious, there must be other provisions to maintain the airway such as an oropharyngeal airway or intubation (advanced level technique).

If at all possible, there should be an attendant accompanying and responsible for the patient during all phases of the evacuation. If this is not possible, then consider transporting the patient on the side. In this case, however, the patient must be thoroughly secured into the litter, and in case of suspected spinal injuries, completely spinal immobilized.

Circulation

Circulation refers to the continual movement of the blood to the tissues of the body and preventing its loss. A major concern is control of bleeding. If this cannot be controlled with a dressing or pressure dressing, direct pressure on the source of the bleeding with a gloved hand is generally the most effective technique to stop bleeding. The use of pressure on the artery at a distance from the wound is usually not effective.

Disability

A number of permanent disabilities to patients--paraplegia and quadriplegia--are the result of mishandling by rescuers. If a spinal injury is suspected, the complete line of the spine, from the head to the hips must be immobilized in line. A cervical collar by itself is not effective. Ambulance personnel traditionally employ rigid backboard, but these are very uncomfortable for long periods and may cause the patient to squirm and counteract the purpose of the immobilization. For long transport periods, a "conforming backboard" such as the KED or the Oregon Spine Splint may provide better patient comfort and, in the long run, better immobilization. However, the conforming backboard must be properly secured with the patient in the litter so there is no movement either lengthwise or side to side.

Loading the Subject into the Litter

The procedures for loading the subject in the litter will depend on his medical condition, but also on where the subject is to be loaded.

Topside

The actual loading of the subject in the litter may be easier if it takes place at the top of the drop. Here there may be more manpower to assist and all members of the rescue team are standing on solid footing.

However, the loaded litter may be more difficult to get over an edge for the litter tender. In this situation, the edge tenders can be of great assistance to the litter tender.

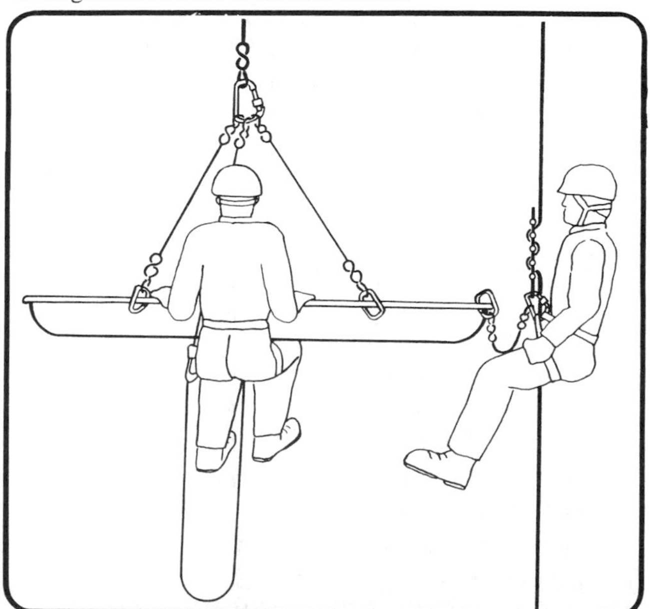

Fig. 14.18 Auxiliary Litter Tender

Partway Down

If the subject has become injured partway down the face of a wall, then it will be necessary to load him in the litter at that point. However, this type of litter loading can be very difficult for the following reasons:

- There is often not much manpower available (there may only be the litter tender).
- With the rescue personnel hanging from their harnesses, they have difficulty getting leverage. If it is a completely free hanging situation (away from the wall), it will be extremely difficult.

The following are some approaches that can assist in this kind of mid-face loading:

- Stop the litter lowering before you get too low. It is better to start the loading attempt with the litter too high, because the brakeman can always let a little bit more rope out to lower some. But if you start out much too low, you may not get another chance. And remember that rope, even static kernmantle, will stretch some. Because it is difficult to lift the subject up to clear the litter rails, the optimum level for the litter is equal to the level of the subject.
- Have the litter positioned for the subject's position (litter head and foot pointing in same direction as subject's head and feet). If at all possible, have the litter in the correct position before starting over the edge. Otherwise, if the litter has to be turned partway down, the belay and main lowering lines will be tangled.
- The subject should have a safety line clipped into him with a seat harness **before** he is moved for loading into the litter. Once in the litter, the subject should have a safety sling run from his seat harness to the carabiners at the top of the spider.
- Of great help in loading the litter partway down will be an **auxiliary tender** (see below).

The Auxiliary Tender *(see Figure 14.18)*

The presence of an auxiliary tender (sometimes referred to also as a "third man") can be very helpful in litter management and in loading the subject.

The auxiliary tender rappels on a separate rope alongside the litter and can be of assistance in several critical ways:

- Responding first before the litter lowering to assess the medical condition of the subject and begin primary treatment.
- Assisting the litter tender in getting the litter over the edge.
- Assisting in loading the subject partway down the wall.
- Helping to maneuver the stretcher around obstructions.

Although the auxiliary tender is on a rappel line separate from the litter lowering system, a tether line running between him and the litter may be helpful in keeping him close to the litter.

ALTERNATIVE APPROACH

The Double Strand Lowering

Sometimes it may be best to employ a double strand lowering with two litter tenders attached to the litter. This approach may be useful in the following situations:

- On an uneven, broken up, vertical face, with obstacles such as overhangs and gullies. These are areas where a single litter tender might have difficulty in managing the litter.

- Where medical considerations or other concerns relating to the rescue subject are too overwhelming for a single litter tender.
- Where it is necessary to change the position of the litter from horizontal to vertical and back again to get it through obstacles on a vertical space or to get through a confined space.

Spiders for Double Strand Lowering

Figure 14.19 illustrates the spider system for a double strand lowering. Note that in this case, there are six legs to the spider. There are three at the head end coming up to meet one lowering rope, and three at the foot end coming up to meet the parallel lowering rope.

Litter Tie-Ins for Double Strand Lowering

As seen in Fig. 14.18, each litter tender has his own tie-in clipped to a separate lowering rope. As with the single strand lowering, each litter tender is also clipped in with a separate safety line to a point on the litter rail.

In the case of double strand lowering, only one litter tender gives the voice communications to the brakeman. This is usually the litter captain, who by tradition is usually stationed at the head of the litter.

Brake Systems for Double Strand Lowering

Both main lowering ropes should run through the same brake device and be controlled by the same brakeman. It is difficult to maintain the same speed for both ropes if the ropes run through different braking devices or are controlled by different brakemen.

Also, both lowering ropes in a double strand lowering should be of the same design and diameter. If they are not, they will have different rates of friction and will consequently run through the brakes at different speeds. Even ropes of different colors may have different rates of friction.

An uneven load on the litter may also cause the ropes to run through the brakes at different speeds. This is often the result of the upper torso of the subject being heavier than the lower torso. So the head end of the litter will often be heavier than the foot end.

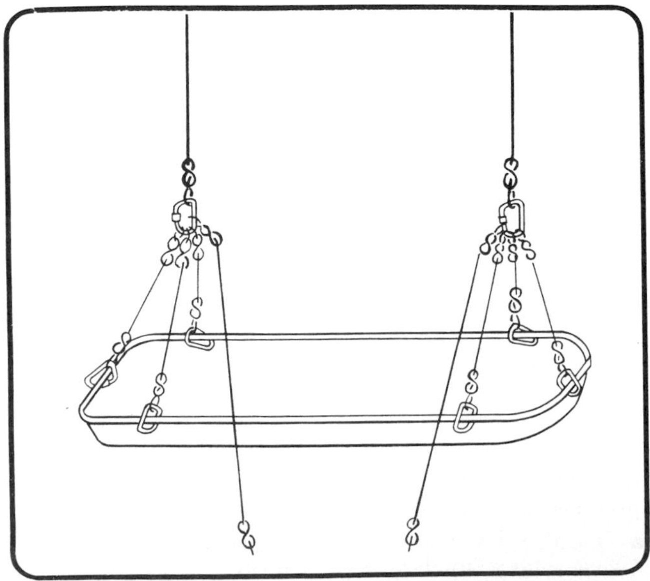

Fig. 14.19 Spiders for Double Strand Lowering

To even out the loading in a two strand lower, the lighter litter tender can sometimes be placed at the head end, and the heavier tender at the foot end.

Changing the Angle of the Litter

The following are steps involved in changing the angle of a litter from horizontal to vertical during a double strand lowering *(see Figure 14.19):*

(The litter has already gone over the edge and is in the midst of being lowered. It is about to reach an area on the wall through which it cannot fit unless it is tilted at a vertical angle.)

1. Litter captain to brakeman: **"Stop!"** (The brakeman stops both ropes from going through the brakes.)
2. Litter captain: **"Down Foot."** (The brakeman allows the rope running to the foot end of the litter to run through the brakes, but he holds the rope that goes to the head end of the litter. As a result, the foot end of the litter goes down, while the head end remains where it is.)
3. When the litter reaches the angle that he desires, the litter captain signals: **"Stop!"** (The brakeman stops the rope going to the foot end of the litter and continues to hold the rope going to the head end. The litter has stopped at a vertical angle with the foot end below the head end. Both litter tenders still hang in the same position, but the litter is now parallel to them. They are still able to reach and tend to the needs of the rescue subject.)
4. Litter captain to brakeman: **"Down Slow."** (The brakeman now allows both ropes to run through the brakes at the same speed. The litter maintains the vertical position it has been set at and continues lowering at that angle.)

To Change the Angle Back to Horizontal

5. Litter captain to brakeman: **"Stop!"** (The brakeman stops all ropes from going through the brakes. The litter stops completely.)
6. Litter captain: **"Down Head."** (The brakeman allows the rope going to the head of the litter to go through the brakes, but holds the rope going to the foot. The head of the litter is lowered while the foot end of the litter remains stationary.)
7. As the head end of the litter becomes even with the foot end, the litter captain says, **"Stop!"** (The brakeman stops the rope to the head end from going through the brakes and holds the rope going to the foot end. The litter is now horizontal and not moving.)
8. Litter captain: **"Down Slow"** (or **"Down Fast"**). The brakeman allows both ropes through the brakes. The litter is lowered in a horizontal position.

Fig. 14.20 Changing Angle/Double Strand Lower

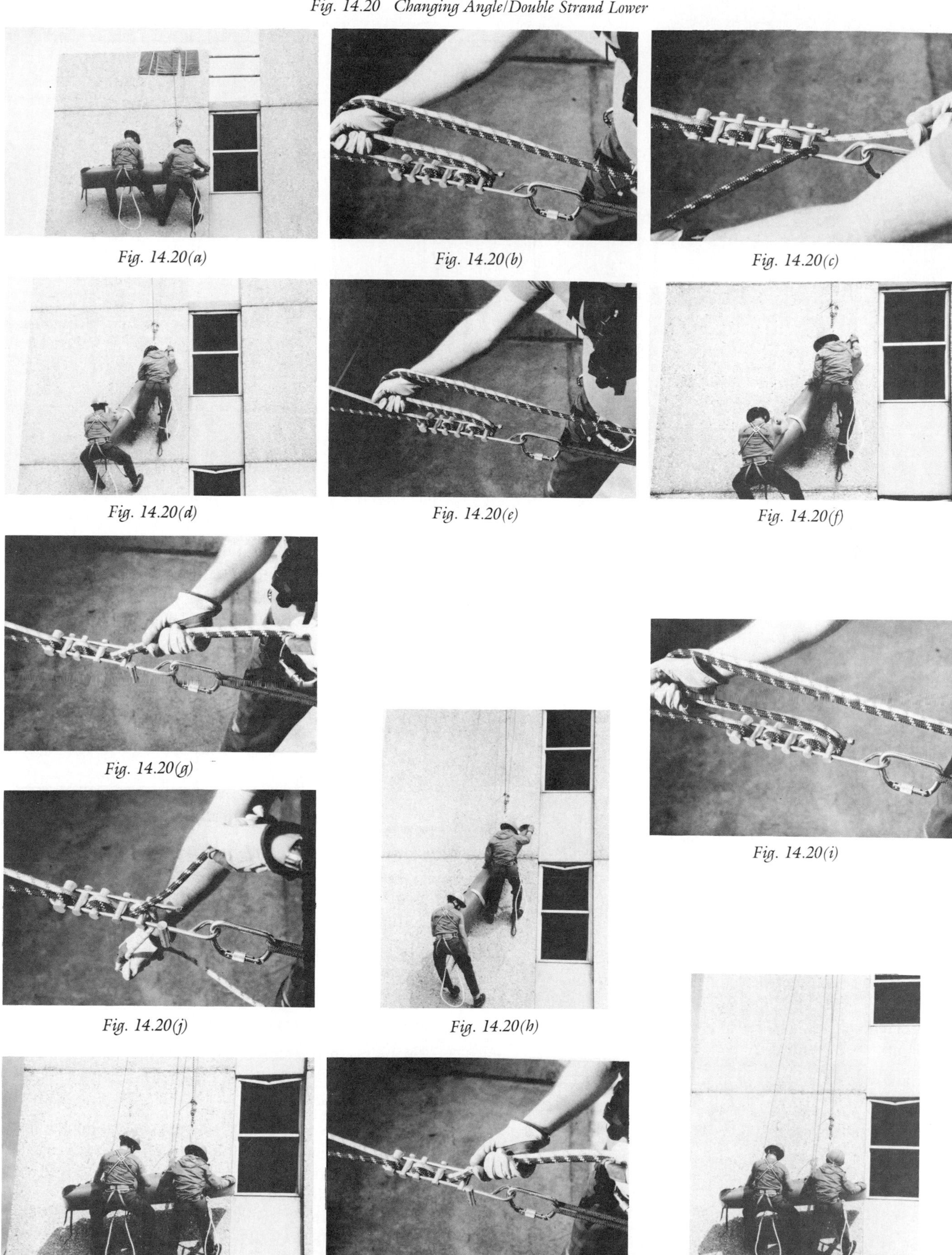

Fig. 14.20(a) *Fig. 14.20(b)* *Fig. 14.20(c)*

Fig. 14.20(d) *Fig. 14.20(e)* *Fig. 14.20(f)*

Fig. 14.20(g)

Fig. 14.20(i)

Fig. 14.20(j) *Fig. 14.20(h)*

Fig. 14.20(k) *Fig. 14.20(l)* *Fig. 14.20(m)*

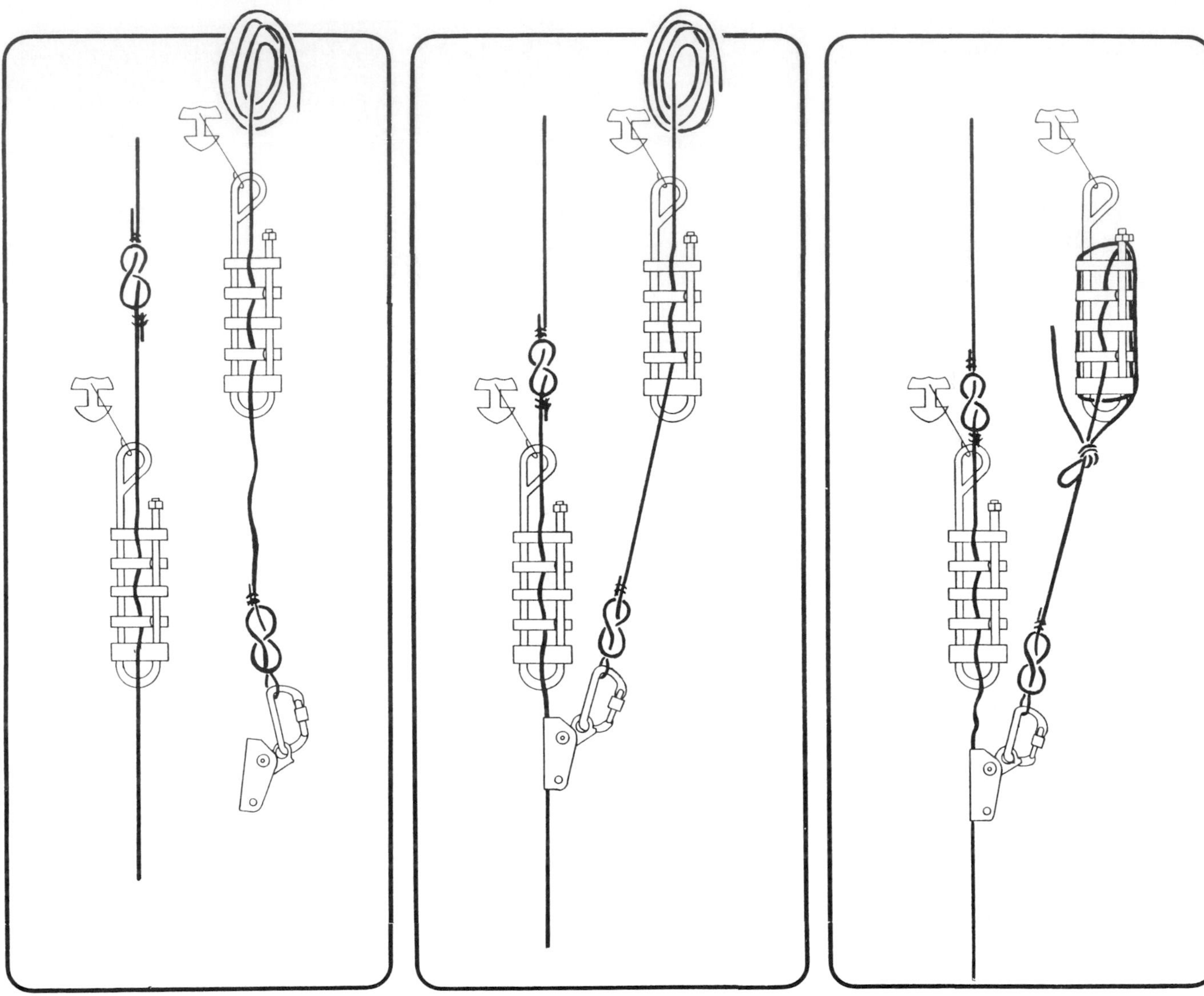

Fig. 14.21(a) Fig. 14.21(b) Fig. 14.21(c)

Fig. 14.21 Passing Knot in Litter Lowering

Passing Knots

If the length of the litter lowering is more than the length of one rope, then it may be necessary to go through a procedure known as **knot pass.** In such a situation, a second length of rope (or more) has to be tied to the first rope in order for the load to reach the bottom. But brake systems will jam if knots enter them. So a bypass procedure has to be used.

Figure 14.21 illustrates a procedure for passing knots.

The following are needed in addition to a regular lowering system:

- A separate anchored braking system.
- A short length of rope (approximately 25 ft.) for interim lowering.
- A cam for each rope in the main lowering system.

1. To avoid delay in the lowering operation, the separate brake system should be rigged and ready before it comes time for the knot pass. The auxiliary braking system is anchored and the short length of rope is rigged into its own brake system. At the end of the short length of rope a Figure 8 on a Bight knot is tied and a locking carabiner clipped into it. The short piece of rope should be adjusted in its brakes so that the cam will reach the main lowering line just below the main brakes.
2. Before less than 20 feet is left on the first rope, the second main rope is tied to it (with a Figure 8 Bend knot or a Grapevine knot *[14.21a]*).
3. Before there is less than 3 feet between the knot and the brakes, the lowering is stopped. **THE KNOT MUST NOT GET ANY CLOSER TO THE BRAKES. IF THE SYSTEM SLIPS AND THE KNOT ENTERS THE BRAKES, IT WILL JAM AND IT WILL TAKE A DIFFICULT HAULING PROCEDURE TO GET IT UNJAMMED.**

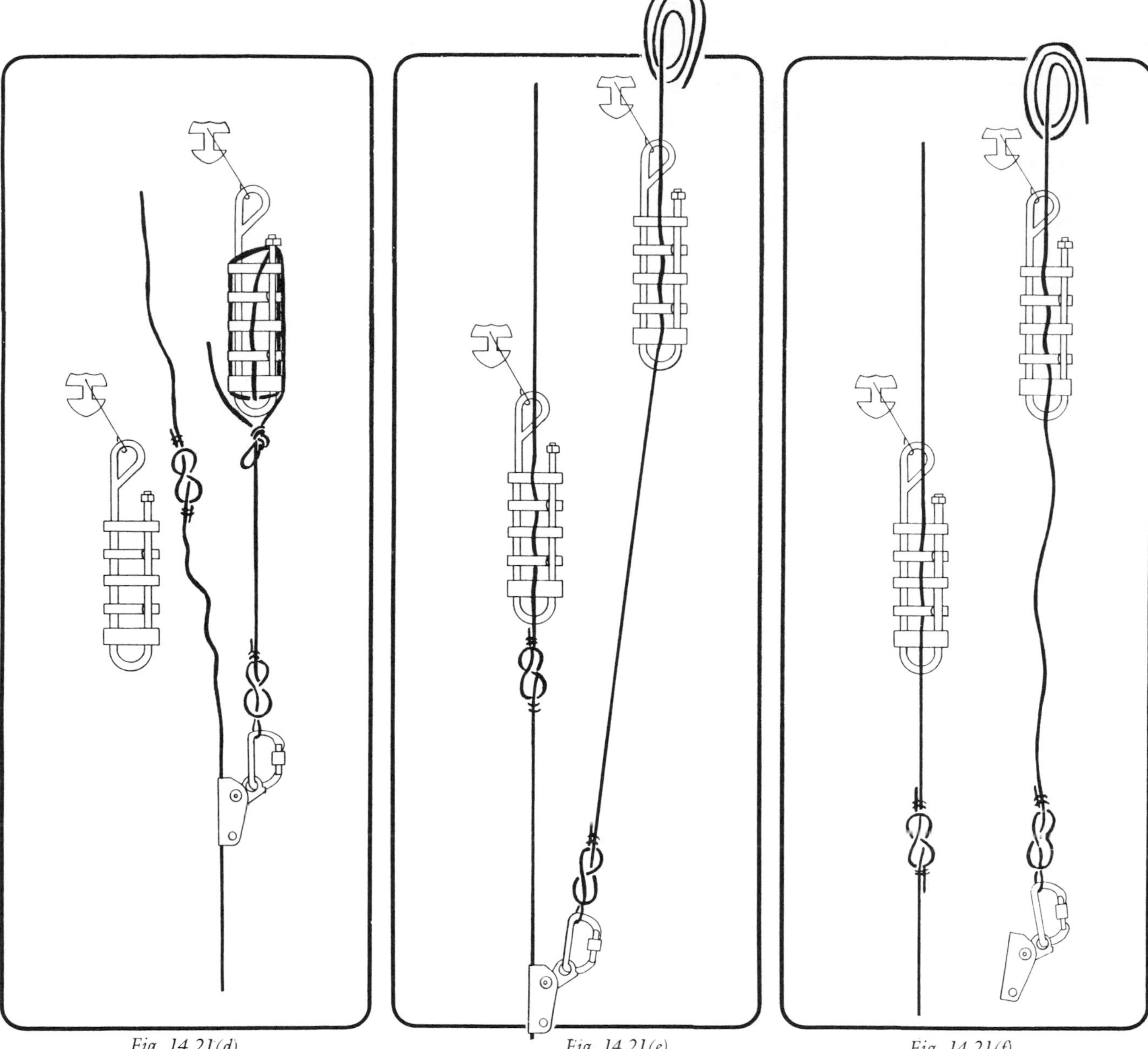

Fig. 14.21(d) *Fig. 14.21(e)* *Fig. 14.21(f)*

4. The cam is placed on the mainline lowering rope just below the main brakes. The arrow on the cam should point to the load (down the drop) so the cam grips the main rope. The cam is locked on the rope *(14.21b).*
5. The auxiliary brakeman holds the auxiliary brakes tight. The main brakeman allows slowly some slack in the main brakes until the load is taken on the auxiliary brake system and the rope through the main brakes becomes slack *(14.21c).*
6. Once the load is fully on the auxiliary brakes, the rope is unlaced from the main brake system *(14.21d).*
7. The auxiliary brakeman begins to lower on his system until the knot on the main lowering line is well past the main brakes.
8. The auxiliary brakeman stops the rope from going through the auxiliary brakes. **THIS MUST BE DONE BEFORE THE CAM(S) ON THE MAIN LOWERING ROPE GET OUT OF REACH OF THE PERSONNEL AT THE TOP.**
9. Once the knot in the main line is past the main brakes, the main line (now into the second rope) is replaced onto the main brakes *(14.21e).*
10. The main brakeman locks off the main brakes. The auxiliary brakeman begins to lower through the auxiliary brakes. The load is taken onto the main line and the auxiliary line goes slack.
11. Once the auxiliary line is slack, the auxiliary brakeman stops rope from going through the auxiliary brakes.
12. The now slack auxiliary line, along with the cam, is removed from the main line *(14.21f).*
13. The main line lowering continues. If an additional knot needs to be passed, then the auxiliary brake system is reset for it.

Questions for Review

1. List three environments in which a high angle lowering system might be used in a rescue.
2. What role does a hip belay have in a rescue lowering system?
3. Repeat from memory and in sequence the voice signals between litter tender and brakeman and between litter tender and belayer during a high angle lowering.
4. In a high angle lowering, the brakeman's dominant hand should be on the slack rope that is feeding into the brakes. This hand is known as the _____ _______. The other hand should be on the rope leading out of the device or, in the case of a Brake Bar Rack, cradling the bars. This hand is known as the _____ _______.
5. Describe the position of the hands and the rope for a "quick stop" position of a Brake Bar Rack during a high angle lowering.
6. When the load goes over the edge during a high angle lowering, what significant change takes place that affects the lowering system and the brakeman's ability to control it?
7. List at least one advantage and one disadvantage for the following types of litter lowerings: a) single strand, b) double strand.
8. In a single strand lowering, there should be a minimum of __________ legs for a spider.
9. List the material that would be needed for a spider used in a single strand lowering of a litter.
10. When a spider is attached to a litter, which direction should the carabiner gates face?
11. Describe the attachment system for litter tender tie-in in that includes the use of ascenders to give adjustable height.
12. List five duties of the litter tender during a high angle evacuation of a rescue subject.
13. List four potential duties for the auxiliary tender during the high angle lowering of a litter containing a rescue subject.
14. Name three conditions that might make a double strand lowering of a litter with a subject preferable to a single strand lowering.
15. How many braking devices and brakeman are employed in a double strand lowering of a litter? Why?
16. List the primary personnel who would be employed in a typical litter lowering.
17. List the things you would check before allowing the litter to go over the edge in a lowering.

Chapter 15

Hauling Systems

PREREQUISITES

Before attempting the activities described in this chapter, you must have demonstrated that you can properly:

1) Use and care for rope.
2) Use and care for other equipment employed in the high angle environment.
3) Tie correctly, confidently and without hesitation the eight knots described in Chapter 6.
4) Apply the principles of anchoring and rig a safe and secure anchor.
5) Apply the principles of belaying and safely and confidently belay another person using either a Munter Hitch or belay plate.
6) Apply the principles of rappeling: rappel safely, confidently, and under control; tie off the rappel device to operate hands free of the rope and then return to a safe and controlled rappel.
7) Apply the principles of ascending: tie correctly, confidently, and without hesitation a Prusik knot and know how to use it; comprehend the uses and limitations of mechanical ascenders; confidently and safely ascend a fixed rope using either friction knots or mechanical ascenders; confidently and safely change over both from rappeling to ascending and from ascending to rappeling; and extricate yourself from a jammed rappel device (or similar problem).
8) Apply the principles of slope evacuation: correctly set the rigging in any of the elements of slope evacuation; and safely and confidently assume the role of litter tender, haul team member, rope handler, brakeman, or belayer.
9) Apply the principles of high angle lowering systems: correctly rig any of the elements of high angle lowering; and safely and confidently assume the role of litter tender, brakeman, belayer, rope handler, or edge attendant.

OBJECTIVES–

At the completion of this chapter, you should be able to:

1. Describe how hauling systems can be used in in rescue and some typical examples of where they might be used.
2. Describe what constitutes the elements of a haul system.
3. Discuss the functions of the following in a rescue haul system: pulleys, haul cam, safety cam, tag line, edge protection, haul team, cam tender, haul captain.
4. Discuss the need for reliable communications in rescue hauling.
5. Describe the principles of a 1:1 haul system, a 2:1 haul system, a 3:1 haul system (Z-Rig), and a 4:1 haul system ("Piggyback Rig").
6. Discuss how to determine mechanical advantage.
7. Define the difference between theoretical and actual mechanical advantage.
8. Discuss the basic criteria for selecting specific haul systems.
9. Repeat from memory the voice communications used in rescue hauling.
10. Select equipment to be used in a rescue hauling system.
11. Describe the primary medical considerations in dealing with rescue subjects in hauling operations.
12. Participate as a member of a haul team.
13. Act as cam tender.
14. Act as haul captain.
15. Act as a belayer for a haul system.
16. Rig a 1:1 hauling system, a 2:1 hauling system, a 3:1 hauling system (Z-Rig), and a 4:1 hauling system ("Piggyback Rig").

TERMS—*relating to rescue hauling systems that a Rope Rescue Technician should know:*

Haul Cam—A cam ascender (or Prusik knot) that grips the rope to provide the "bite" in hauling.

Mechanical Advantage—The relationship of how much load can be moved to the amount of force it takes to move it.

Piggyback System ("Pig-Rig")—Common name given to a specific type of 4:1 hauling system in which one 2:1 system is attached to ("piggybacked onto") another 2:1 system.

Ratchet Cam—The cam ascender (or Prusik knot) in a hauling system that holds the rope to prevent the load from slipping while the haul team resets the system to get another "bite" on the rope. In some hauling systems, the ratchet cam may be the same as the safety cam.

Rescue Hauling—Techniques for using rope, pulleys, cams, and other equipment to give mechanical advantage, convenience, or added safety for raising a rescue load.

Safety Cam—The cam ascender (or Prusik knot) in a hauling system that prevents the rope and the load from accidentally slipping should a mishap occur to the haul system. In some hauling systems, the safety cam may be the same as the ratchet cam.

Tag Line—A line attached to a load that can be used to maneuver the load and prevent it from snagging, and to hold it away from a vertical face.

Theoretical Mechanical Advantage (TMA)—Mechanical advantage without allowance for friction and other losses of advantage.

Traveling Pulley—A moving pulley that is attached to a load or to a haul cam and which adds to the mechanical advantage.

Z-Rig—Common name given to a specific type of 3:1 hauling system. The name is taken from the general shape that the rope makes as it runs through the system.

Rescue Hauling Systems

While the knowledge of lowering systems is essential for competent rope rescue personnel, not all accidents take place on high places. In fact, depending on the location, many rescues will involve the raising of rescue subjects out of a lower place. Some examples might include:

- Silos.
- River gorges, canyons, escarpments.
- Grain elevators.
- Sewers.
- Tank cars.
- Basins.
- Utility vaults.
- Industrial storage bins.
- Fuel tanks.
- Air vents.
- Mine shafts.
- Caves.
- Tunnels.

Purposes of Hauling Systems

In general, there are two basic reasons for rescuers to employ hauling systems:

- To make the raising more convenient and safer.
- To make the raising easier.

Convenience and Safety

The knowledge of hauling systems and the ability to properly use the equipment could mean that rescuers can establish the location of a raising in a place that is more convenient and safer for them.

They could, for example, establish a hauling system that was:

- Closer to vehicles/roadway.
- Away from rockfall.
- Away from potential hostile activity.
- Away from bystanders.
- Out of view of the media.
- At a shorter drop.

Easier

Hauling systems make the job of raising a load easier by spreading the work over (rope) distance. It is the same principle as using a long lever to move a heavy rock.

How Hauling Systems Work

By using hauling systems, you can spread the weight of the load over distance, in this case, over greater rope length. The following examples illustrate how hauling systems work.

Note that in Figure 15.1(a) a rope goes down a drop of 10 feet and is connected to a load that weighs 100 pounds. The rope is directly on the load, and it is a straight haul. So, to bring the 100-pound load to the top requires a 100 pounds of force. It will also take 10 feet of rope to do it.

The relationship of how much load can be moved to the force it takes to move it is referred to as the **Mechanical Advantage**, or MA.

In reality, rescuers never get the full amount of MA out of any hauling system they rig because some of the force they exert in the hauling is lost through such things as friction in the pulleys, rope abrasion, wind resistance, etc. The resulting MA, which we work with in the field, is known as the **Actual Mechanical Advantage**. The MA that we talk about without consideration of the factors such as friction and abrasion is known as the **Theoretical Mechanical Advantage**. But to simplify matters, in this manual, we will talk about hauling systems in terms of the Theoretical Mechanical Advantage, and simply refer to it as the MA You can assume that you will never get the full MA out of a haul system. The amount you do get will depend on how efficiently you can rig the system. There will be several points throughout this chapter on how to avoid loss of advantage in hauling.

In the simple example in Figure 15.1(a), we can calculate the MA as follows:

Weight of Load	to:	Force It Takes to Move Load
100 lbs.	**to:**	**100 lbs.**

So the ratio, or MA, is 1:1

Now note that in Figure 15.1(b) there is the same load (100 pounds) that needs to be moved up the vertical drop. In this case, the rescuers have attached a pulley on the load (this is known as a traveling pulley since it travels with the load). The rope is anchored at the top of the drop. It then runs down through the pulley and back up to the top where the end of the rope is held by the rescuers. So there is now 20 feet of rope and two strands of rope that are moving. As the rescuers pull up the rope, they will pull 20 feet of rope to move the load 10 feet. As a result of having half the load on the anchor and half the load on the side they are hauling, they will pull with a force of 50 pounds.

Calculating the MA in this second example looks like this:

Weight of Load	to:	Force It Takes to Move Load
100 lbs.	**to:**	**50 lbs.**

So the MA is 2:1

To add to the convenience of hauling, consider Figure 15.1(c). Note that the rescuers have now added a second pulley at the top, so they can pull the rope horizontally instead of vertically. This second pulley is stationary and is attached to a sling that is suspended from a tripod, a strong tree, or other similar anchor. This pulley does not add any additional mechanical advantage. As in the second example, there is still 20 feet of rope being pulled, while the load moves 10 feet. It still takes 50 pounds of force to move the 100 pound load. So, the calculation is still the same as in the previous example, with an MA of 2:1.

Although the rescuers have not added any MA, they may have made things better for themselves by making things more **convenient** and possibly **safer**. Now they no longer have to stand bunched together on the edge of the drop awkwardly pulling up on the rope and fighting gravity. With the addition of this stationary pulley (which also acts as a directional), they can now walk back in a line away from the drop as they pull the rope and raise the load.

This illustration is just one example of how convenience and safety must be considered, along with MA, in rigging rescue haul systems.

WAYS OF ADDING TO THE MECHANICAL ADVANTAGE

The creation of additional MA can be achieved in either of two ways: a) by increments, or b) by multiplying.

Incremental Addition of MA

Figure 15.1(d) illustrates a way of moving from a 2:1 MA to a 3:1 MA by changing the system. This particular type of system is commonly called a "Z-Rig" because of the general shape taken by the rope as it runs through the system.

Note that in the 3:1 system at point "A" there is an anchor to which a pulley is attached. This will be a stationary pulley. Through it runs the main line hauling rope. At point "C" there is also a pulley through which the rope runs. But this is a traveling pulley. To this traveling pulley is attached a cam ascender with a carabiner. The cam in turn attaches to the main strand of the same rope. But the pulley and the cam will move as they are pulled by the top end of the rope. The cam will grasp the rope that goes to the load and pull up the load.

Fig. 15.1 Calculating MA

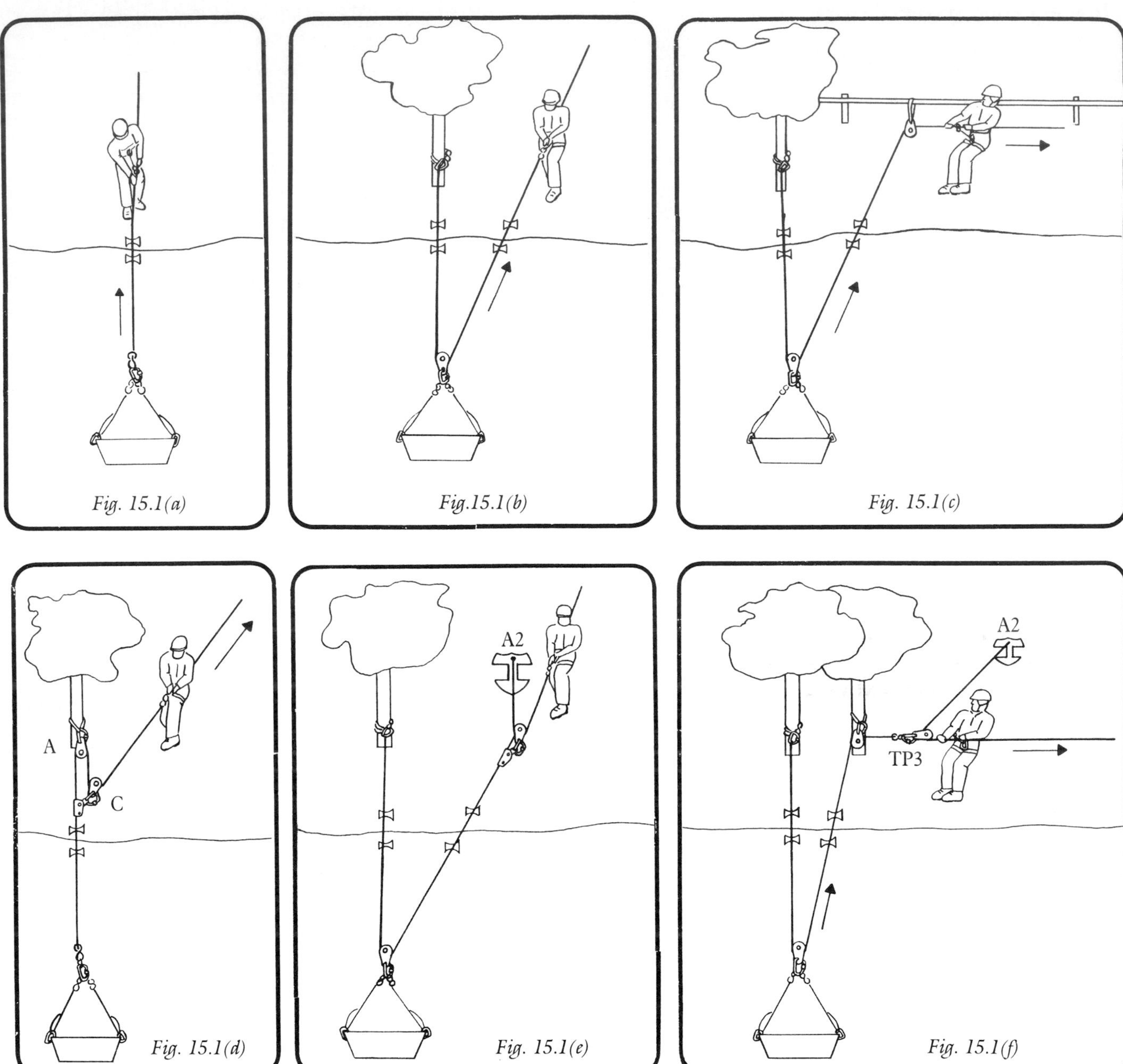

Fig. 15.1(a) *Fig. 15.1(b)* *Fig. 15.1(c)* *Fig. 15.1(d)* *Fig. 15.1(e)* *Fig. 15.1(f)*

In this system, there are three stands of moving rope. For every foot that the load moves, the haul team will have to pull 3 feet of rope. The load is supported by three moving ropes, and each rope is supporting approximately one third of the load. To move the 100 pound load, the haul team will have to exert a force of approximately 33.3 pounds:

Weight of Load	to:	Force It Takes to Move Load
100 lbs.	to:	33.3 lbs.

So the MA is 3:1

Multiplying the MA

In some cases, the MA of a rescue hauling system can be increased by multiplying it. These are called compound systems.

The general rule of thumb is this: when two rescue hauling systems are joined together, the resulting MA is obtained by multiplying the two MAs.

For example:

If a 2:1 MA hauling system is joined to a 2:1, the result is 4:1.

If a 2:1 MA hauling system is joined to a 3:1, the result is 6:1.

And so on.

NOTE: This is a general rule as applied to the basic rescue hauling systems that are examined in this chapter. There are some exceptions to the multiplying rule. In certain compound block and tackle systems, for example, the rule may not apply. The most reliable way to determine an MA is to measure the length of rope pulled against the distance the load moves.

A 4:1 MA System

To create one type of a 4:1 hauling system, take the 2:1 hauling system from Figure 15.1(b) and add another 2:1 system. Start by adding a second pulley at the end of the rope where before the rescuers were hauling (*see Figure 15.1[e]*). Now take a second rope and anchor one end of it at point A2. Run the rope through the second pulley at the end of the first rope and bring the second rope back parallel with itself towards the anchor A2. The haul team will now pull on the second rope at its free end. As the haul progresses, there will be four strands of rope moving. To move the load up 10 feet, 40 feet of rope will have to be pulled by the rescuers. To move the 100-pound load, it will take a pulling force of 25 pounds.

Weight of Load	to:	Force It Takes to Move Load
100 lbs.	to:	25 lbs.

So the MA is 4:1

If this were a real situation, then the rescuers could not rig and operate the system as shown in Figure 15.1(e). The haul team could not pull up while standing in space, just as the rope could not be anchored in space. But a 4:1 system could still be achieved in this situation if the rescuers were to first add a directional at the edge of the drop as they did in Figure 15.1(f). After they rigged the directional pulley, they could rig a second 2:1 MA system as shown in Figure 15.1(f).

First, they would set anchor A2 a way back from the drop. Next they would pull the top end of the first rope through the directional (and stationary) pulley. Then they would attach the second traveling pulley to the end of the first rope *(point TP3)*. After that they would pull the second rope back from the anchor, and thread it through the second traveling pulley. They would then pull the end of the second rope back from the edge of the drop and haul on it.

Again, the stationary pulley at the top of the drop would serve only as a **directional** and add no mechanical advantage. But it would provide for greater convenience and safety in the hauling system. The rescuers would now be able to haul by walking back away from the drop. The situation for MA in Figure 15.1(f) would still be the same as in Figure 15.1(e):

Weight of Load	to:	Force It Takes to Move Load
100 lbs.	to:	25 lbs.

The MA would still be 4:1

ELEMENTS OF HAULING SYSTEMS

The Role of Cams in Hauling Systems

In hauling systems, devices referred to as cams are used to grip the rope. These devices are usually either one of two types:

a) Cam ascenders, or

b) Prusik knots

WARNING NOTE

Handled ascenders must not be used in hauling systems. Handled ascenders are designed only for one person's body weight. They are not designed for the high stresses resulting from the multiplication of forces that take place in hauling systems. The use of handled ascenders in hauling systems can result in tearing of the rope or in the structural failure of the handled ascender. Either one of these may result in the failure of the entire system.

Depending on its position in the hauling system, a cam will serve the following purposes:

- To grasp the rope so it can be pulled by the hauling system. This is known as the **haul cam.** An example of a haul cam is at point "C" in the 3:1 hauling system in Figure 15.1(d).
- To hold the rope while the haul team resets itself to get another bite on the rope. This is known as the **ratchet cam.**
- To act as a safety by grabbing the rope in case the haul team slipped and let go of the rope, or another mishap took place. This is known as the **safety cam.**

In many hauling systems, the ratchet cam and the safety cam can be the same. For simplicity, the hauling systems described in this chapter will have them be the same.

The Nature of The Cam

In the United States, most rescuers use cam ascenders for both the safety and haul cams. Some people, however, prefer the use of Prusik knots. The following is a comparison of the major advantages and disadvantages of cam ascenders vs. Prusik knots.

Cam Ascenders

Advantages:

- Easy to use.
- Can be rigged to reset "automatically."
- Can be manipulated quickly, even when used as a safety cam.

Disadvantages:

- Can cut through rope when under extreme stress.

Prusik Knot

Advantages:

- If correctly used can have a good margin of safety.

Disadvantages:

- Difficult to manipulate, particularly when used as a safety cam.
- Slippage under load can cause friction heat, resulting in melting and failure.
- Wide variation in the types of Prusik material used and in the ability of personnel to use it.

ALTERNATIVE APPROACH

This manual focuses on hard cams in hauling systems because they are more commonly used in the United States and are easier for most rescuers to correctly use than are Prusik knots. **No camming device or technique is perfect for every rescue situation; every camming device has some drawbacks.** Any device, including hard cams, can be made to fail if they are improperly used or if the system is stressed beyond what it is designed for.

Because hard cams can be made to fail under high shock loading, some rescue teams prefer to use the **Tandem Prusik** system as an alternative camming device for the main line in hauling systems *(see Figure 15-A)*. During some tests, the Tandem Prusik system was shown to have greater energy absorption capability than a number of other methods for belaying large loads.

In a low impact fall, Tandem Prusiks are designed to catch the load and spread the force of absorbing the fall over a greater area of the rope while not harming the rope.

In a more severe fall, the Prusiks may grab the rope and the surface of the Prusik material melts, creating a "clutch" effect that can absorb the energy of the fall.

In a very severe fall, it is possible for the Prusiks to melt through completely and fail.

Most of the data indicating the advantages of the Tandem Prusik system have been gathered in tests dropping a rigid weight for 3 meters (9 feet) in free air. Some rescuers feel that these specific test situations are beyond what most rescuers are going to experience in real rescue situations. In these specific tests, there are no edges, faces or directionals to help absorb the energy of the fall, nor is there the elasticity that might exist in anchor systems using slings or rope.

The Tandem Prusik system has not yet been in use long enough in the field, in differing environments, and by different kinds of rescuers to indicate that it works in all field applications as well as it does in the laboratory. But for those persons who may consider using the Tandem Prusik system, the following information is included.

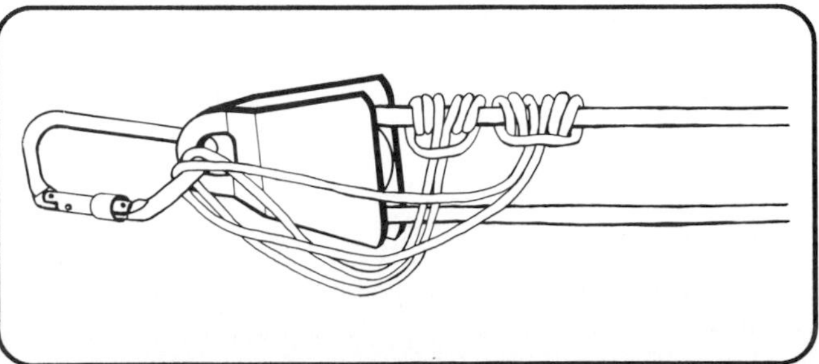

Fig. 15-A

Considerations for Using the Tandem Prusik System

- The system requires meticulous care in rigging. For example, the Prusiks must be tightened just the right amount or they will not grab, the diameter of the Prusik material must be just right in relation to the rope diameter, and the distance between the Prusiks must be just right or they will not grab together.
- It is a more complex system that requires more frequent training sessions. Rescuers with less frequent training may forget how to rig the tandem prusik system.
- It performs best under special conditions. For example, the Prusiks grab best when the rope comes in at an angle, and do not grab as well when the rope comes in from above.
- Its performance is very sensitive to the brand of Prusik material used, flexibility, the diameter, construction, rope coating and age of Prusik material.
- The successful tests of Tandem Prusik systems have been specific to tandem, equalized, triple-wrapped Prusiks of a specific diameter. You should use caution when using single, double, or non-equalized Prusiks.

Positioning of the Safety Cam

For the safety cam to offer greatest protection against failure in other elements of the hauling system, it should be positioned as far forward of the hauling system (toward the load) as possible, while still being safely in reach of the rescuers.

For example, if the rescuers are hauling a load up a vertical drop, the safety cam should be close to the edge, but not over it and out of reach.

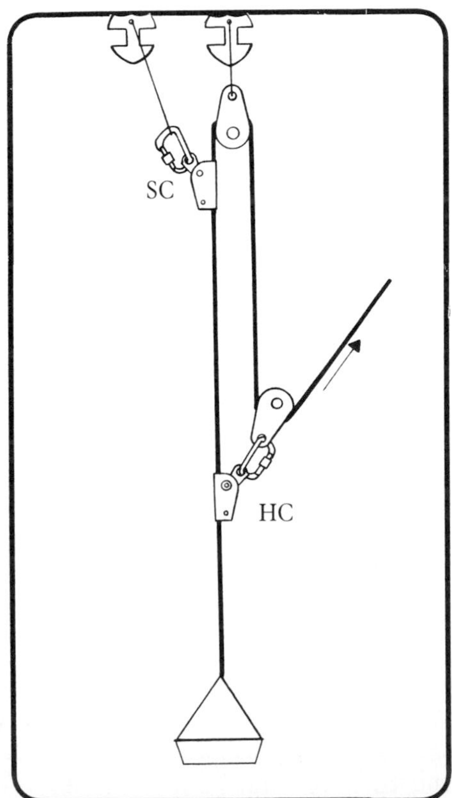

Fig. 15.2 Alternative Position of Safety Cam

It is important to remember that should anything else in the hauling system fail—anchors, hauling team, haul cam, pulleys, etc. —the safety cam is the last hope to grab the main line rope and prevent the system and the rescue load from falling. For this reason, **THE SAFETY CAM SHOULD BE ON A VERY GOOD ANCHOR.**

ALTERNATIVE APPROACH

In certain cases, it may not be possible to place the safety cam forward of the hauling system. One such situation would be where there are no anchors available for doing so. It is possible when using a Z-Rig (3:1 MA) to place a cam **back of** the hauling system *(see Figure 15.2).* However, in these cases, rescuers must keep in mind it would act more as a **rachet** cam in that position and would not be a **safety** for much of the hauling system. It would offer little protection should there be a failure anywhere in the system in front of (towards the load) from the safety cam. If, for example, the cam were back of the hauling system and the haul cam cut through the rope, then the system could fail.

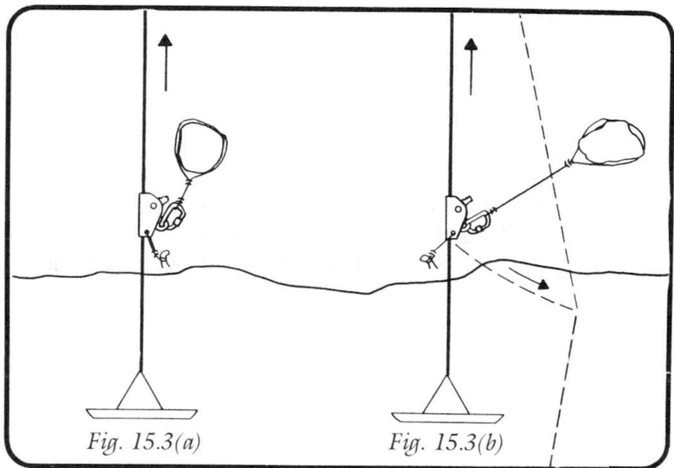

Fig. 15.3 Alignment of Safety Cam

Rigging the Cam in Hauling Systems

(Review material on cams in Chapter 5, "Basic Vertical Equipment" and in Chapter 13, "Slope Evacuation.")

When rigging a safety cam, it should be set so that its anchor and anchor sling are as much in line as possible with the rope it is camming. This should be done without interfering with or jamming the rope *(see Figure 15.3[a]).* If the cam is rigged too far off to the side of the rope it is protecting, there will be too much slack in the cam's anchor sling. This will result in two problems *(see Figure 15.3[b]):* a) it can result in dangerous shock loading of the cam, its sling, and its anchor, and b) the hauling team will lose some of the purchase it has gained when it sets the cam to reposition for another bite.

Setting of a Safety Cam

When rigging a safety cam, there are two primary concerns:

- The cam must clamp on the rope should a safety be needed.
- The cam must not ride up the rope as the rope moves. This could result in dangerous shock loading.

How the cam is specifically rigged depends in part on the specific type of cam used and in part on the specific circumstances of the rigging. *(See Figures 15.4[a] through 15.4[c].)*

Free Running *(Not Spring-Loaded)* Cam

Figure 15.4(a) shows one method of ensuring that the cam stays in place and clamps on the rope when needed. This technique uses the services of a person known as the **cam tender.** As the main line hauling rope moves up, the cam tender makes certain that the cam does not travel up the rope. He does this by holding the back side of the shell with the palm and with fingers extended out of the way of the cam. He holds the shell of the cam in this manner so that fingers or gloves do not get caught if the cam suddenly shock loads. A finger caught by the cam could get injured, and a glove caught in the cam could prevent it from clamping the rope.

While the use of a cam tender can be effective, it does require extra manpower. And there is always the potential for human failure or inattention.

A second technique for using a free-running cam as a safety is shown in Figure 15.4(b). A bungee (elastic) cord is clipped into the empty hole usually found in the "point" of the arrowhead. The other end of the bungee cord is anchored securely to a convenient spot towards the load. A great deal of tension is not required on the bungee cord. But there should be enough to:

a) Keep the cam from riding up the rope, and

b) Keep the cam clamped on the rope.

NOTE: Tie the bungee cord **only to the shell,** and not to the cam itself. If the bungee is tied to the cam, the cam might not close when you want it to.

A third alternative is shown in Figure 15.4(c). This method uses a short piece of cord. It is attached to a weight and the weight is hung so that it will pull the shell of the cam towards the load.

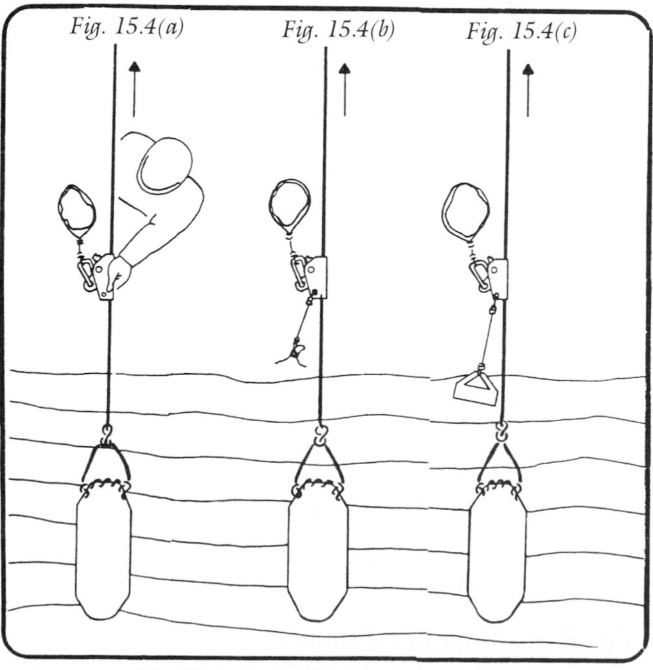

Fig. 15.4 Setting Safety Cam on Automatic

Avoiding Edge Friction in Hauling Systems

One of the most common problems in the rigging of hauling systems is one that often goes unnoticed until it causes problems. This is rope friction over the edge of the drop and elsewhere in the system. Because ropes in hauling systems are often highly loaded, this can result in two significant problems:

- It can result in a tremendous increase in load for the haul team.
- It can result in severe damage to the rope and other equipment.

Possible solutions to edge friction include:

- **Edge Rollers**

  Edge rollers may often be the most efficient solution to edge friction. However, they must be centered under the rope and securely anchored at each side. Otherwise, they can easily be turned over as the rope moves slightly from side to side.

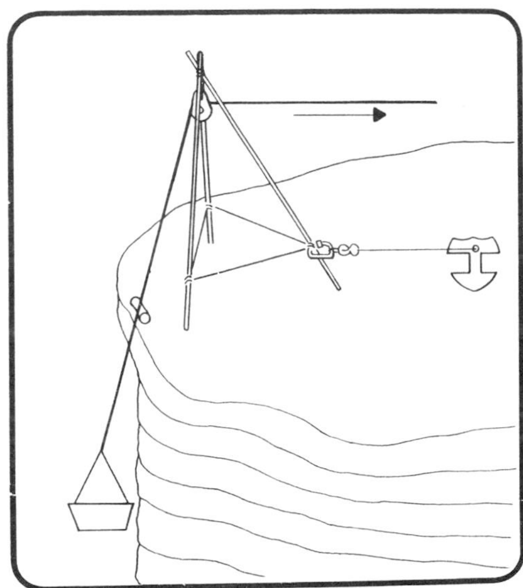

Fig. 15.5 Tripod Used as Directional

- **Directionals**

  To protect against edge friction, a directional pulley must hold the rope above the edge. Two ways of doing this are with a tripod *(see Figure 15.5)* or, if in a natural area, with a very strong tree.

- **Changing the Position of the Haul Rope**

  Either move the haul system to where the edge is less sharp, or move the rope at a higher angle above the edge so there is less friction.

- **Rope Padding**

  This may only slightly reduce the friction problem. But it could significantly reduce the damage to the rope (see Chapter 4, "Care and Use of Rope and Related Equipment," for tips on padding ropes).

The Role of the Haul Team

See page 156, Chapter 13, "Slope Evacuation," for information on the haul team.

Tag Lines

A tag line can be used for controlling the movement of a load and preventing it from hanging up during a hauling operation. A tag line is particularly helpful in hauling litters, where the line can be attached to the foot of the litter *(see Figure 15.6).*

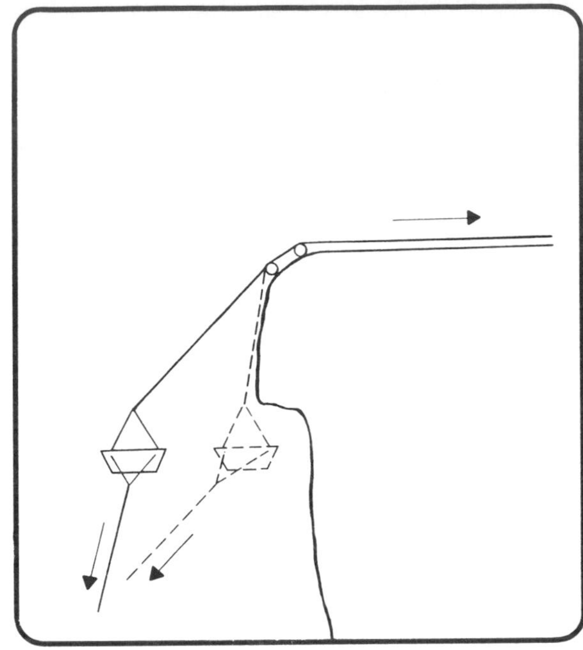

Fig. 15.6 Tag Line on Litter

Among the ways that a tag line can be of assistance are:

- To position a litter when it is hauled through a confined space.
- To keep a litter from getting snagged on an overhanging edge.
- To hold a litter away from a wall to ensure a smoother ride for the rescue subject in the litter.
- To prevent the litter from spinning on free-hanging hauls.

Getting Over the Edge

Getting a litter over an edge when hauling can be very difficult. It also should be done slowly and carefully, since a hung up litter can over stress elements of a hauling system and cause failure.

1. Before starting a haul, tie short tag lines on both the foot and the head of the litter.
2. At the top, station Edge Attendants **securely tied in.**
3. Make certain the Litter Attendant is in the best position for getting his feet under the litter and against the face *(see Figure 14.16).*
4. Before the litter makes contact under the edge, stop the haul.
5. Have the Litter Attendant hand a tag line to each of the Edge Attendants, one at the foot and one at the head of the litter.
6. Restart the haul **very slowly** and be prepared to stop instantly should the litter hang up. If the litter gets hung up and cannot be freed, you may have to slack off on the system to free it.
7. As the litter attendant pushes his feet against the face and pulls the litter rail opposite him and away from the face, the two Edge Attendants help to pull the litter up and over the edge.

Communications

See page 160, Chapter 13, "Slope Evacuation," for information on communications used in hauling systems.

The Rigging and Use of Hauling Systems in Rescue

WARNING NOTE

Hauling systems can create tremendous stress on rope, hardware, and anchors. Hauling team members and other rescuers must keep in mind that forces exerted by the hauling team will be multiplied on portions of the system. If, for example, a haul team exerts a force of 1,000 pounds into a 4:1 system, the resulting force on portions of the system could be around 4,000. Higher MA systems could result in even higher forces.

These forces can develop quickly. Should there be a mishap, such as a jammed knot or tangled equipment, system failure could quickly occur.

All rescuers must be aware of the forces they are creating and be constantly on guard against problems that are developing.

If a haul suddenly becomes difficult, do not simply pull harder. The system may have become jammed, and harder pulling may cause a failure. Stop the haul and examine the system for problems.

A 1:1 MA Hauling System

Figure 15.7 illustrates a 1:1 hauling system for raising a load up a vertical drop.

Minimum Equipment Requirements

(1) Main line rope.
(1) Anchor sling.
(2) Locking carabiners.
(1) Cam (safety).

The Following Steps Are Employed In Rigging The 1:1 System

A1. At the top of the drop, establish a secure anchor point well back from the edge. Into this anchor point securely attach an anchor sling. Into the end of the anchor sling, clip a locking carabiner. Into the carabiner clip a pulley. This pulley is the directional through which the rope will run. This directional should be located so that the haul team will be able to move the rope in a convenient direction.

A2. Thread the main line hauling rope into the pulley. So that the rope will not slip completely through the pulley and fall over the drop while you are doing other rigging, tie a "stopper knot" in the end of the rope. This could be a simple Figure 8 Overhand Knot. Or you can secure the rope by temporarily tying off the upper end.

A3. Drop the lower end of the rope over the edge. The lower end should be close to where the load is to be hauled.

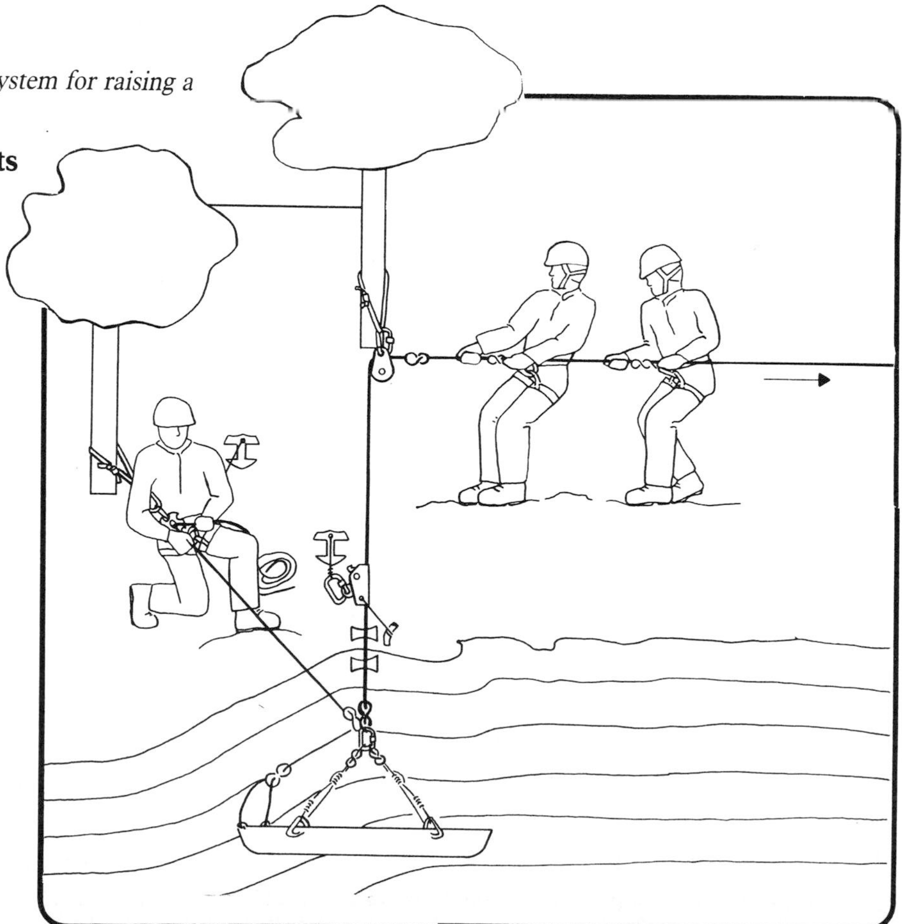

Fig. 15.7 1:1 Hauling System

A4. Establish a separate anchor point for the safety cam. This anchor should be very strong. It also should be located close enough to where the main line rope will run, but slightly off to the side of it so that the cam and its anchor sling do not tangle with the rope. Attach an anchor sling securely onto the anchor point. Into the end of the anchor sling attach a locking carabiner. Into the carabiner clip a cam ascender (or Prusik).

A5. Thread the safety cam onto the rope. If it is a cam ascender, it should have its arrow pointing towards the load. Be certain that after placement on the rope and after the rope is loaded, the cam creates very little bend in the rope. If there is a great deal of bend in the rope created by the cam, it means that the cam is creating a great deal of rope friction that will increase the load for the haul team.

A6. Examine the area where the rope will run over the edge. If there is edge friction, place an edge roller under the rope or rig a suspended directional. If neither one of these is available, carefully pad the rope.

A7. Establish a belay system on a separate anchor system.

A8. Attach the lower end of the main line rope to the load. Remove slack in the main line rope between the safety cam and the load.

A9. Attach the lower end of the belay rope to the load. Make certain that there is the proper amount of slack between the load and the belay device.

A10. Position the haul captain so that he has, if possible, a field of view of both the load and the haul team. If he is near the edge, he should be tied in to a safety line.

A11. Position a cam tender to make certain that the cam will not travel up the rope, and that it will set on the rope when it is supposed to. (Or: set the cam for "automatic" with a bungee cord or weighted line. If this is done, then someone, such as the haul captain, must have the cam in view to make certain that it sets when it is supposed to.)

A12. Position a belayer with the belay system ready to establish the load on belay.

A13. Position the haul team with the rope in their hands, ready to haul.

Beginning The Haul

A14. When he is ready to be hauled, the person on the load (practice rescuer or litter tender) initiates the belay cycle.

A15. When everyone is ready, the person on the load says, **"Haul Slow"** (or **"Haul Fast"**). The haul captain then relays the communication to the haul team: **"Haul Slow"** (or **"Haul Fast"**).

A16. When the haul team goes as far as it can, or the system needs to be reset (and the load has not reached the top), the haul captain says, **"Set."** The haul team immediately stops hauling and slowly eases back so the safety cam catches. The cam tender makes certain that the cam will set (if the cam is rigged for "automatic," the haul captain or other assigned person makes certain it will set).

A17. The haul captain then says, **"Slack."** The haul team instantly reverses direction and moves back up the rope to take another "bite."

A18. The haul captain then says, **"Haul,"** and the cycle continues until the load reaches the top.

Getting the Load over the Edge

One of the most critical and difficult parts of a haul sequence is getting the load over the edge. As in rappeling and lowering, getting over the edge can be particularly difficult if it is sharp and/or undercut. But in hauling, the problem can be even more difficult and potentially dangerous. By the time the load is near the top, the main line rope tends to be pulling at a low angle that forces the load against the edge. This often increases the stresses that are already present in a haul system.

As in rappeling or lowering, the one factor that significantly affects the degree of difficulty of bringing the load over the edge is the angle the rope makes from the load to the anchor point when the load is at the edge. This will range from the most difficult for a horizontal angle (the anchor on the same level as the load when it is at the edge) to the easiest for a vertical angle (the anchor above the load when it is at the edge). There are various ways of increasing this angle including raising the level of the anchor to establishing a high directional. But remember **hauling systems tend to create greater stress on anchors and other elements of the system, so any elevated anchors or directionals must be able to take this stress.**

The use of a tag line (see above) can be very helpful in these situations. As the load reaches the top, the tag line can be handed over from persons below to an edge attendant. The edge attendant can then use the tag line to help pull the load over the top.

In addition, the following procedures can be helpful in getting the load over the edge, while avoiding equipment failure and potential serious injury:

- As the load approaches the edge, the haul captain says, **"Haul very slowly."** The haul team moves very slowly and is prepared to stop instantly.
- If a size-up of the situation at the edge is needed, the haul captain says, **"Stop!"** The entire hauling process stops while the haul captain evaluates the situation.

- All personnel should be on alert to problems or stresses developing at the edge. Anyone who sees such a situation can call, **"Stop!"**
- The use of edge attendants can be very helpful in getting the load over the edge. This is particularly true with litters, which are very prone to getting snagged. (All edge attendants must be securely attached to safety lines.)

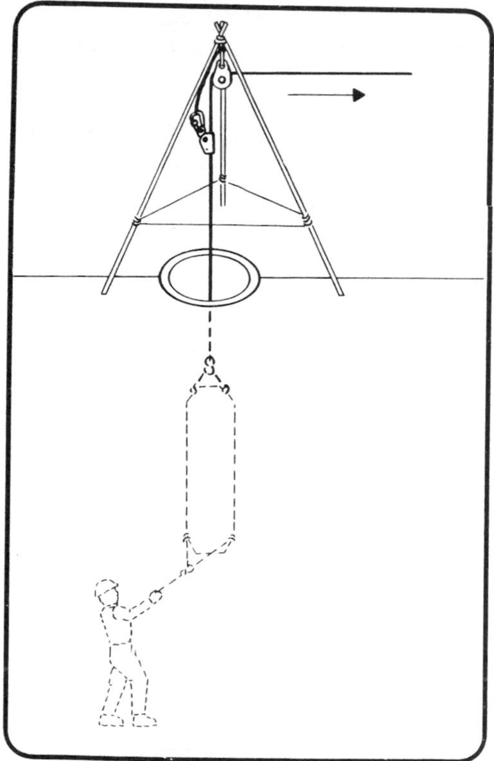

Fig. 15.8 1:1 Haul From Confined Space

Hauling from a Confined Space

Figure 15.8 illustrates the use of a 1:1 hauling system to perform a rescue from a vertical confined space. This system employs an "A" Frame constructed from two ladders lashed together near their tops. A large tripod may also be used. The addition of spreaders near the base of the ladders or the tripod will give added stability.

Additional features of this 1:1 haul system include the following:

- The ladder "A" frame must be securely anchored, particularly in a direction **opposite** to the direction of haul.
- A pulley (directional) is attached via a sling anchored to the apex of the "A" frame.
- A safety cam is also attached to a separate sling that is anchored at the apex of the frame. The safety cam should be lower down and closer to the ground. But it should be high enough so that it does not interfere with getting the load out of the hole.
- A tag line should be attached to the load. Initially this is used by the rescuers below to position the load so that it does not snag on the edge. Once the load is out of the hole, however, a rescuer at the top will need to control the tag line. He will use the tag line to swing the load from over the hole to where it can be safely detached.
- Once the load is at the top, the haul team will have to give slack, and the cam tender will have to release the cam so the load can be moved from over the hole.

A 2:1 Hauling System Without the "Diminishing V"

The problem with a conventional hauling system with a traveling pulley on the load *(see Figure 15.1[b])* is that it creates a "Diminishing V" by having the rope run around the traveling pulley. This "Diminishing V" tends to get easily snagged on brush, rock, building projections, and anything else that it can. Another problem is that there are two strands of rope going over an edge to create additional friction.

One solution to the problem of the "Diminishing V," that still has a 2:1 MA is shown in Figure 15.9.

Note that there is only a single haul line attachment going to the load, as would be the case in a 1:1 hauling system.

However, after the rope has gotten to the top, a short 2:1 hauling system is attached to it with a haul cam. By keeping the "Diminishing V" smaller and at the top, the system is less likely to snag. And if it does, the rescuers are able to reach it to free it.

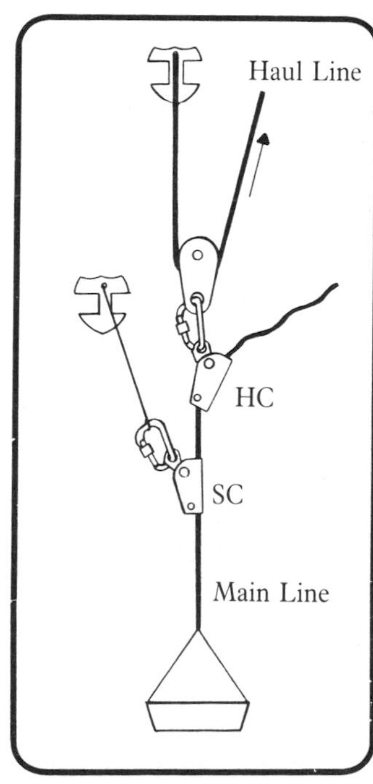

Fig. 15.9 2:1 Hauling System

Setting Up a 2:1 System

B1. Follow steps A1 through A9 above for setting up a simple 1:1 MA hauling system.

B2. At the top, and where rescuers can reach it, attach a haul cam onto the main line rope. This should be placed above the safety cam (between the safety cam and the anchor). To get the greatest amount of bite possible, the haul cam should be close to the safety cam. But it should be far enough away from it so that the two cams will not jam during hauling operations. Onto the cam clip a locking carabiner. Onto the carabiner clip a pulley.

B3. Take a shorter second rope (the haul rope) and fold it in half. Attach one free end of the haul rope to an anchor that is in the direction to be pulled *(point A1 in Figure 15.9).*

B4. Thread the haul rope through the pulley that is attached to the haul cam.

B5. Pull the second strand of the haul rope back parallel with the first strand that goes to the anchor. The haul team will pull on the second strand of the haul rope.

Hauling with the 2:1 System Attached to the Main Line

B6. If there is a shortage of personnel for this operation, the haul captain may be stationed at the haul cam to manipulate it. He should have a good field of view of the operation. If enough personnel are available, a person should be assigned to be haul cam tender.

B7. Follow steps A11 through A16 above.

B8. When the haul captain says, **"Slack,"** the haul team instantly reverses direction and moves back up the rope to take another bite. As they do this, the person attending the haul cam resets the system. He does this by grasping the carabiner attached to the haul cam and pulling it back down the rope towards the load. With his other hand, he should pull on the mainline rope so no slack develops on the main line rope between the haul cam and the load.

B9. When the haul cam reaches its original position, the team is ready to begin another haul cycle.

Rigging A 3:1 Haul System (Z-Rig)

Figure 15.10 illustrates a 3:1 haul system. This particular 3:1 system is also commonly known as a "Z-Rig" because of the approximate shape the rope takes as it goes through the system.

Minimum Equipment Requirements

(1) Main line rope

(3) Locking carabiners

(2) Pulleys

(2) Cams (1 for hauling, one for safety)

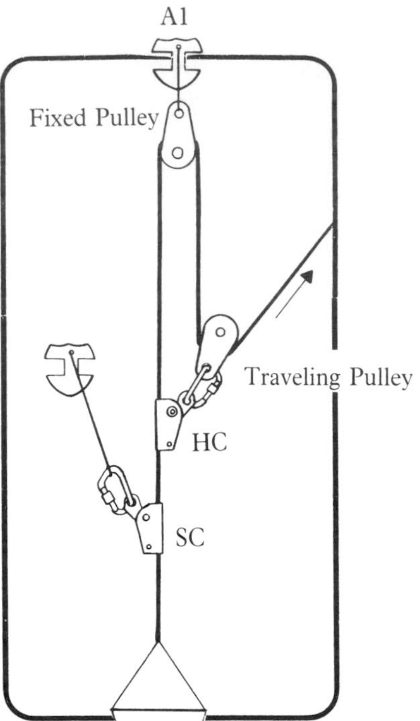

Fig. 15.10 3:1 Hauling System

The Following Steps Are Employed in Rigging the 3:1 System

(See Figure 15.10).

C1. Establish an anchor point that is above the load *(point A1 in Figure 15.10).* Into this anchor point, securely attach a sling. Into the end of the anchor sling, clip a locking carabiner. To this locking carabiner, clip a pulley. This anchor system should be well back from the edge of the drop to allow space for setting up the haul system.

C2. Thread the main line rope onto the pulley that has been anchored.

C3. Lower an end of the main line rope to the load.

C4. Establish a separate anchor point for the safety cam. This should be close enough to where the main rope will run, but slightly off to the side of it so the safety cam and its sling do not tangle with the rope. Attach an anchor sling onto the anchor point. Into the end of the anchor sling, attach a locking carabiner. Into the carabiner, clip a cam ascender (or Prusik).

C5. Thread the safety cam onto the rope. If it is a cam ascender, it should have its arrow pointing toward the load. Be certain that after placement on the rope and after the rope is loaded, the cam does not create much of a bend in the main line rope. If there is a great deal of bend created by the cam on the rope, it means a much higher load for the haul team.

C6. At a point just above the safety cam, place the haul cam (HC in Figure 15.10). The haul cam arrow should point towards the load. Clip a locking carabiner into this haul cam. Into the locking carabiner, clip a pulley.

C7. Bring the upper end of the main rope back in the direction of the load and run it through the pulley that is on the haul cam.

C8. Now again reverse direction with the upper end of the rope and bring it back towards the first anchor and parallel with the other two strands. This will be the hauling end of the rope. Remove all the rope slack from the hauling system.

C9. In this configuration, the haul must be back towards Anchor 1 and parallel with the other strands of the rope. Otherwise, if it is off to the side, some of the advantage will be lost. If, because of limited space or some other reason, the haul team must go off in another direction, then set a directional, using a stationary pulley so they can. In this way, the only advantage lost will be the small amount due to the friction of the directional pulley.

C10. Follow steps A11 through A15 on page 202.

C11. Before the two pulleys at A1 and HC come together, the haul captain says, **"Set."** The haul team immediately stops hauling. The safety cam tender makes certain that the cam will set. (If the cam is rigged for "automatic," the haul captain or other assigned person makes certain it will set.)

C12. The haul captain says, **"Slack."** The haul team instantly reverses direction and moves back toward the load with the rope still in their hands. As they do this, the person attending the haul cam resets the system. He does this by grasping the carabiner attached to the haul cam and pulling it back down on the rope until it reaches the safety cam.

C13. When the haul cam reaches its original position, the team is ready to begin another cycle.

A 4:1 Hauling System *(Piggyback System)*

Minimum Equipment Required:

(1) Main line rope

(1) Hauling rope (50-to 100-feet long, depending on space available for the haul

(3) Locking carabiners

(2) Pulleys

(2) Cams (1 for safety, 1 for hauling)

Rigging The Piggyback System *(see Figure 15.11)*

D1. Attach the main line rope to the load. Secure the upper end so that it does not accidentally slip over the edge.

D2. Establish a strong anchor point for the safety cam. This should be close enough to where the main rope will run, but slightly off to the side of it so that the safety cam and its anchor sling do not tangle with the rope. Attach an anchor sling onto the anchor point. Into the end of the anchor sling, attach a locking carabiner. Into the carabiner, clip a cam ascender (or Prusik).

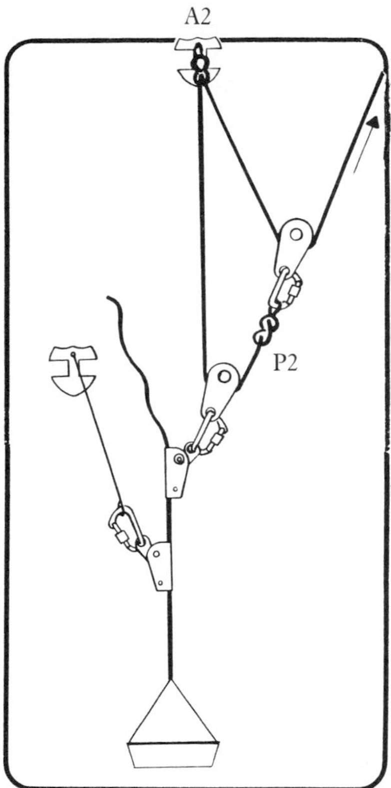

Fig. 15.11 4:1 Hauling System

D3. Establish an anchor point for the hauling system *(point A2 in Figure 15.11)*. This anchor point should be in line with the load and well back from the edge to allow room for the haul system. Attach a sling to the anchor point. In the end of the sling, attach a locking carabiner.

D4. Take the haul rope (the shorter line). Find the center. At the center of the rope, tie a Figure Eight Overhand knot. Clip the Figure 8 knot into the locking carabiner on the anchor sling.

D5. After tying the Figure 8 Overhand knot, you have two strands of the hauling rope. Take one leg of the haul rope back in the direction of the load. At about 2/3 of the way down this leg of the haul rope create a bight. Put a pulley on the rope at this bight. Into the pulley, clip a locking carabiner. Into the carabiner, clip a cam (or Prusik). Attach the cam or Prusik onto the main line rope. It must be attached on the main line rope above the safety cam (between the safety cam and the top end of the main line rope). The arrow of the cam should point towards the load. This is the point where the hauling system "piggybacks" onto the main line rope.

D6. On the same leg of the haul rope where you have attached the cam, find the free end of the rope. In this end, tie a Figure 8 Overhand knot *(point P2 in Figure 15.11)*.

D7. Into this Figure 8 Overhand knot, clip a locking carabiner. Into the carabiner, clip a pulley.

D8. Take the second leg of the haul rope. Bring it down to the pulley you have just clipped into the first leg of the haul rope. Thread the second strand through the pulley.

D9. Now pull the end of the second strand back toward the anchor for the hauling system *(point A2 in Figure 15.11)*. This is the end that the haul team will pull on.

D10. Follow steps A11 through A15 on page 202.

D11. Before the two pulleys at points P2 and A2 in Figure 15.11 come together, the haul captain says, **"Set."** The haul team immediately stops hauling. The safety cam tender makes certain that the cam will set. (If the cam is rigged for "automatic," the haul captain or other assigned person makes certain it will set.)

D12. The haul captain says, **"Slack."** The haul team instantly reverses direction and moves back towards the load with the rope in their hands. As they do this, the person attending the haul cam resets the system. He does this by grasping the carabiner attached to the haul cam and pulling it back down on the rope until it reaches the safety cam.

D13. When the haul cam reaches its original position, the team is ready to begin another cycle.

General Considerations for Rescue Hauling Systems

As with any rope rescue technique, hauling systems must be adapted to the rescue circumstances and environment, and not the other way around. While thorough knowledge of equipment and techniques is good, you should not necessarily always rig the most complex system that you know. In fact, the simplest system that is workable under the circumstances will often mean an expeditious and successful rescue.

The following are some of the considerations to be made when deciding on a hauling system:

- **Necessity for Speed**

  A simple system will often mean:

  a) It can be set up quickly, and

  b) The haul itself can be done quickly.

- **Few persons available for haul team**

  May necessitate a higher MA, and therefore more complex system.

- **Small amount of gear available for rigging**

  Usually means a simpler hauling system.

- **Cluttered area for hauling**

  Hauling systems more likely to get snagged. A simpler system may be less likely to get jammed.

- **Higher loads**

  May necessitate a higher MA (more complex system).

- **Lighter loads**

  Possibly use a smaller MA (less complex system).

QUESTIONS for REVIEW

1. Name ten situations in which rescue haul systems might be used.
2. What are the two basic reasons that rescuers employ hauling systems?
3. Define Mechanical Advantage.
4. How does Theoretical Mechanical Advantage differ from Actual Mechanical Advantage?
5. Assuming a load of 400 pounds, what would be the force needed for a haul team to move it in a 1:1 hauling system? In a 2:1 hauling system? A 3:1? A 4:1?
6. Which of the following can be used to create additional mechanical advantage in some hauling systems: a) stationary pulley b) traveling pulley.
7. In a basic rescue hauling system, if a 2:1 haul system were added to another 2:1 system what would likely be the resulting mechanical advantage? What would be the MA if a 2:1 were added to a 3:1?
8. Why should handled ascenders not be used in hauling systems?
9. What should be the position of the safety cam in relationship to the other elements of the haul system?
10. What might be the possible consequences of rigging a safety cam so far to the side of a main line rope that it creates a significant bend in the rope?
11. What are two consequences of edge friction on a rope in hauling systems?
12. What are two ways of reducing such edge friction?
13. What technique can a rescue team use to help control the movement of a litter that is being hauled and to help prevent it from becoming snagged?
14. List the minimum equipment required to rig a 1:1 hauling system. A 2:1 hauling system. A 3:1 (Z-Rig) hauling system. A 4:1 ("Piggyback") system.
15. What can be a significant problem with a 2:1 hauling system that uses a traveling pulley attached to the load?

★ ACTIVITIES ★

1. Draw a 2:1 hauling system using a traveling pulley on the load. Label the parts and the forces acting on the system.
2. Draw a 3:1 hauling system using a "Z-rig." Label the parts and the forces acting on the system.
3. Draw a 4:1 hauling system using a "Piggy back" rig. Label the parts and the forces acting on the system.

Chapter 16

Highlines

PREREQUISITES

Before attempting the activities described in this chapter, you must have demonstrated that you can properly:

1) Use and care for rope.
2) Use and care for other equipment employed in the high angle environment.
3) Tie correctly, confidently, and without hesitation the eight knots described in Chapter 6.
4) Apply the principles of anchoring and rig a safe and secure anchor.
5) Apply the principles of belaying and safely and confidently belay another person using either a Munter Hitch or belay plate.
6) Apply the principles of rappeling: rappel safely, confidently, and under control; tie off the rappel device to operate hands free of the rope and then return to a safe and controlled rappel.
7) Apply the principles of ascending: tie correctly, confidently, and without hesitation a Prusik knot and know how to use it; comprehend the uses and limitations of mechanical ascenders; confidently and safely ascend a fixed rope using either friction knots or mechanical ascenders; confidently and safely change over both from rappeling to ascending and from ascending to rappeling; and extricate yourself from a jammed rappel device (or similar problem).
8) Apply the principles of slope evacuation: correctly set the rigging in any of the elements of slope evacuation; and safely and confidently assume the role of litter tender, haul team member, rope handler, brakeman, or belayer.
9) Apply the principles of high angle lowering systems: correctly rig any of the elements of high angle lowering; and safely and confidently assume the role of litter tender, brakeman, belayer, rope handler, or edge tender.
10) Apply the principles of hauling systems: can determine the mechanical advantage required for a hauling system; can correctly rig any of the elements of a hauling system; can rig for a 1:1 MA hauling system, a 2:1 MA hauling system, a 3:1 MA hauling system (Z-Rig), and a 4:1 MA hauling system ("Piggyback Rig"); and can safely and confidently assume the role of cam tender, haul captain, haul team member, or belayer.

OBJECTIVES–

At the completion of this chapter, you should be able to:

1. Describe how highlines can be used in rescue and some typical examples of where they might be used.
2. Describe the basic elements of a rescue highline system.
3. Describe the major considerations in determining if a highline is a feasible rescue technique for a specific situation.
4. Describe the possible consequences of over-tensioning a highline.
5. Discuss the role of rope sag in highlines.
6. Apply the Ten Percent Rule to a variety of highline spans and load weights.
7. Describe the functions of the following: main line rope, load, pulleys, lowering/belay line, and tag line.
8. Select equipment to be used in highline systems.
9. Rig a highline system using a single main line, a lowering/belay line and a tag line.
10. Describe under what circumstances a highline constructed of parallel main lines would be preferable to a highline constructed of a single main line.
11. Rig a litter and litter tender attachments for a highline.

TERMS– *relating to highline systems that a Rope Rescue Technician should know:*

Highline—A system of using a rope suspended from between two points to move persons or equipment over an area that is a barrier to the rescue operation.

Horizontal Highline—A highline in which the two suspension points are close to being on the same level.

Lowering/Belay Line—The line that is attached to a load on a highline and is used to control the load from the near-side point.

Steep Angle Highline—A highline in which one of the suspension points is considerably higher than the other.

Tag Line—The line that is attached to a load on a highline and is used to control the load from the far-side point.

Tadem Pulleys—Two in-line pulleys on the same rope. In highlines, tandem pulleys are employed to stabilize a load and to spread its weight out along the rope.

Telpher—Another name for a highline.

Tyrolean—Another name for a highline.

Highlines in Rescue

Highlines are also sometimes referred to as **tyroleans** or **telphers.** In all cases, these terms refer to a rope linc suspended from one point to another between which people, rescue subjects, or equipment can be moved.

Horizontal Highlines are suspended from two points that are close to the same level, as in Figure 16.1(a). **Steep angle highlines** are suspended between two points in which one is at a much higher level than the other, as in Figure 16.1(b). Highlines may also be suspended from points that result in an angle anywhere between a steep one and a horizontal one.

Uses of Highlines

Highlines are used to transport rescuers, rescue subjects, and/or equipment across an area that is a barrier to rescuers. Some typical uses of highlines might be:

- **To Bypass an Obstacle**
  Highlines could be used to cross a deep canyon or gorge.
- **To Avoid Hazardous Terrain**
  An example of this would be a situation where a highline would be used to bridge a swiftly flowing river.
- **To Avoid Difficult Terrain**
  A highline might be suspended over an area that contained large boulders or thick debris through which it would be very difficult and time-consuming to move a litter containing a rescue subject.
- **For Emergency Evacuation**
  Highlines might be used to evacuate persons from a hazard in which there is the threat of injury or death, and where there is no other practical or expeditious means of evacuation. An example would be a fire in a building.
- **In Tactical Situations**
  Highlines might be used to evacuate civilians and/or to insert tactical personnel in a barricade or hostage situation.

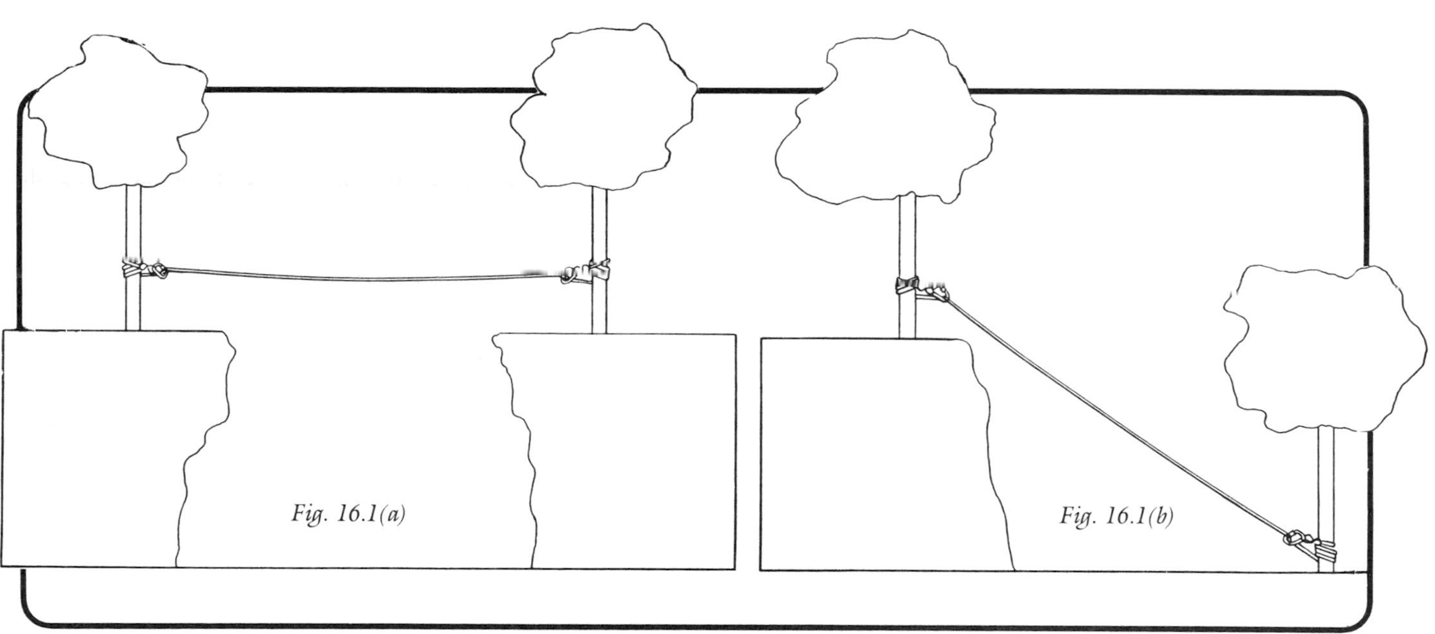

Fig. 16.1 Horizontal/High Angle Highline

Problems With Highlines

Potential Stress and Failure of Equipment

Perhaps more than any other rope rescue technique, highlines have the potential for overstressing rope, equipment, and anchors, thus causing the failure of the system. The rigging and use of highlines require thorough knowledge of the potential forces involved.

Time-Consuming

Highlines require a great deal of teamwork and communications to rig. When first attempted, teams find this rigging to be very time-consuming. Even rescue teams experienced with highlines often find other techniques preferable when time is crucial.

Getting Initial Personnel and Rope Across

One of the most difficult parts in rigging a horizontal highline is getting the initial personnel to the second point and getting the rope across to them.

Elements of a Highline

Figure 16.2 illustrates the basic elements of a highline system that might be used in a rescue. In this operation, a rescue team is moving a litter containing a rescue subject from left to right in a horizontal highline. In many cases, it would be desirable to include a litter tender. But for simplicity, this is not shown in this illustration. (A discussion of litter tenders for highlines is included in a later section of this chapter.)

- **Main Line Rope**
  This is the line that supports the major portion of the weight of the load in the highline. In many cases, this can be a single line. But under circumstances of high loading, a double line may be desirable. The rope should be of a low-stretch design, such as static kernmantle. Otherwise there could be a great deal of uncontrollable stretch in the system. There must be an appropriate amount of sag in the main line rope to prevent overstressing it, other equipment, and the anchors.

- **Near-Side Anchor**
  This is the anchor to which the main line rope is initially anchored. Because of the stresses generated in highlines, the main line anchors must be extremely reliable. In a horizontal highline, both main line anchors will be subjected to similar stresses. In steep angle highlines, the upper anchor will be the one most subjected to stress, much as an upper anchor would be in a lowering system.

- **Far-Side Anchor**
  This is the anchor to which the main line rope is attached once rope has gotten across to the far-side point. This anchor must be as strong as the near-side anchor. In a steep angle highline, the far-side anchor will receive less stress while the load remains near the top *(see Figure 16.3)*. However, as the load approaches the bottom, the far-side anchor will be subjected to greater stress.

- **Load**
  Where there is a seriously injured subject to be transported, the load on the highline would be a litter with the subject and possibly a litter tender. In other cases, the load on a highline could be one person, a rescue subject attached to a rescuer, or equipment.

- **Pulleys**
  The load is connected to a pulley (or pulleys) that travels along the main line rope. With higher loads, tandem pulleys are preferable to single pulleys because tandem pulleys create less of a bend in the rope and spread the load along the rope. It is important that pulleys be set so they travel in a straight line along the rope. Otherwise, they may torque and create drag. It is conceivable that if a pulley were not available, a large locking carabiner might be used in its place. This, however, would mean greater friction and the potential for wear and damage to the carabiner.

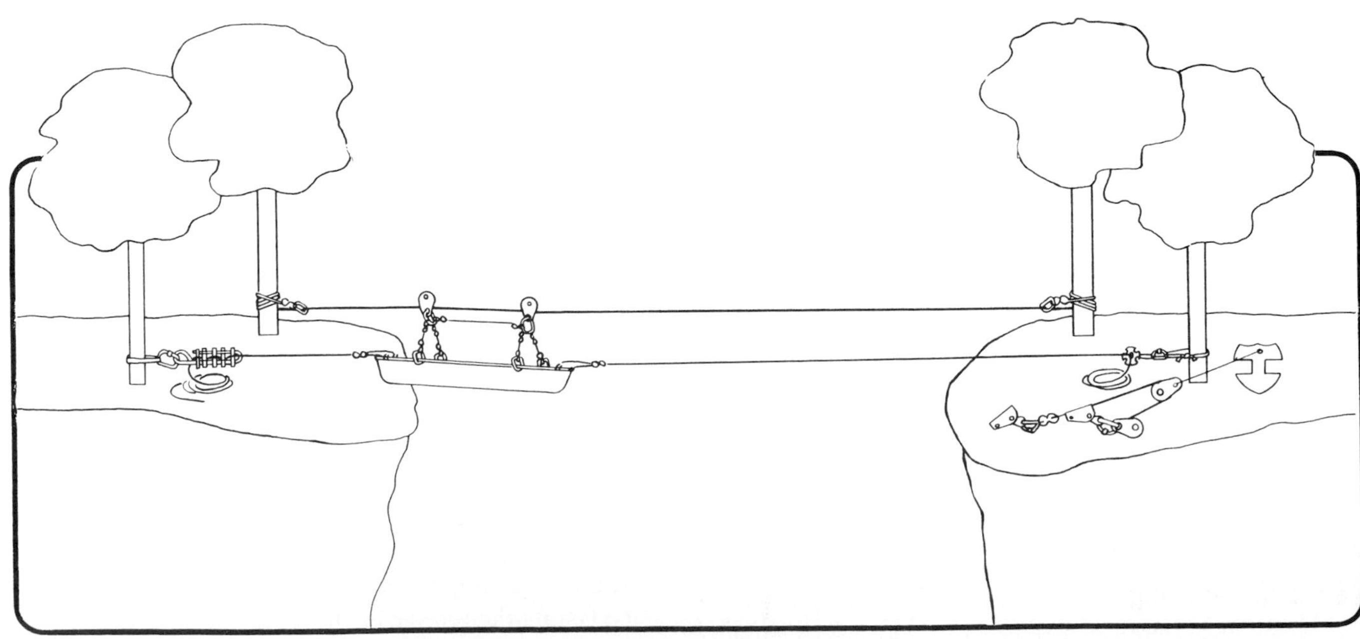

Fig. 16.2 Elements of a Highline

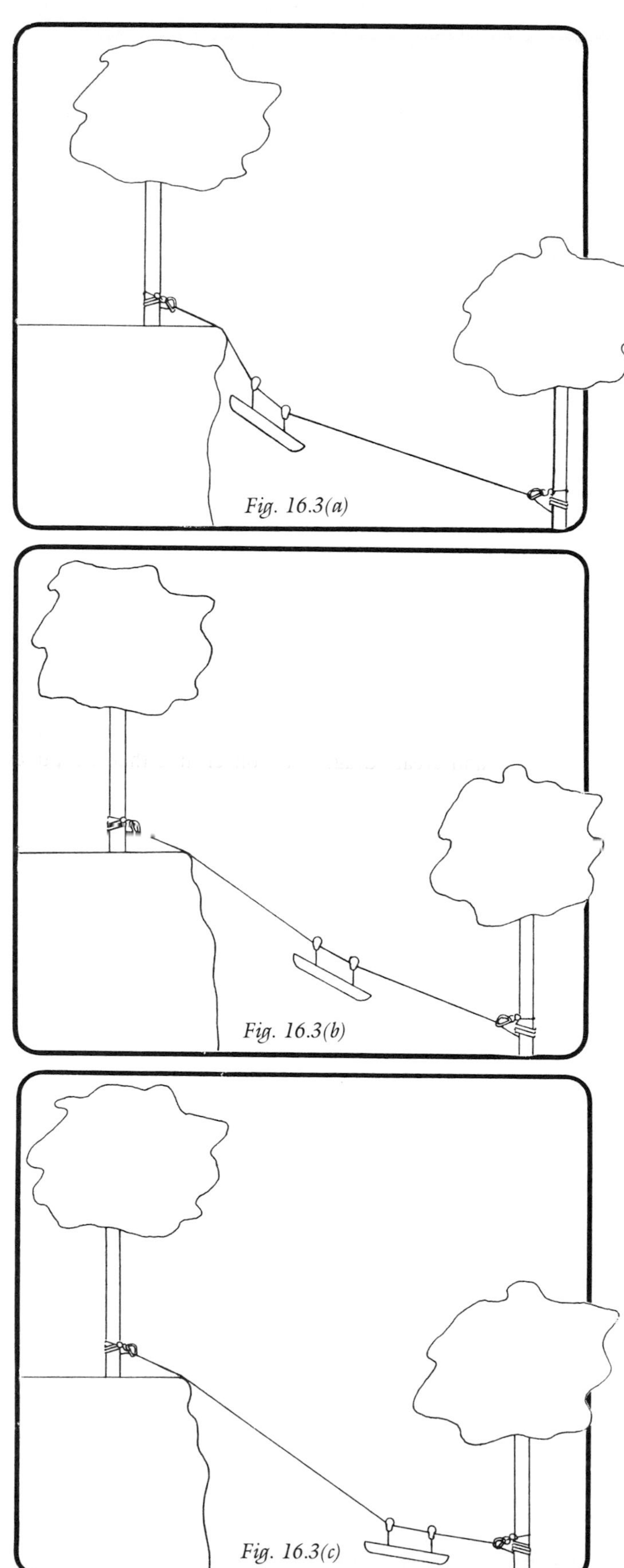

Fig. 16.3(a)
Fig. 16.3(b)
Fig. 16.3(c)

Fig. 16.3 Forces on High Angle Highline

■ Lowering/Belay Line

The lowering/belay line runs from the near-side point and is connected to the load. It serves two purposes.

First, it is used to control the speed of the load. This control is maintained by running the lowering/belay rope through a lowering device, such as a large Figure 8 descender or a Brake Bar Rack anchored on the near-side. In a horizontal highline this lowering effect will usually only be necessary until the load reaches the center of the main line *(see Figure 16.4)*. At that point, because of sag and stretch in the rope, the load will start "uphill" towards the far side. After this, the lowering/belay line will act more in its function of a belay.

However, in a steep angle highline, the lowering/belay line will function slightly differently. In a steep angle highline, if the load is traveling from the upper point to the lower one, the line will act in the lowering function for the most part until the load nears the bottom. In such a steep angle situation, the lowering/belay line will also require a lowering device with a great deal of friction and control. (If in a steep angle highline, the load is going from the lower point to a higher one, then a hauling system will be involved.)

■ Tag Line

The tag line runs from the far side point and is connected to the load. It can also serve two purposes.

First, it is used to belay the load. Consequently, it should be run through a belay device that is attached to a secure anchor on the far side. The belayer will be taking in rope continually through the movement of the load from the near side to him, so he must have a belay device that can easily take in rope. Such belay devices include belay plates and Munter Hitches.

The tag line is also used to haul the load. Once the load reaches the center of the main line, the personnel on the far side will actually have to pull the load to their side with the tag line. If there is significant sag and stretch, this could be an effort.

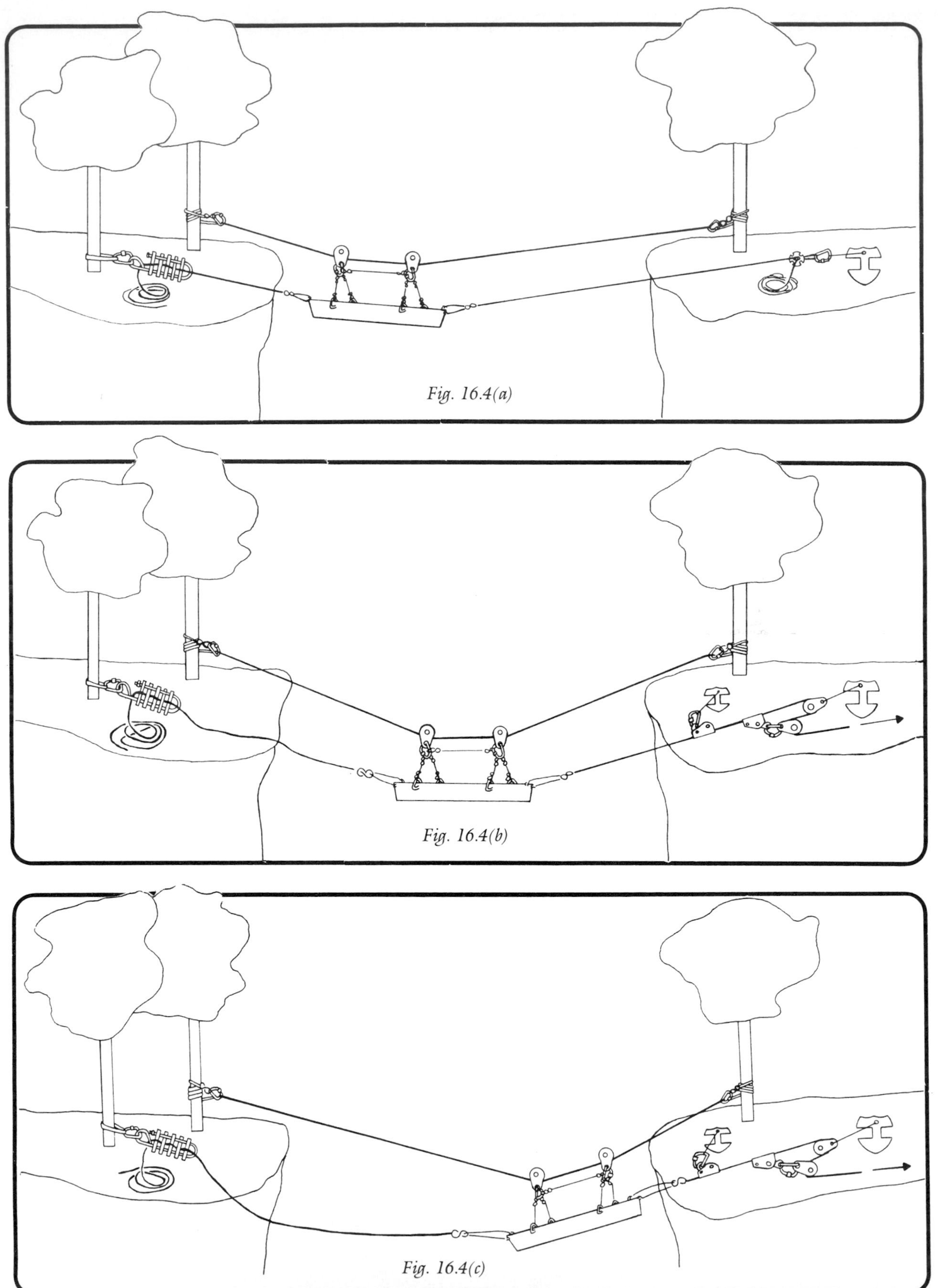

Fig. 16.4 Progress of Load on Horizontal Line

Highline Loads

One-Person Loads

Figure 16.5 illustrates a highline load with only one person. The essential elements of this are:

- A support sling which runs up to the pulley. A large carabiner clips it into the seat harness front tie-in point. At the top end it is clipped into a large locking carabiner. The Figure 8 is clipped into the pulley with a large locking carabiner. The Figure 8 is used as a central tie-in point because it can be pulled from either direction without twisting the system. Also clipped into the Figure 8 are:
  - Lowering belay line (with a carabiner).
  - Tag line (with a carabiner).
- Note also that the person acting as load is wearing gloves. This is to prevent rope burns should he grab a rope while he is moving and to prevent injuries to the fingers in the moving pulley. If the person being transported on the highline is a rescue subject, it may be wise to make the support sling long enough so that he cannot reach the highline and cause injury to his hands.

Fig. 16.5 One Person Load on Highline

Two-Person Loads

If the person on the line is rescuing another person, then the rescue subject may be clipped in through various possible means:

- Into the same support sling to the pulley as the rescuer. To manage this, the subject should be connected by a short sling to his own harness. This is clipped directly into the support sling. (***Not* into the rescuer's seat harness.**)
- One alternative is to connect the rescue subject to the pulley with his own support sling. If this is done, there should be an additional short sling between the rescuer's seat harness and the subject's seat harness. In this way, the rescuer will have some control over the position of the subject.
- An additional alternative is to have the rescue subject connected with his own support sling into a separate pulley on the main line. If this is done, there should be a sling connecting the seat harnesses of the rescuer and the subject. In addition, if possible, there should be a spreader between the two pulleys. This would ensure that the two pulleys would have a smoother ride down the rope.

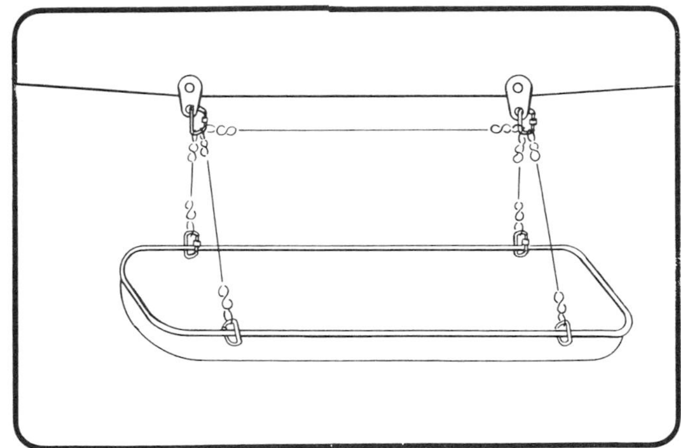

Fig. 16.6 Litter Load on Highline

Litter Loads

Figure 16.6 illustrates the rigging for a litter attached to a highline. It consists of the following elements:

- **Spiders**

  The litter system for a highline uses two spiders, each with at least two legs. One set is clipped into the litter rail at the head end, while the other set is clipped in at the foot end. Each set of spider legs comes up to a large locking carabiner that is clipped into a pulley on the high line.

- **Litter Tender Attachment**

  Figure 16.7 illustrates an attachment system for a litter tender on a highline. It consists of the following elements:

  - ***Pig Tail***

    The main attachment to the litter system is a "pig tail" made from approximately 12 feet of rope. The top end of the pig tail is attached with a Figure 8 Overhand knot to the pulley carabiner over the head end of the litter. To prevent the tender's ascender attachments from accidentally slipping off the end of the pig tail, the lower end of the pig tail is clipped into the tender's harness carabiner with a Figure 8 Overhand knot.

  - ***Ascender Attachments***

    So that the litter tender can adjust his height in relation to the litter, he uses two spring-loaded ascenders that clip into the pig tail. One of the ascenders is attached with a sling to the litter tender's seat harness. The other ascender is attached with a sling to the tenders foot.

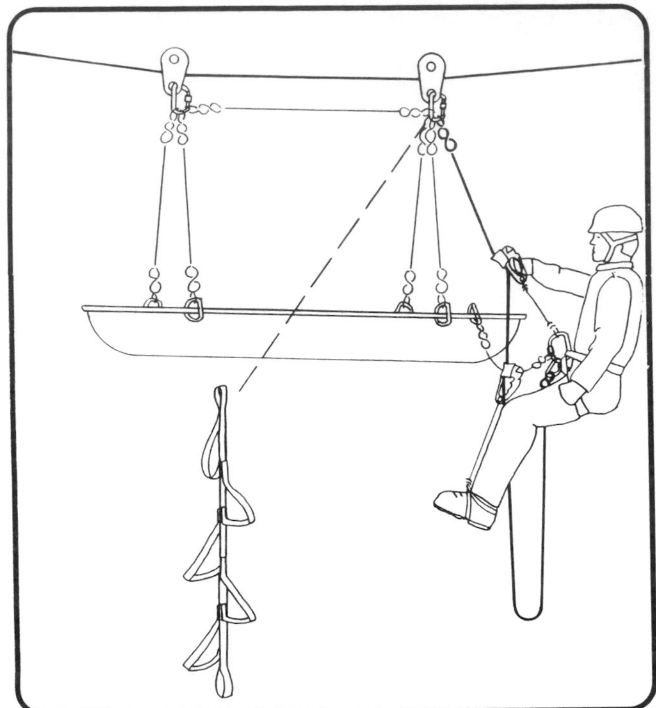

Fig. 16.7 Tender Attachment for Highline

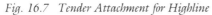

ALTERNATIVE APPROACH

An alternative litter tender tie-in is a daisy chain. The end of the daisy chain is clipped into the carabiner at the top of the spider, and the litter tender clips into the appropriate "pocket" on the daisy chain to give the proper position on the litter.

The daisy chain tie-in offers the following advantages:

- It is simple to use.
- It can be used by personnel who have no experience with ascenders.

But offers the following disadvantages:

- Once the litter tender is hanging in the daisy chain and the litter is over the side, the tender cannot easily and safely change his position on the daisy chain.
- The litter tender lacks the mobility to move below the litter to clear obstructions or perform other tasks.

- ***Etrier*** *(Optional)*

  One optional piece of gear that assist the litter tender in moving up and down on the litter is an etrier. This is a sort of short ladder made of webbing. The etrier may be used **in place of** the ascenders or **in addition to** them. The etrier best serves the litter tender if it is extra long. The standard etrier is normally too short for this purpose, since it requires the tender to raise his foot very high to get it into the bottom step. Two standard etriers clipped together are often too long and easily become snagged or entangled.

- ***Safety Tie-In***

  A safety sling runs from the litter tender's seat harness and is clipped into the rail of the litter near the head end.

■ **Litter Tender Position**

The litter tender position is similar to that used in a litter lowering: sitting comfortably in the seat harness, with the litter slightly above his lap. The tender normally keeps both hands on the litter rail nearest him in order to help stabilize the litter.

■ **Litter/Subject Position on Highline**

When rigging a litter onto a horizontal highline, it should be set with the head to go across first. This means that during the initial movement of the litter from the near-side to the center of the sag, the subject's head will be down. But then, as the litter moves "uphill" from the sag to the far side, the head will be up. In most cases, it will take longer to move the litter from the center of the sag to the far side, than from the near side to the center of the sag, because the far side team will be fighting gravity in moving the litter up from middle of the sag and will have to be hauling the load. So, by setting the litter with the head across first, it will minimize the time that the subject's head will be down.

The situation is different, however, in a steep angle highline. In this situation, the litter will stay in essentially the same angle until it nears the end of its travel at the bottom. Thus, in a steep angle highline, the litter should be rigged with the head up.

■ **Spreader**

The spreader serves to keep the two pulleys on the main line a proper distance apart. It also may serve as a stabilizer should one of the pulleys fail.

The spreader is created from webbing or a short piece of rope. It runs between two carabiner pulleys.

■ **Attachments for Lowering/Belay Line and Tag Line**

The lowering/belay line and the tag line are attached to the litter head and foot rail in a way that is similar to that used for slope evacuation. (See Chapter 13, "Slope Evacuation.")

The difference is that after tying the Figure 8 Follow Through knot, there should be enough tail to bring the end of the rope back up to the nearest pulley on the high line. At that point, a Figure 8 Overhand knot is tied in the end of the rope and this is clipped into the pulley carabiner.

This procedure is designed not only as a safety factor, but to keep the pulley in place when the litter is at an angle on the highline.

Determining the Amount of Sag in the Highline

One of the most critical activities in rigging a rope for a highline is determining the proper amount of sag in the main line rope.

WARNING NOTE

A highline system *must never be stretched tight and then loaded.* This could result in overstressing and failure of the rope, other equipment, or the anchors. The reasons for this are explored below.

Figure 16.8(a) diagrams the forces present in a highline system when the rope has been stretched tight horizontally with no visible sag. The following example shows that this is an unsafe condition.

In Figure 16.8(b) a load of 200 pounds has been placed in the middle of the rope. Note that while there is a force of 200 pounds in a downward direction on the rope, the resulting forces off to the side at a right angle are multiplied tremendously. This creates enormous stress on the rope, anchors and hardware. It easily could result in complete failure of the entire system.

Now note in Figure 16.8(c) what occurs using the same 200-pound load on the same 100-foot length of rope. The rope now has been slacked so that there is a sag of 10 feet in the center of the rope. As a result, the forces at a right angle have been reduced to within tolerable limits. This basic principle of physics must be kept in mind whenever rigging a highline.

Among some technical people there has been considerable debate on exactly how much slack should be allowed in a highline. As a result, there are available complex mathematical equations for working this out. But rather than carrying a computer with them every time they rig a highline, many people simply adhere to what is known as the "The Ten Percent Rule."

For a detailed discussion on the forces created in highlines, see *Wilderness Search and Rescue* by Tim Setnicka.

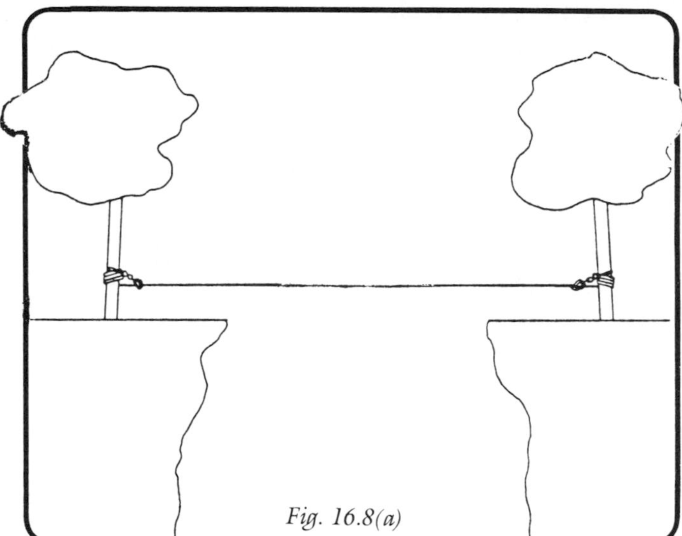

Fig. 16.8(a)

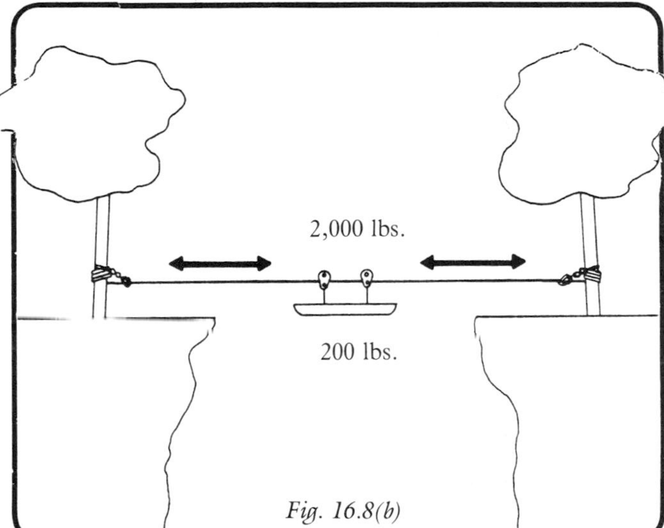

Fig. 16.8(b)

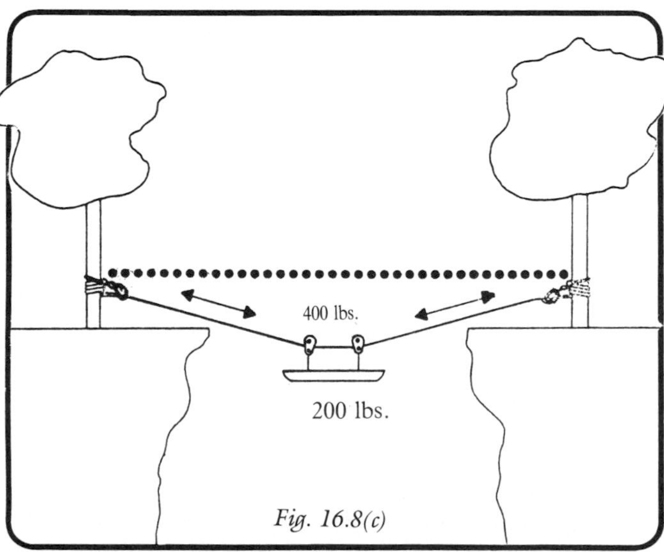

Fig. 16.8(c)

Fig. 16.8 Forces on Highline

The Ten Percent Rule

The Ten Percent Rule means that for every load of 200 pounds, with 100 feet of span in the rope, there should be a sag of 10 percent.

REMEMBER: This means *total weight of the load* (not just the weight of the person[s] on the load) and the *total length of span between the two supports*, such as the anchors (not just the width of the gap it bridges).

The *rope sag* is the visible amount of sag in the main line *before the load is applied.*

There are two variables in the Ten Percent Rule: a) the weight of the load and b) the length of the rope span. If either one of these variables changes, then the amount of sag has to change. For example:

200 pound load (1L) on a 100 foot span-
1L x 100' x .10 = 10 foot sag required.

200 pound load (1L) on a 200 foot span-
1L x 200' x .10 = 20 foot sag required.

400 pound load (2L) on a 200 foot span-
2L x 200' x .10 = 40 foot sag required.

Steps in Rigging a Highline

Before proceeding, have all equipment in order:

(1) **Mainline Rope.** The total length will include:
- Length of gap to be bridged, plus
- Length from edge to anchor **on each side**, plus
- Length needed to tie onto anchors on **each side**.

(1) **Lowering/Belay Rope.** The total length will include:
- Distance from lowering device on near side to point on far side where load will be derigged, plus
- Amount of rope needed to tie to the load, plus
- 20 feet to spare.

(1) **Tag Line.** The total length will include:
- Distance from where load is rigged on near side to belay device on far side, plus
- Length needed to attach to load, plus
- 20 feet to spare.

(13) **Carabiners** *(locking)*
- For anchoring (single line system) . . . **4** *(minimum)*
- For attaching (litter) load to highline **2**
- For attaching pig tail from lowering/belay and tag lines to pulley carabiners **2**
- For attaching litter spiders to litter **4**
- For attaching tender to litter **1**

Total . . . . . . 13 *(minimum)*

Anchor Materials

Litter Rigging

(2 sets) Litter Spiders

(1) spreader
(1) litter
(2) pulleys for main line
(1) litter tender pig tail
(2) ascenders plus slings for tender
(1) safety sling for tender
edge protection for both sides.

1. **Select an Appropriate Site**
   Considerations would include:
   - As narrow a span as possible.
   - Room on both sides to rig/derig load and for personnel to get on and off.
   - Availability of anchors—
     Must be very strong and secure.
     Must be high enough so that load can get over the edge without dragging.
2. **Get Second Team to Far Side**
   All personnel must be thoroughly briefed beforehand on steps to be followed and on communications. Radios will be very helpful.
3. **Get the Far-Side End of the Rope Over**
   Depending on the physical circumstances, this may be one of the more difficult parts of the operation. If it is a horizontal span, one of the following approaches may be necessary:
   - A line gun may have to be used. This usually involves first shooting over a smaller diameter line. Once this line is over, then the end of the main line rope is attached and pulled over.
   - The personnel who go across to the far side may have to trail the rope behind themselves.

- If in an urban situation and a line gun is not feasible, then:
  a) Lower the far-side end of the main line rope to the ground.
  b) Have someone pull the end to the base of the structure on the other side.
  c) Have the far-side team drop a haul line to the ground. The two ends are tied together and then the team pulls up the main line rope.

  If it is a steep angle highline, then follow steps (a) and (b) above.

4. Have the team on the far side anchor their end of the main line rope.
5. The near-side team pulls on the main line rope until the proper amount of slack is achieved. They then anchor their side of the main line rope.
6. Get the tag line to the far side. (This may be done at the same time as the main line.) On the near side, secure the end of the tag line so it does not slip over the edge.
7. Have the far-side team establish an anchor for the tag line belay device.
8. Have the far-side team attach a belay device to the anchor system.
9. Have the far-side team thread the tag line through the belay device.
10. On the near side, establish the anchor for the lowering/belay rope.
11. Attach the lowering system (Figure 8 descender or Brake Bar Rack) to the anchor on the near side.
12. Attach the lowering/belay rope to the load.
13. Thread the lowering/belay rope into the lowering system and lock the rope off.
14. Rig the litter or the single person that is going to be the load.
15. Set the load on the main line. Make certain that the belayer is in control of the lowering/belay line.
16. Attach the tag line to the litter system. Have the tag line belayer on the far side remove slack from tag line.
17. Have the litter tender attach himself to the litter system.

Beginning Movement of the Load

18. Before beginning movement, recheck rigging, including anchors on both sides.
19. Make certain all persons are ready, including:
    a) The person attached to the load.
    b) The brakeman on lowering/belay line.
    c) The belayer on far-side tag line.
    d) The edge tenders.
20. When all personnel are ready, the litter tender or person attached to load says: **"On Belay."**

    Lowering/belay line brakeman: **"Belay On."**

    Tag line belayer: **"Belay On."**

    Litter tender: **"Down Slow."**

    The lowering/belay line brakeman begins allowing rope through the brakes on the near side, while the belayer on the far side tag line begins taking in slack.
21. Once the load reaches the far side and is in a secure position, then the litter tender or person attached to the load, says, **"Stop!"** The lowering/belay line brakeman stops feeding rope.
22. When the litter tender or person attached to the load is in a secure position on the far side, he then says, **"Off Belay."**

    Lowering/belay line brakeman: **"Off Belay."**

    Tag line belayer: **"Off Belay."**

QUESTIONS for REVIEW

1. Name four situations in which highlines might be useful to rescuers.
2. Name three common problems that rescuers encounter in rigging highlines.
3. Name seven basic elements of a highline system.
4. The litter system for a highline consists of ________ spiders sets, each with ________ legs.
5. Why should two pulleys in tandem be used for supporting a litter system on a highline?
6. Name two purposes of a spreader for a highline litter system.
7. Name the ways in which a litter tender is attached to the litter in a highline system.
8. Assuming a horizontal highline, should the litter be moved across head or feet first? Why?
9. Assuming a steep angle highline, should the litter be moved head or feet first? Why?
10. Describe the functions of the following in a highline system:
    a) Lowering/belay system.
    b) Tag line.
11. Describe how the lowering/belay line and tag line should be attached to the litter system in a highline.
12. In a horizontal highline, what kind of device should be used to control the lowering/belay line? The tag line?
13. Describe what occurs if a highline is stretched tight and then loaded.
14. The Ten Percent Rule says that for every load of about ________ pounds and for every ____________ feet of rope in a highline span, there should be a sag of ____________ percent.
15. The Ten Percent Rule contains two variables. If either one of these changes, then the amount of sag should change. What are the two variables?
16. Estimate the amount of visible sag needed in a highline under the following conditions:
    a) 200-pound load on 100-foot span.
    b) 200-pound load on 200-foot span.
    c) 400-pound load on 100-foot span.
    d) 100-pound load on 200-foot span.
17. List the minimum equipment that would be required in rigging a highline with a 100-foot horizontal span, and using a litter system with tender for a load.
18. Describe the criteria in selecting an appropriate site for a highline.
19. List three possible techniques for getting the rope across the gap when establishing a highline.
20. Describe the precautions that should be taken before beginning movement on a highline.
21. List the steps taken when moving a load across a highline.

Chapter 17

Helicopter Operations

NOTE: Helicopters are being used more and more in rescue, tactical, and aeromedical operations. The following chapter is presented to familiarize you with helicopter operations and safety. Helicopter operations vary, depending on local training, on the local environment, and on the particular aircraft in use. For that reason, specific skills for helicopter operations can be gained only by actual local use.

If your organization does not have helicopters, you should strive to develop an association with helicopter units. In this way, you may work for mutual training and understanding to develop confidence in one another's abilities.

One thing that helps the survival of people in a risky business, such as helicopter piloting, is distrust of the unknown, in particular, the unknown abilities of unknown people.

Some Things You Should Know About Helicopters

(Before You Get In, Around, Under, or Anywhere Near Them)

How Helicopters Fly

Helicopters have some very special aerodynamic characteristics that enable them to perform in the special way they do. But these characteristics also make helicopters vulnerable in other ways.

- **Collective**
  Each main rotor blade acts like an airfoil, and when whirling through the air, creates lift. The pilot can vary the pitch of these blades and their rate of rpms to create more or less lift. Under the right conditions, this causes the helicopter to rise or drop.

- **Cyclic**
  The main rotor can be tilted to create a direction of flight. If it is tilted forward, then the helicopter, under the right conditions, will head in that direction.

- **Anti-Torque**
  A helicopter with only a single main rotor would spin helplessly in the direction opposite to the turn of the main rotor. To solve this problem, the tail rotor provides thrust in a direction opposite to the spin of the main rotor in order to stabilize the aircraft. The more power that goes to the main rotor, the more power must go to the tail rotor for counter thrust.

- **Ground Effect**
  When helicopters are close to the ground their hovering is assisted by a cushion of air they compress below themselves. But this only works well in stable situations. If the cushion of air is disturbed, then the helicopter becomes unstable.

- **Power Margin**
  The ability of a helicopter to lift itself is affected by the density of the air. If it is a hot day, or if the altitude is high, the air is less dense and the helicopter has less ability to lift itself. A particularly dangerous combination is a hot day at high altitude. When the margin is close to the line, a small change, such as the additional weight of one person can make the difference between a helicopter lifting off successfully or crashing.

Roles and Responsibilities

The Pilot

You must realize that the pilot is in command of his ship. He is the one who decides on things that affect the safety of the machine and the people in and around it. Such things include if and when conditions are right for take off, who boards and when, where people sit on the aircraft, whether gear will be taken aboard and how it will be stowed on the aircraft, and if the mission is to be competed or aborted. If he is uncertain about the people or other conditions in a situation, it is not only his right, but his DUTY to avoid involvement of his aircraft and crew.

IN SHORT: THE PILOT HAS THE FINAL WORD. He is the one who knows best the limitations of his aircraft, the flight conditions and himself.

No one should approach the helicopter without the permission of the pilot. This is often given through a visual sign such as the "thumbs up."

Crew Chief

On larger aircraft, the crew chief performs many of these duties that might otherwise be done by the pilot.

Copilot

On larger helicopters, a copilot may be involved with navigation, communication and crew coordination. During hovering operations, the pilot may have his attention to things outside the aircraft, so the copilot will be responsible for monitoring the aircraft's engine and performance instruments.

On smaller aircraft, the copilot may also operate a hoist or tend a belay line.

Observer

On smaller aircraft, the person commonly referred to as the observer may sit to the left of the pilot and take on many important responsibilities to relieve the pilot of the burden.

- Directing the pilot into the target, and advising him of rotor clearances.
- At the landing zone stationing himself to keep spectators and other persons from helicopter danger zones.

Helicopter Safety

Each year, there continue to be a large number of helicopter mishaps that result in injury and death. While some of these mishaps relate to mechanical failure, the majority of them relate to human error. These errors occur both among inexperienced personnel and among personnel who have long experience with the aircraft, but who have become so casual about aircraft operations that they have dangerous lapses in attention. Consequently, those persons who work around helicopters must not only educate themselves about the dangers, but also continually remind themselves about the dangers.

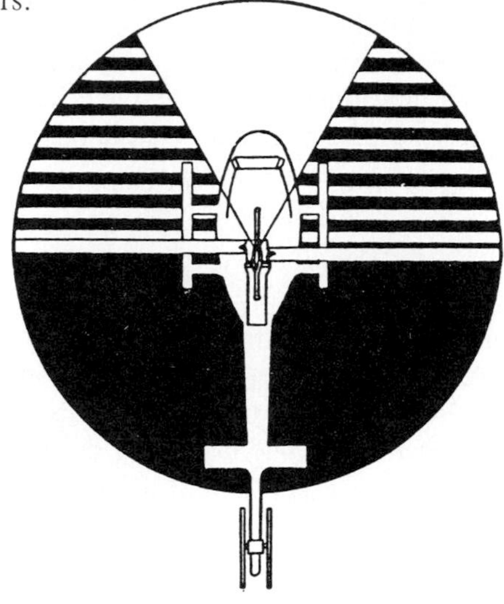

Fig. 17.1 Danger Zones of a Helicopter

Danger Zones

Figure 17.1 illustrates the common danger zones of a helicopter. This illustrates the helicopter danger zones as you would look directly down onto the aircraft from above. The 12 o'clock position is to the front of the aircraft, 6 o'clock is to the rear, 9 o'clock is on the left axis, and 3 o'clock is on the right axis.

Helicopter danger zones relate both to the physical hazards on the aircraft and to the pilot's restricted field of view. Note that the shaded area behind the 9 o'clock/3 o'clock axis in Figure 17.1. This is the **Pilot's Blind Area** and no personnel should approach the aircraft from this direction.

With full power on, the threat from the main rotor may not be obvious. But, as they slow down, many main rotors will droop closer to the ground, as low as head high in some models.

The other danger relates to the main rotor on sloping ground *(see Figure 17.2)*. When a helicopter sets on sloping ground, anyone on the up hill side will be automatically placed closer to the rotor blades. Consequently, when on sloping ground, always approach a helicopter **from downhill.**

Tail Rotor

Another primary hazard is the tail rotor. It poses a great danger for two reasons:

- The tail rotor in operation is often whirling at such a high speed that in the noise and excitement the rotor is unnoticed and people walk into it.
- A sudden wind gust moves the helicopter on the ground, or the pilot makes a quick take off and shifts direction, mowing down a person or persons on the ground.

  Consequently, to avoid contact with the tail rotor follow these rules strictly:

  Never approach a helicopter from the rear and never go around the rear of a helicopter to get to the other side.

Unsecured Objects and Debris

Take special care with objects, such as radio antenna, litters, weapons, skis, ice axes, etc., that protrude above head level and could make contact with the main rotor. This could cause serious injury or death to you and cause the rotor to fragment, resulting in serious injury or death to other persons. Objects such as these should be carried low and parallel to the ground.

Objects, such as hats and helmets, that could be caught by rotor wash and be sucked into the rotor must be thoroughly secured.

Rotor Wash

As the helicopter approaches for a landing, the rotor wash will blow all lightweight objects away to either be sucked into the rotor or tail rotor, or take flight causing them to be chased cross country. Such things include caps, tents, tarps, loose clothing, empty packs, sleeping bags, parkas, and helmets.

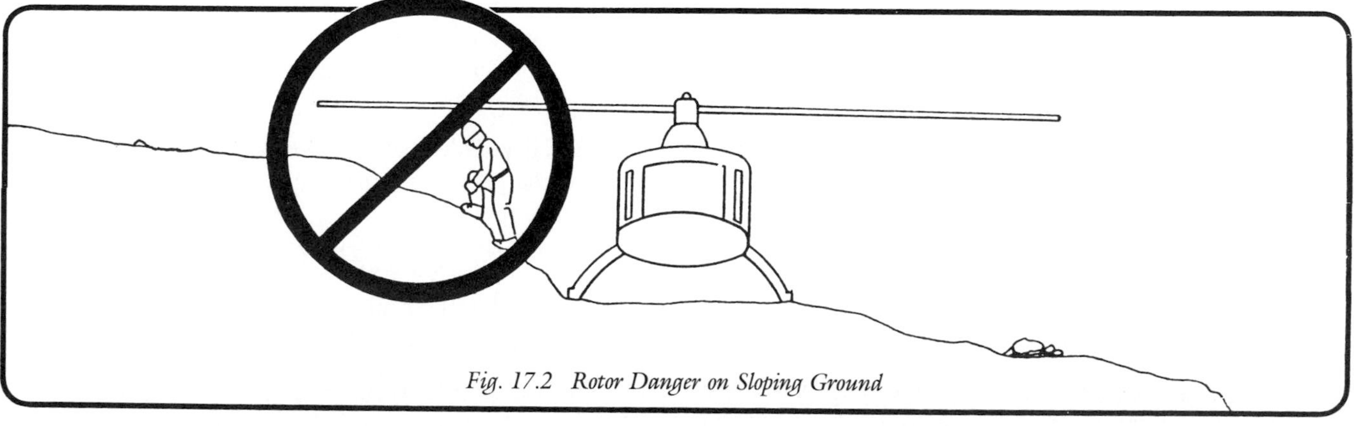

Fig. 17.2 Rotor Danger on Sloping Ground

Additional Helicopter Safety Rules

- Do not smoke within 100 feet of a helicopter.
- Always buckle up when inside the aircraft.
- Never stand under a helicopter or in its landing/takeoff zone, unless you are authorized to attach sling loads.
- When working in a helicopter LZ, wear eye protection against blowing dust and sand.
- After you exit a helicopter always make certain that seat harness buckles are secured inside the aircraft and the door is firmly closed.

Mountain Flying

Among the greatest tolls on rescue helicopters is mountain flying. Some veteran pilots have described rescue mountain flying in North America as being scarier than combat flying in Vietnam.

There are two primary reasons for mountain flying's being so treacherous:

- Aircraft performance limitations at altitude

  As you ascend in altitude the air becomes less dense, so the helicopter rotor has less to "bite" into, reducing its capability to lift the aircraft. During warm weather, the air also becomes less dense, meaning even less lift for the aircraft. Thus, a hot summer day at high altitude is a dangerous environment for a helicopter.

  An additional problem is that there is less lift for a helicopter in the air over water. There have been a number of tragedies in which helicopters at high altitude have taken off or approached for landing over a mountain lake and crashed into the water.

  Use caution in judging a helicopter ceiling from manufacturer specifications. These are often the ideal, calculated from stripped down models flying only with a pilot. If the same helicopter has rescuers aboard, and its full complement of gear and equipment, the ceiling may be considerably lower.

- Down drafts

  Mountains tend to make their own weather, which may be unpredictable at lower altitudes. Veteran pilots who have been flying for years in the mountains are often able to predict what these conditions may be. But it is often the veteran pilot who gets caught in them.

Landing Zones

One thing that helicopter pilots tend to be distrustful about is landing zones (LZs) with which they have had no previous experience. The primary reason for this is that unless they know the ground personnel well and trust their judgement, there is no way they can be sure those persons setting up the LZ are fully aware of the hazards.The following are some general considerations for the creation of LZs.

Selection of Landing Zones

- Terrain

  Helicopters are not at their most controllable when landing and, in particular, when taking off at a complete vertical. They operate better and safer when landing and taking off at a slope and, as with fixed-wing aircraft, into the wind.

  Figure 17.3 illustrates an ideal terrain for a helicopter LZ. Note that the terrain slopes down both on the approach and takeoff sides, and the landing and takeoff are into the wind. Obviously, there must be adequate space at the top for the landing and maneuvering of the aircraft.

- Debris

  Natural debris such as pine needles, leaves, dust, and light snow can obstruct vision and create hazards causing potential mechanical failure.

- Obstructions

  The landing zone should be free of vertical obstructions. Any such obstructions that cannot be removed must be reported to the pilot, and, if possible, marked or lighted. Particularly dangerous are utility lines, which are often invisible to the pilot.

- Meadows with tall grass

  Tall grass dissipates a helicopter's ground cushion, creating dangers in landing, takeoff, and hovering. Tall grass can also obscure hazards such as rocks, stumps, and bogs.

- Snow

  Loose, dry snow may be kicked up by the rotor wash and blind the pilot. This may cause dangers both in takeoff and landing.

  Wet snow may cause dangers by adhering to the skid or landing gear. This may require more power for takeoff and when the snow adhesion to the undercarriage breaks unevenly, the aircraft can roll on takeoff.

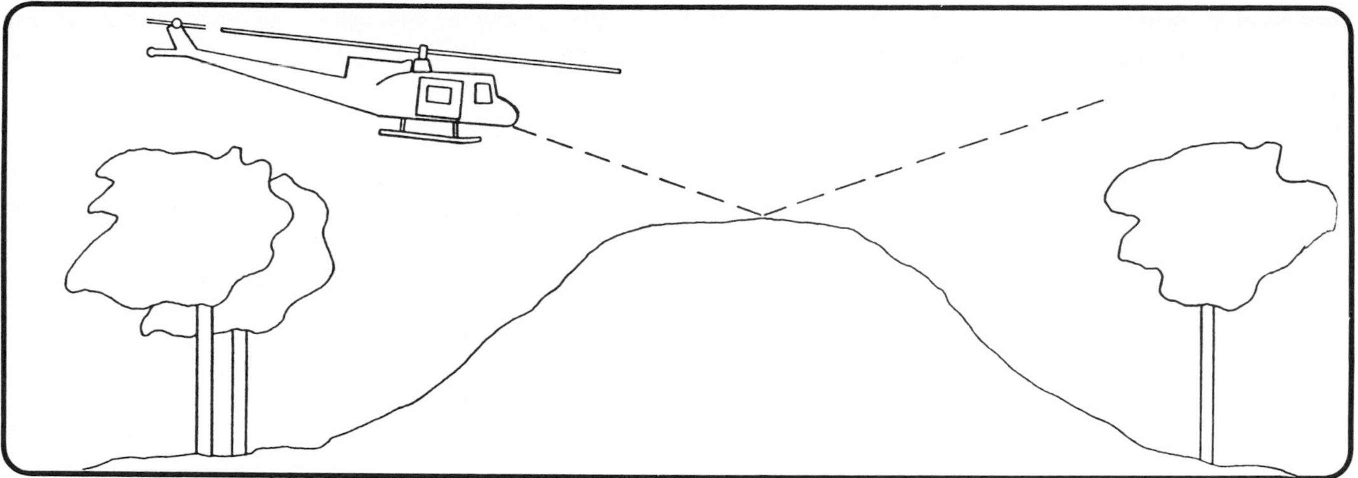

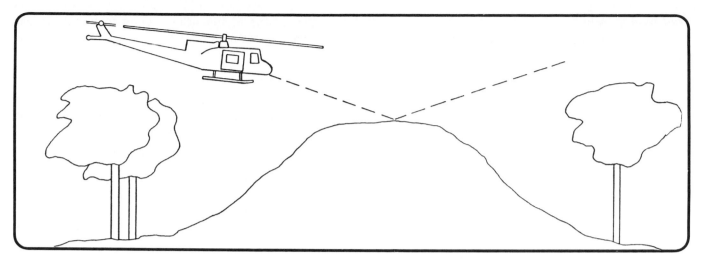

Fig. 17.3 Ideal Terrain for Helicopter LZ

Preparation of Landing Zones

- Remove debris

  Remove or secure all debris or loose equipment.

- Remove non-essential personnel

  Remove all persons except those who are completely essential for the landing, loading, and take off of the helicopter. Those who are loading must not approach the aircraft until the pilot orcrew cheif signals them to do so.

- Wind indicators

  Pilots will need to know the wind direction both on landing and takeoff. Many services use smoke grenades for wind indicators, but some pilots do not like them because they feel smoke obscures their view of an area. Other possibilities for wind indicators include the throwing of dirt, pine needles, leaves, or snow into the air.

When arriving at an unfamiliar LZ, pilots will commonly make one or more reconnaissance passes to determine the best approach pass and if a landing can be made safely.

Night Landings

The objective of night landing lights is to provide the pilot with adequate orientation without blinding him.

- For long distance orientation

  Strobes or emergency vehicle lights may stand out among other distracting lights and orient the pilot from a distance. As the pilot begins his approach, these and other bright lights should be extinguished because they may blind him.

- LZ marker lights

  For marking the LZ perimeter, pilots prefer low-intensity lights, such as chemical light sticks. Avoid flares since they are too bright and are a fire danger.

Helicopter Insertion/Evacuation Techniques

Full Landing

Where the terrain permits a suitable LZ, and where the weather allows for it, the full landing is usually the most desirable for the insertion or evacuation of personnel.

After the landing, the personnel must await signal from the pilot or crew chief to exit the aircraft. As in entering the aircraft, they must take the same precautions: secure all equipment, do not project any equipment into the rotor, keep heads down, stay in view of the pilot, and **move forward in the pilot's view away from the aircraft and avoid the tail section.**

Hot loading or **unloading** is a condition where the pilot keeps the rotor turning during loading or unloading. It may be necessary for a variety of reasons, most commonly the following:

- Speed

  Example: A critically injured rescue subject, forest fire, etc.

- Questionable LZ

  The pilot keeps the rotor rpms up to avoid the skid slipping off rock.

- Tactical reasons

  Example: The LZ may be exposed to weapons fire.

One Skid Landing

One skid (**"Toe-In"**) landings are sometimes used where the terrain is unsuitable for a full landing. Personnel enter and leave the aircraft by moving carefully and deliberately across the skid in use so they do not upset the equilibrium of the aircraft.

The basic sequence for unloading is as follows:

1. The pilot slows the ship and touches the skid on the passenger side of ship. He maintains the engine rpms to keep the ship level and makes certain that he has stabilized the ship.
2. The pilot signals that he has stabilized the ship. The passengers unfasten their seat belts.
3. One at a time, the passengers move carefully onto the skid.
4. Carefully, they step off and beyond (never between the skid and the helicopter).
5. They remain in a crouch on the ground until the helicopter clears the area.

The basic sequence for boarding is as follows:

1. Stay crouched until the pilot has steadied the ship.
2. Move close to the skid while watching the pilot. Do not touch the skid until the pilot signals.
3. When the pilot signals, place one foot on the skid and grasp the door frame. Your movement must be slow and deliberate. Do not switch feet on the skid. Move slowly and deliberately through the door placing your other foot inside the ship.
4. Follow through by moving your first foot from the skid. Take your seat and fasten the safety belt.

Hovering Tactics

WARNING NOTE

Hovering is one of the least desirable maneuvers for the helicopter pilot because it places the ship in a very vulnerable and unstable situation. Several factors create this situation:

- **In a hover, the helicopter is being kept in the air only by the collective (angle of the rotor blades) and engine rpm.**
- **In a hover, the aircraft lacks the forward momentum that could add to its lift and carry it away both from its own destruction and the danger to people below it.**
- **In a hover, the helicopter may be endangered by its own rotor wash which may be recirculated to create an unstable air mass.**

For these reasons, helicopter hover time must be kept to a minimum.

Hover - Jumps

A hover - jump is often used when weather conditions, such as cross winds, prevent even a one skid landing. In a hover - jump, the personnel exit the aircraft by moving carefully and deliberately onto the skid and then dropping onto the ground. It must be done only in the best of conditions and where the terrain is known.

Hoist Systems

Hoists are based on either hydraulic or electric winch systems and are usually found in larger aircraft, such as the Huey series. While they add convenience to some helicopter operations, winches require a well trained and experienced operator.

Their failure is usually due to one of the following reasons:

- Cable break

  The cable has been overstressed previously and the defect gone undetected. When loaded, the cable fails catastrophically.
- Failure in the Winch Mechanism

  This can occur when the winch mechanism fails catastrophically as the cable is being wound, or the cable has been completely extended, but the mechanism is unable to draw it in.

WARNING NOTE

When attaching a person to a helicopter lift system, whether it be hoist or static line, use only hardware that is designed and rated for life support. NEVER use hardware, such as cargo hooks, that is designed and rated only for lifting material.

Belayed Hoists

As a matter of course, some services belay the hoist with a separate line. The separate line is usually a nylon rope that would be the same as is used in a high angle environment.

The rope is run through a conventional belay device, such as a belay plate, which is anchored to a structural part of the aircraft.

> **WARNING NOTE**
>
> **Helicopters build up large charges of static electricity due to the action of the rotors against the air. This static electricity will be discharged to the ground through the helicopter or metal components, such as cable. Therefore, whenever a component, such as sling or stretcher is being lowered by a helicopter, always allow it to touch ground before reaching for it.**

Horse Collar

The horse collar is one traditional method used by the military for extricating personnel. *(see Figure 17.4)*. During the past several years there have been numerous incidents in which persons have slipped out of the horse collar, resulting in severe injury or death. As a result, some military search and rescue units do not use the horse collar with untrained civilian personnel.

Fig. 17.4 Horse Collar

The most common cause of the personnel slipping out of the horse collar is misuse of it. The device is designed to be used as follows *(see Figure 17.4)*:

1. Bring horse collar over the head and shoulders.
2. Place the collar under the armpits with the hoist cable in front.
3. Grasp the horse collar just below the cable hook to prevent the hook from snapping into your face when tension comes onto the cable.
4. When the cable becomes taut, the cable hook should remain at just above eye level.

   NOTE: Some versions of the horse collar have "retainer straps" to help secure a person in the horse collar. These are located in pockets on each side of the collar and are usually marked "pull." The straps are run around the back of the person and clip together with a "V" ring and a quick ejector fitting.

> **WARNING NOTE**
>
> **a) The horse collar can only be used on a fully conscious individual.**
>
> **b) Even an alert, well-trained individual may fall from a horse collar. The collar may impinge on nerves and result in the individual's loss of sensation in the arms. He may then lose his grip and fall from the harness.**

Jungle Penetrator

The Jungle Penetrator is a metal device with a tapered end. It has arms that swing down on which persons can ride, and a webbing loop to hold the user onto the device *(see Figure 17.5)*.

Full Body Fishnet *("Screamer Suit")*

The Full Body Fishnet *(see Figure 17.6)* is a net-like device constructed of mesh and webbing. It is designed to encompass rescue subjects who are disabled or unconscious to prevent them from falling out. Because the Full Body Fishnet requires no knowledge of it by the rescue subject, or any previous experience with a hoist, it is appropriate for civilian rescues.

It is also very simple to use by the rescuer, and is attached via a single large locking carabiner.

Fig. 17.5(a) Fig. 17.5 *Jungle Penetrator* 17.5(b)

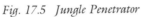

Some services require a litter spider to meet certain specifications before they will hoist it. One service, for example, requires that the spider be made of cable. When it is feasible, many services will have a crewman leave the helicopter to inspect the spider before they attempt to hoist it.

Critical time often can be saved when you know if the service requires specifications for the litter spider.

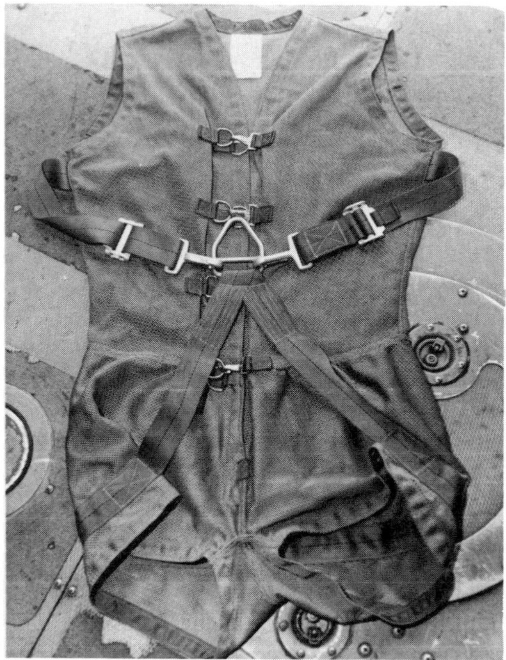

Fig. 17.6 Screamer Suit

MEDICAL CONSIDERATIONS

Even when they are in their best physical and mental condition, people may find helicopters intimidating. When a person is seriously injured and in a state of anxiety, a helicopter may be very frightening and could increase the medical problems. Before transporting a patient by helicopter you must:

1. Thoroughly evaluate his medical condition and his state of mind.
2. Explain to the subject the nature of the steps that are to take place.
3. Reassure him.

Litter Hoists

The hoisting of a litter to a helicopter is commonly carried out by some military services. There are some primary considerations for such litter hoists.

The Spider

The litter spider must have a low enough profile so that when hoisted to the helicopter, the litter does not hang so low that the air crew is unable to bring it inside. Figure 17.7 illustrates the maximum height for a litter spider to be hoisted to a Huey model helicopter.

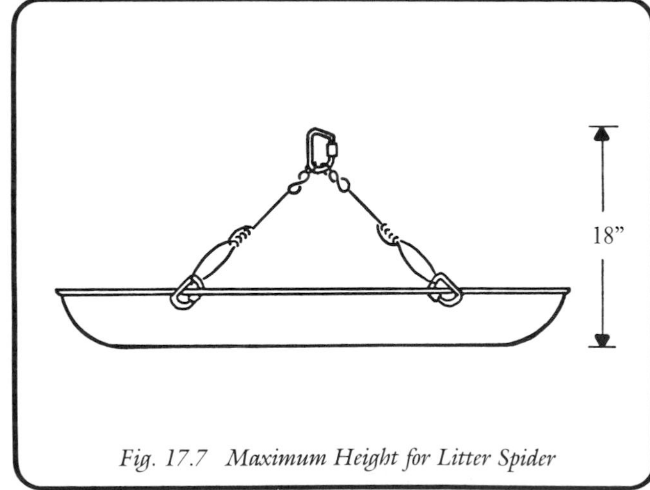

Fig. 17.7 Maximum Height for Litter Spider

Tag Lines

One problem with hoisting a litter is that it tends to spin and pendulum. One way to reduce these problems is to have a tag line tied to the foot of the litter and steadied by the ground crew as the litter is being hoisted. The tag line should have a simple attachment to the foot of the litter, such as a carabiner. This is so when the litter reaches the aircraft, a crew member can quickly detach the tag line and drop it to the ground so it does not get sucked into the tail rotor.

Litter Movement

Complete all necessary litter movement before attaching it to the cable. If the litter has been lowered, remove it from the cable before moving it. These actions will free the helicopter for any maneuvering it needs to make, and should a wind gust come or the wind change direction, will prevent the helicopter from dragging the litter along the ground.

Patient Protection

The patient must be protected against chill and wetness. His face and eyes must be protected from blowing debris.

WARNING NOTE

Any litter used in the water must have a flotation kit. This kit, which includes a ballast bar on the foot end of the litter, supports the patient and litter in the water, keeps the patient's face out of the water, and prevents the litter from floating face-down.

Fixed Line Flyaways

A fixed line flyaway is a helicopter extrication technique in which a person, or persons, are connected to a rope attached to a helicopter and the aircraft lifts off with the person carried below it. The exact length of the line depends on the specific needs of the operation.

Fixed line flyaways are commonly used in smaller helicopters where winches are not available. There are a number of specific procedures that can be used with the fixed line flyaway technique.

Nets

A well-designed net has the advantages of being able to quickly evacuate personnel, to lift out disabled people, and to lift off several people at once. There are, however, some design considerations for nets that should be considered:

- The net should have some rigid components so that the net does not close in and entrap the subject or create a feeling of claustrophobia.
- If used in the water, it should have proper flotation so that the net does not drag the subject under the water.
- The net should create a user friendly atmosphere so that the subject desires to climb into it and does not try to hang onto the outside.

Attached Person Flyaways

One of the simplest forms of flyaway is the attached person flyaway. This is often used in tactical operations where personnel have to be removed hurriedly. One common tactic when using larger aircraft such as the Huey series is to have four tactical personnel on the ground ready and wearing seat harnesses. The helicopter approaches with four lines attached, each with Figure 8 knots in the bottom end. The four personnel then quickly attach themselves to the Figure 8 knots and the helicopter flies away with them.

A major problem with a multi person flyaway is independent penduluming—the swinging away from one another. This can cause discomfort and injury to the persons involved, but the major problem is that such wide penduluming and sudden closing together may cause such large weight shifts that it results in control problems for the pilot.

One way to avoid this kind of penduluming is for the persons to link themselves together with their arms *(see Figure 17.8)*.

Fig. 17.8 Linking Arms to Prevent Pendulum

Also used are one-and two-person insertion/extrication techniques using a fixed line flyaway. Though the pendulum problem may still be a possibility, there are not other people with which to collide. One technique for rescue is for the rescuer to be attached to a fixed line under the aircraft and to come in and pick off a rescue subject. This technique is essentially the same as a one person rescue technique (see Chapter 12, "One-Person Rescue Techniques"). As with one-person rescue techniques, the subject **should not be attached to the rescuer's seat harness, but directly to his vertical system, such as the rescuer's rappel device, his main attachment carabiner, or the main line rope.**

WARNING NOTE

Any person involved in a helicopter flyaway will be vulnerable to a high degree of heat loss from the rotor wash and environmental wind. In cold weather and where water is involved, this heat loss could be substantial and could result in dangerous chilling both to rescuers and rescue subjects. Consequently, there must be protection from wind and soaking for all persons involved in a flyaway.

Litter Flyaways

Where helicopter hoists are not available for litters, litter flyaway procedures are sometimes used. It is best to have a litter tender attached to the litter system to fly along with the patient. The main reason for this relates to medical considerations for the patient, primarily to ensure an open and a clear airway.

There may, however, be circumstances in which the use of a litter tender on a flyaway is not practical. These often relate to altitude where the aircraft's performance is so limited that the additional weight of the tender cannot be tolerated. In these cases, the actual time of the flyaway must be kept as short as possible.

MEDICAL CONSIDERATIONS

When an injured patient is transported without a litter tender, one of the primary dangers is blockage of the airway, which will quickly result in death. Airway blockage is often associated with the following conditions:

- Unconscious patient

  When a person is lying supine (on the back) the tongue can fall back and block the airway.

- Material that is the result of the injury (such as blood or vomitus) can be aspirated into the lungs.

If a patient cannot be attended during transport, then he should be transported in as short a time as possible and transported in the "coma position" (on the side).

WARNING NOTE

Avoid Litter Spin

Due to down draft and pendulum effect, litters in a flyaway tend to spin uncontrollably. This can complicate existing injuries and create injuries on its own. The litter should be rigged with a spider so that it does not spin.

Helicopter Rappeling

Helicopter rappeling is one of the most spectacular and exhilarating of the high angle techniques. It also can be used to insert personnel in otherwise unavailable areas for such purposes as rescue, fire fighting, and tactical operations. However, helicopter rappeling has some severe constraints and hazards and requires some very special training techniques if it is to be done successfully.

Aircraft Type

One of the most important factors affecting how rappels are conducted from a helicopter, and indeed if they are conducted at all, is the aircraft type.

It is very important that the aircraft have a great deal of axial stability so that the weight of the rappeler coming onto the side of the aircraft does not cause disequilibrium.

The axial stability of a helicopter will depend on such factors a engine power, surface area of rotors, and number of rotors. Aircraft such as the Huey series, or the Blackhawk, for example, have greater stability, while such aircraft as the Long Ranger and the Hughes 500 have less axial stability. This does not necessarily mean that rappeling cannot be done from the latter craft. It does mean that such rappeling does require greater preparation and care.

Pilot and Crew Experience.

The air crew and pilot must have worked with rappeling and have the confidence and experience to control the aircraft while the rappeling is occurring. If they do not have the confidence and experience, the rappel could endanger the aircraft or injure the rappeler.

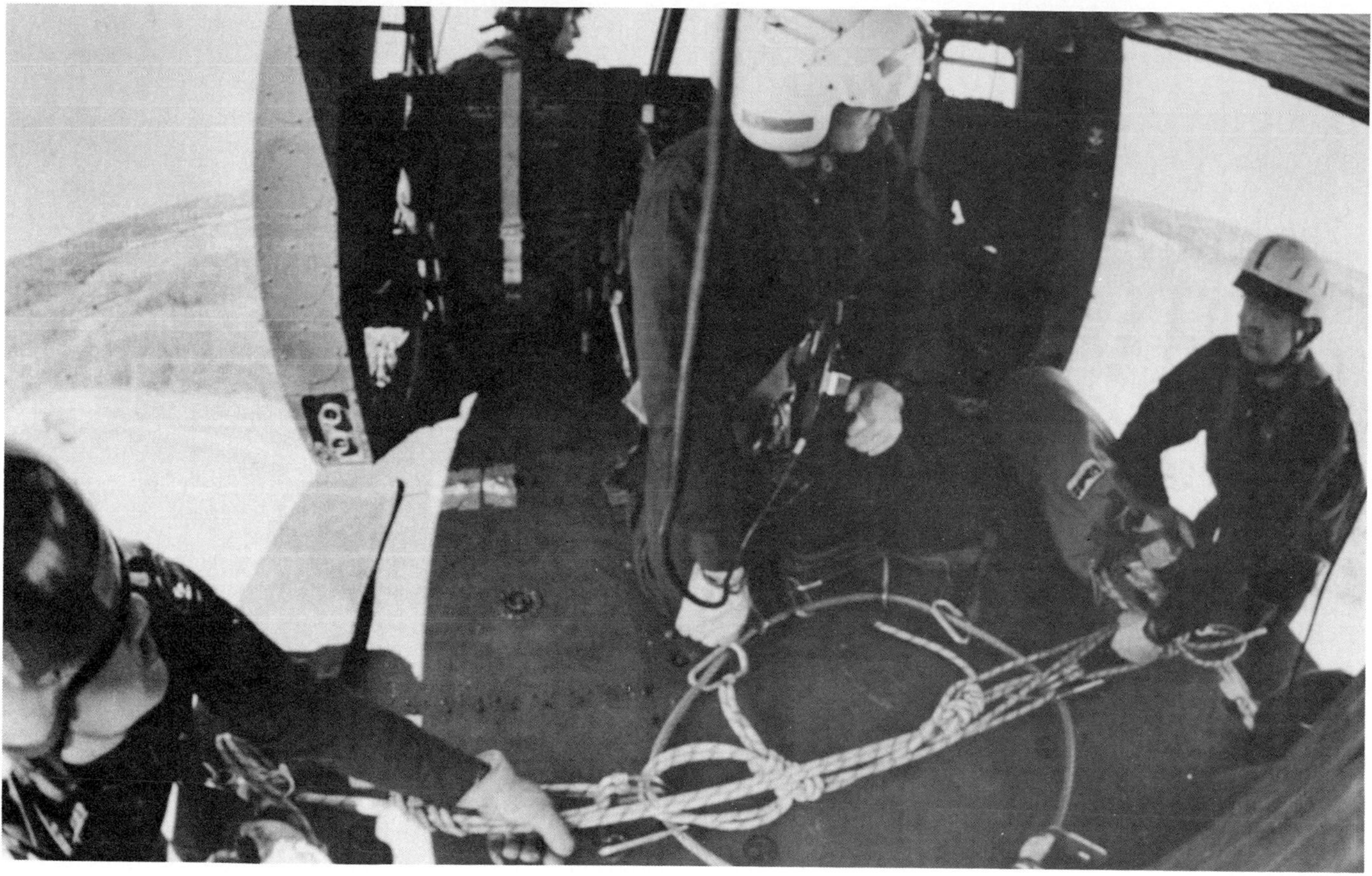

Fig. 17.9 Deck Anchor for Rappeling

Rappeler Experience

The person performing the rappel must have a great deal of experience and practice at rappeling and, most importantly, the ability to remain in control of the rappel. A sloppy, jerky rappel can cause the pilot to experience difficulty in controlling the aircraft and potential injury to the rappeler.

Most persons with the experience agree that their first helicopter rappel was an intimidating one. The helicopter rappels must be practiced with real aircraft and the proficiency kept up before one attempts an actual rappel during a rescue or tactical operation.

Interior Anchoring for Helicopter Rappels

One of the most critical elements of a helicopter rappel system is the establishment of a secure interior anchor system in the aircraft. Because helicopters are constructed of lightweight materials, with few structural members, it may be difficult to establish secure anchors inside the aircraft.

- Ceiling anchors

  Anchors for rappeling established in the ceiling are usually the most convenient for exiting the aircraft. This is because such an anchor creates a favorable angle on the rope (see Chapter 9, "Rappeling"). This type of anchor can be used only in a limited number of helicopter designs, usually those that have a large amount of ceiling space, such as the early model Hueys.

- Deck anchors

  An alternative for rappel anchors is in the helicopter deck. Unfortunately, most helicopter decks are not designed as load bearing, so the anchor must be attached at several points (in effect a multi point anchor). Figure 17.9 illustrates a typical deck anchor system.

- Struts

  In many of the older helicopters still in use, there may be doubts about the structural integrity of the floor. A third alternative is to tie the anchor system into the structural struts of the helicopter.

Types of Rappel Devices

Descenders used for helicopter rappeling should have all the positive characteristics of regular rappel devices (see Chapter 9, "Rappeling"). But they should also have the additional advantages of being able to take them off the rope and put them on the rope quickly, and correctly and while under stress.

Two of the most commonly used rappel devices for helicopter rappeling are the Sky Genie™ and the Figure 8 with Ears.

Ropes for Helicopter Rappeling

For helicopter rappeling, kernmantle construction rope is preferable to laid rope. Laid rope tends to spin when a person is free hanging on it, as in helicopter rappeling, and it also tends to kink. Kinks in a rope can cause rappel devices to jam, stranding the rappeler on the rope. While a difficult problem in an ordinary situation, in a helicopter rappel situation, this can have some extremely serious consequences.

Having a jammed descender in a rescue situation might mean you have to abort the rescue. In a tactical situation it might mean that you are there struggling with both hands to free the descender while someone is shooting at you. Not a nice situation to be in.

The Sky Genie™ device takes a rope specially manufactured for it and **it must be used only with that rope.**

If using one of the other types of rappel devices, such as the Figure 8 with Ears, then the decision is whether to use dynamic or static kernmantle ropes.

One theory is that dynamic rope is preferable for helicopter rappeling because its shock absorbing qualities dampen the effect on the aircraft caused by a rappeler suddenly loading the rope and by pendulums. However, this theory does not seem to noticeably bear itself out in practice and the trend now seems to be towards using static kernmantle ropes for helicopter rappeling. One advantage of static ropes is that they are more resistant to abrasion and cutting, factors that should be of great concern to anyone involved in helicopter rope operations (see paragraph below on padding of ropes).

As in any other rope handling situation, if severe shock loading is expected, then dynamic ropes should be used.

Rope Length

Many people experienced in helicopter rappel operations prefer to rappel on their ropes doubled, with the center of the rope anchored in the aircraft. The doubled rope has more friction (and therefore more control) for rappelling than a single rope and has more gripping surface for the brake hand.

If dynamic ropes are used, then there may be few options for length, since most dynamic ropes come in standard lengths such as 150 or 165 feet.

If static ropes are used the options will depend on local needs. Theoretically, the rope could be any length. You could in theory have a helicopter rappeling rope that was a thousand feet long. But this may not be reasonable in practical terms. A rope that long would cause some severe problems for the rappeler in terms of controlling friction on the rappel device and in bagging it.

The important thing is to settle on a standard length for the rappel rope, such as 200 feet overall, so that the rappelers and air crew know what to expect from it. If the air crew is aware of the length, then they will not attempt to hover too far above the ground for the rope length.

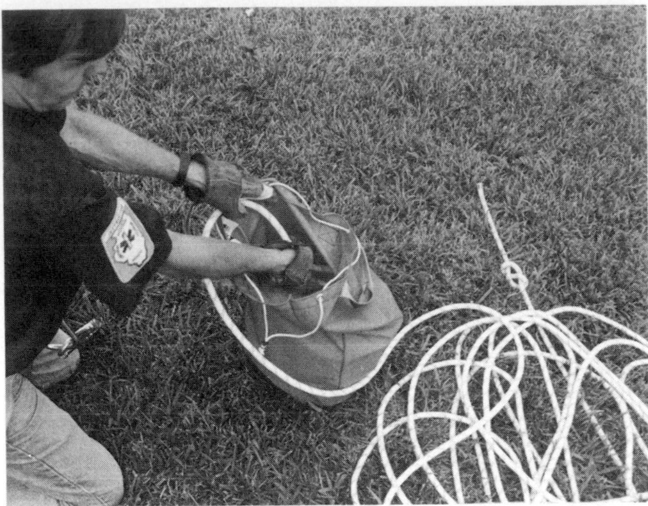

Fig. 17.10 Bagging Rope

Padding Ropes on Helicopters

Any place the rope runs across the aircraft that is at all potentially rough or sharp should be padded. An inspection for rough or sharp points should be made both visually and with an ungloved hand. Be alert for small items such as bolts and rivets that can snag and cut the sheath of a loaded rope.

It is generally more difficult to place effective rope pads on a helicopter than it is in a natural or building environment because the rotor wash and slipstream blow away padding that is not completely secured and, once in flight, it is difficult or impossible to reach some areas of the aircraft.

One padding technique is the use of PVC pipe that can be easily cut to fit shapes such as the skid and the edges of doors and ramps.

Rope Control

Rope control during helicopter rappeling operations is essential. The alternatives could lead to disaster. One possibility would be snagging the rope on an object on the ground. For this possibility, every helicopter conducting rope operations must carry a knife to cut snagged lines. One possibility that is even worse is getting the rope in the tail rotor, resulting in the aircraft's coming to the ground rather quickly.

In addition, ropes need to be carefully controlled when they are in the aircraft. If not under control, ropes can get snagged on many things, which can easily interrupt flight operations.

Bagging

One way to help control rope is bagging *(see Figure 17.10)*. Rope bags for helicopter rappeling should have a large grommet in the bottom or some other means of securing the bottom end of the rope to the bag. This will prevent rappeling off the end of the rope and ensure that the bag will not be lost when it is deployed.

WARNING NOTE

It is difficult to accurately estimate the length of rope when it is in a bag. This increases the danger of rappeling off the end of a bagged rope during helicopter operations, particularly when the bag is attached to the rappeler.

Bagged rope should always either be attached to the bag or have a large Figure 8 knot tied in the bottom end to prevent rappeling off the end.

Deployment/Recovery of Rope

Ground Practice for Helicopter Rappeling

There are many elements to the environment of helicopter rappeling that make it a very new and intimidating experience for the first time user. There is noise and vibration from the engines, the wind from the rotor wash and slip stream, the inability to hear the spoken word, the fact that you are about to leap off a moving platform, and the increased hazards of such things as the tail rotor.

Also intimidating for the first time user may be the fact that this rappel will be **completely free in space.** This means two problems for those who have never rappelled from a helicopter skid.

- They have to become accustomed to a controlled rappel with no place to put their feet.
- They need to get off the skid without banging their faces or other body parts against it.

To learn this technique, the initial training should be **on the ground** in a controlled atmosphere where there is good communication with a qualified instructor.

A Practice Skid System

A practice skid system consists of the following elements:

- A mockup of a helicopter skid. This can be made from a real skid or created of pipe made from the exact dimensions of the skid of the helicopter to be used in the rappel.
- A practice site for controlled conditions. The skid should be easily accessible for the trainee and an instructor should be close by. A fire training tower can often be used for a practice site.
- The practice skid must be rigged in such a manner so that the rappeler does not hit the side of the structure as he swings forward after taking his feet off the skid. If the practice skid is on the side of a building, it may have to be set so that the rappeler swings through a window.

WARNING NOTE

The following practice procedure must be conducted only under the direct supervision of a qualified and experienced instructor.

Until the trainee displays the confidence and ability to securely control his rappel, he should have a top belay.

Procedure for Rappeling off a Skid

Rappeling from a Huey Type helicopter is similar to rappeling from a severe cliff overhang. It means planting the feet firmly on the skid, lowering the upper body until it is lower than the feet, and then taking the feet off the skid in a controlled manner so there is very little pendulum. Everything about this procedure must be under control which means that the person **must be very experienced in rappeling and be using a rappel device that gives him sufficient control.** *(See Figure 17.11 on page 232.)*

Fig.17.11(a)

Fig. 17.11(b)

Fig. 17.11 Rappeling off a Skid

1. Before stepping onto the skid, the rappeler clips his rappel device into the rope, onto his seat harness, and locks off the rappel device.
2. The instructor inspects the rappeler's rigging, including the anchoring system to the aircraft, the lacing of the rappel device to the rope, the attachment of the rappel device to the rappeler's seat harness carabiner and the rappeler's seat harness.
3. The instructor signals the rappeler to unlock his rappel device.
4. The rappeler unlocks his rappel device.
5. The instructor signals the rappeler to take the rappel stance.
6. The rappeler leans back in the rappel stance and waits.
7. The instructor signals the rappeler to begin his rappel.
8. With the use of the rappel device, the rappeler leans back until his feet, still planted on the skid, are higher than the rest of his body.
9. When he is certain that all of his body will clear the skid, he drops his feet by flexing his knees and pendulums forward.
10. He gently drops his feet until he is in the normal rappeling position.
11. He continues by smoothly rappeling to the ground.

Communications with the Pilot

It is absolutely essential that during rappeling operations there be continuous hardwire communications between the pilot and the observer, crew chief, or whoever is supervising the operation. This is necessary because many actions will be occurring that the pilot will be unable to see, but which he must take into account to properly conduct his ship. Among these are:

- Direction to drop zone.
- Notification when the ship is directly over the drop zone.
- Warning of any hazards.
- When a rappeler is about to leave the ship.
- When a rappeler is about to drop off the skid.
- When a rappeler has left the skid.
- When a rappeler is about to reach the ground.
- When a rappeler has reached the ground and detached.
- When the ship is free to leave the area.

APPENDIX I

GLOSSARY

Abrasion—The damaging wear on rope and other gear caused by their rubbing against hard material.

Anchor—In the high angle environment the means of attaching the rope and all other portions of the system to something secure.

Anchors—The means of attaching the high angle system to secure points so that the system will not fall.

Anchor Point—A single secure connection for an anchor. It will range in size from a piece of hardware wedged in the crack of a rock to a large tree or rock.

Anchor System—Multiple anchor points rigged in such a way that together they provide a "bombproof" anchor.

Arm Rappel—(Guide's Rappel) A type of rappel in which the rope wraps around both outstreched arms and across the person's back. The technique is better suited for slopping terrain than for vertical situations.

Artificial Anchors—The use of specifically designed hardware to create anchors where no good natural anchors exist.

Ascender—A mechanical device, or a friction knot, that is used in ascending a fixed rope. They are secured to the rope and attached via attachments (slings) to the person using them.

Ascender Sling—Attachments of webbing or rope that connect a person to his ascenders.

Ascending—A means of traveling up a fixed rope with the use of either mechanical devices or friction knots that are attached with slings to the user's body.

Auxiliary Tender—A person who rappels alongside the litter as it is being lowered in order to assist in the rescue. Duties may include medical assessment and/or primary treatment of the rescue subject, assistance in getting the litter over the edge, assistance in handling the litter on the vertical face, and assistance in loading the subject into the litter.

Backing Up—The creation of an additional independent anchor, or anchors, to sustain the high angle system should initial anchors fail. Backing up may be to the same anchor point if it is very solid, or to additional anchor points.

Back-Up-Knot—A knot used to secure the tail of another knot. Also known as a **Safety** or **Keeper Knot.**

Belay—The securing of a person with a rope to keep him from falling a long enough distance to cause harm.

Belayer—The one who performs the belay.

Belay Plate—A simple metal plate containing one or more slots for rope, and used to create rope friction with a carabiner. It is commonly used in belaying.

Bend—A knot that joins two ropes.

Bight—The open loop in a rope formed when it is doubled back on itself.

Body Rappel—(**Dulfersitz Rappel**)—A type of rappel that uses the body as friction by running the rope through the legs, across one hip, over the opposite shoulder, and to a braking hand. Because of the discomfort involved, and potential damage to body parts, the technique has largely been supplanted by other techniques.

Bolts—Metal devices used to create permanent anchors in rock. Because they permanently deface the rock and take time to place, they have limited application in high angle activities.

Bombproof—An anchor that will not fail.

Brake Bar Rack—A descending device consisting of a "U"-shaped metal bar to which are attached several metal bars that create friction on the rope. Some "racks" are restricted to use for personal rappeling, while others may also be used for rescuer lowering. Also commonly known as a **Rappel Rack**.

Brake Hand—The hand, usually the dominant one, that grasps the rope to help control the speed of descent during a rappel.

Brakeman—Person who operates the braking device that controls the rate of descent of a load in lowering operations such as slope evacuation and high angle lowering.

Cams— (**1**) A generic term for ascenders that grip the rope through pressure. Some cams are spring-loaded to assist in this function.

(**2**) Devices used in climbing for protection or in anchoring and which lodge in a rock crack through offset cam action.

Carabiners—Metal snap links used to connect elements of a high angle system. Sometimes (outside of the United States) spelled **Karabiner**. Also known as **biners** or **crabs.**

Carabiner Wrap—A rappel technique that uses several rope wraps around a seat harness carabiner to create friction and control the descent. It is generally not considered a safe and secure technique for rappeling.

Changeover—To transfer from an ascending mode to a rappeling mode, or from a rappeling mode to an ascending mode.

Chest Harness—A type of harness worn around the chest for upper body support. In the high angle environment, it should never be used as the only source of support, but always be used in combination with a seat harness.

Chicken Loop—A safety loop that fits around the ankle to secure the ascender sling and prevent the foot from slipping out of the sling should an upper connection fail and the person ascending falls over backwards.

Counter Balance Hauling System—A procedure for hauling that uses a 1:1 ratio and a haul team that moves in a direction opposite to the load;

Descender—A rappel device that creates friction by a rope running through it, and is attached to a rappeler to control descent on a rope. Most descenders can also be used as a fixed brake lowering device.

Descenders—Metal device that, through friction with the rope, creates braking action for a controlled rappel or lowering.

Directional—A technique for repositioning a rope at a more favorable angle than would exist using only its anchor.

Double Strand Lowering—The use of two ropes attached to the litter in a lowering rigged so that they may be operated independently to change the angle of the litter. Double strand lowerings usually involve two litter tenders.

Dulfersitz—See **Body Rappel**.

Dynamic Rope—A type of rope designed for high stretch to reduce the shock on the climber and anchor system. Usually employed in rock climbing and mountaineering.

Edge Rollers—In-line, free-turning rollers that are anchored at an edge of a cliff or building to reduce rope friction.

Edge Tender—A person connected to a safety attachment who works at the edge of a drop in a high angle lowering. His duties include assistance in getting the litter over the edge, reducing edge abrasion to the rope, and, when necessary, relaying communications between the litter tender and the brakeman.

Emergency Seat Harness—A temporary, tied harness to be used when a manufactured, sewn seat harness is not available.

Fall Factor—A calculation used to estimate the impact forces on a rope when it is subjected to stopping a falling person. It is expressed as a number to indicate the relation between rope length and the distance the person falls.

Foundation Knot—A simple knot that is tied as the first step in tying a more complicated knot. Examples of foundation knots include the Overhand and the Simple Figure 8.

Figure 8 Descender—A device used for rappeling and, in some cases, for lowering. It is in the general shape of an "8," with a large ring to create friction on the rope and a smaller ring for attaching to a seat harness.

Full Body Harness—A type of harness that offers both pelvic and upper body support as one unit.

Guide Hand—The hand, usually not the dominant one, that cradles the rope to help in balancing the rappeler.

Handled Ascenders—Ascenders with frames large enough to accommodate built-in handles that can be comfortably gripped with the hands. Most handled ascenders have toothed cams that grip the rope.

Haul Cam—A cam ascender (or Prusik knot) that grips the rope to provide the "bite" in hauling.

Haul Team—The group of persons who provide the power to raise the load.

Helmet—Head covering that protects against head injury both from falling objects and from head impact. When used in this manual, "helmet" indicates head protection specifically designed for high angle work.

High Angle—The high angle environment in which one must be secured with rope and other equipment to keep from falling.

Highline—A system of using a rope suspended from between two points to move persons or equipment over an area that is a barrier to the rescue operation.

Horizontal Highline—A highline in which the two suspension points are close to being on the same level.

Kernmantle—A rope design consisting of two elements: an interior core (kern) which supports the major portion of the load on the rope, and an outer sheath (mantle) which serves primarily to protect the core and also supports a minor portion of the load.

Kevlar™—Trade name for a type of Aramid fiber manufactured by the Dupont Corporation and which has high tensile strength, low elongation and high resistance to heat.

Laid Rope—A rope design that consists of fiber bundles twisted around one another.

Litter Captain—The person in slope evacuation who manages the litter team and coordinates the litter movement with other members of the rescue team.

Litter Tender—A person who physically manages the litter in slope evacuation.

Load—That which is being lowered or raised by rope in a high angle system. Some examples include a rescue subject, a rescuer and subjects in a litter with an attached litter tender.

Locking Carabiner—A carabiner with a locking sleeve on its gate side that secures the gate shut.

Locking Off—The technique of jamming a rope into a descender or tying off securely so that the rappeler can stop the descent or lowering and operate hands free of the rope.

Lowering/Belay Line—The line that is attached to a load on a highline and is used to control the load from the near-side point.

Manner of Function—The method in which a particular piece of equipment was designed to be used.

Mechanical Advantage—The relationship of how much load can be moved to the amount of force it takes to move it.

Mechanical Ascenders—Rope grab devices used by individuals to ascend a fixed rope or, with specific types of ascenders, used in the creation of hauling systems. There are two categories of ascenders: (1) **handled ascenders** normally used for no more than one person's body weight and (2) **cams** which are used both as personal ascenders and for hauling systems.

Mountaineering—The use of skills such as climbing, snow and ice travel, and camping to ascend a mountain.

Multi-Pitch—More than one rope length.

Munter Hitch—A type of running knot that slips around a carabiner to create friction against itself. It is commonly used in belaying.

NFPA—National Fire Protection Association. A national organization that sets safety standards, among them life safety equipment for fire fighters.

Non-Locking Carabiner—A carabiner without a means of securing its gate shut.

Nylon 6—A type of nylon used in rope manufacturing. One trade name for this type is **Perlon.**

Nylon 6,6—A type of nylon used in rope manufacturing. It is manufactured by Dupont and Monsanto in North America.

1:1 Hauling System—A procedure for hauling where the force needed to haul is roughly the same as the load being hauled, i.e., without mechanical advantage

Packaging—The placing of a rescue subject in a litter so that the primary medical considerations are cared for and the subject is physically stabilized in the litter.

Pendulum—To swing on a rope.

Perlon—A trade name for one type of nylon type 6.

Pig Tail—A short piece of rope with which the litter tender attaches to the litter system.

Piggyback System ("Pig-Rig")—Common name given to a specific type of 4:1 hauling system in which one 2:1 system is attached to ("piggybacked onto") another 2:1 system.

Pitch—One rope length.

Piton—A slender metal wedge, with an eye for attachment, that is driven into a rock crack for climbing protection or for anchoring.

Polyester—A type of fiber used in some rope manufacturing. Also known by the trade name Dacron™.

Polyolefins—A group of fiber types used in manufacturing ropes that are often used in water applications. In this group are polypropylene and polyethylene.

Prusik—A type of friction knot used in ascending. It has also come to be used by some individuals as a term synonymous with **ascending**, even when mechanical devices are used, i.e., "to Prusik."

Prusik Loop—A continuous loop of rope in which a Prusik knot is tied.

Pulley—A device with a free-turning, grooved metal wheel (sheave) used to reduce rope friction, and which has side plates to which a carabiner may be attached.

Rappel Rack—See **Brake Bar Rack.**

Rappeling—The controlled descent of a rope using the friction of the rope against one's body or through a descender.

Ratchet Cam—The cam ascender (or Prusik knot) in a hauling system that holds the rope to prevent the load from slipping while the haul team resets the system to get another "bite" on the rope. In some hauling systems, the ratchet cam may be the same as the **safety cam.**

Rescue Hauling—Techniques for using rope, pulleys, cams, and other equipment to give mechanical advantage, convenience, or added safety for raising a rescue load.

Rock Climbing—Ascending while making direct contact with the rock and commonly using rope and other equipment for safety should one fall.

Rope Handler—The person in a litter lowering operation who assists the brakeman with rope management.

Rope Rescue—The performing of rescue in an high angle environment where the use of rope and related equipment is necessary.

Safety Belt—A belt-like harness worn around the waist to prevent falls from elevated positions. Should never be used as sole means of suspension.

Safety Cam—The cam ascender (or Prusik knot) in a hauling system that prevents the rope and the load from accidentally slipping should a mishap occur to the haul system. In some hauling systems, the safety cam may be the same as the **ratchet cam.**

Safety Factor—The ratio between the maximum load expected on a rope and the rope's breaking strength. The larger the ratio, the greater the safety factor.

Safety Knot—See **Back-Up Knot**

Seat Harness—A system of nylon or polyester webbing that wraps and supports the pelvic region to attach the wearer to the rope or other protection in the high angle environment.

Self-Equalizing Anchor—An anchor system established from two or more anchor points that: a) maintains near equal loading on the anchor points despite direction changes on the main line rope, and b) reestablishes equal loading on remaining anchor points if any one of them fails. Also known as the **SEA.**

Single Rope Techniques (SRT)—Ascending and descending directly on the rope without direct aid by contact with the rock.

Single Strand Lowering—The use of one main lowering rope with a belay in litter lowering.

Slope Evacuation—The movement of a rescue subject over terrain that is so rugged or angled that it requires the litter to be safetied with a rope and its descent controlled by a braking device or its ascent assisted with a hauling system.

Software—A category of high angle equipment that is not hardware. In this category are rope and webbing.

Spider—The system of attaching a lowering rope to a litter. A spider usually has four or more legs that connect to various points of a litter to equalize loading.

Static Rope—A type of rope designed for low stretch. It is used in applications such as rescue, rappeling, and ascending where high stretch would be a disadvantage and where no falls, or very short falls, are expected before being caught by the rope.

Steep Angle Highline—A highline in which one of the suspension points is considerably higher than the other.

Stopper Knot—A knot that helps provide security in rope work. Examples would be a Simple Figure 8 tied in the bottom end of a rope to prevent a person from rappeling off the end or tied in an end of the rope to prevent it from accidentally slipping through equipment.

System—The combination of all the various elements, including rope, hardware, anchors, etc., used in the high angle environment.

Tag Line (1)—The line that is attached to a load on a highline and is used to control the load from the far-side point.

Tag Line (2)—A line attached to a load that can be used to maneuver the load and prevent it from snagging, and to hold it away from a vertical face.

Tandem Pulleys—Two in-line pulleys on the same rope. In highlines, tandem pulleys are employed to stabilize a load and to spread its weight out along the rope.

Technical Rescue—See **Rope Rescue**

Telpher—Another name for a highline.

Tensile Strength—A measurement of the greatest lengthwise stress that a rope or piece of equipment can resist without failure.

Theoretical Mechanical Advantage (TMA)—Mechanical advantage without allowance for friction and other losses of advantage.

Traveling Pulley—A moving pulley that is attached to a load or to a haul cam and which adds to the mechanical advantage.

Tree Wrap—A technique of running a rope around a tree trunk to create friction for a braking effect in a litter lowering.

Tying Off Short—A safety technique which creates an extra point of attachment during ascending by tying the person directly into the main line rope.

Tyrolean—Another name for a highline.

UIAA—The Union of International Alpine Associations. An organization that sets performance standards for ropes, harnesses, ice axes, helmets, and carabiners to be used by climbers and mountaineers.

Vertical Caving—The travel through caves that have vertical, or near vertical, sections that require the use of rope and ascending and descending equipment.

Z-Rig—Common name given to a specific type of 3:1 hauling system. The name is taken from the general shape that the rope makes as it runs through the system.

APPENDIX II

SKILLS CHECK LIST

High Angle Skills Checklist

KNOTS

Instructions

As each student completes a skill check, the evaluator notes the date in the evaluator column. If the student demonstrates the prescribed level of competence for the skill, the instructor signs his initials; if the student fails, the instructor places a 'U' (Unsatisfactory). If successful, the student signs his initials in the adjacent column to the right.

Student:

Evaluator(s): ______________________

____

____

	1st try		2nd try	
	EVALUATOR	STUDENT	EVALUATOR	STUDENT
Tie a Simple Overhand Knot.				
Tie a Simple Figure 8 Knot.				
Tie a Figure 8 on a Bight Knot.				
Tie a Figure 8 Follow Through Knot.				
Tie a Figure 8 Bend.				
Tie a Water Knot in Webbing.				
Tie a Barrel Knot.				
Tie a Double Fisherman's (Grapevine) Knot.				

High Angle Skills Checklist

ANCHORING

Instructions

As each student completes a skill check, the evaluator notes the date in the evaluator column. If the student demonstrates the prescribed level of competence for the skill, the instructor signs his initials; if the student fails, the instructor places a 'U' (Unsatisfactory). If successful, the student signs his initials in the adjacent column to the right.

Student:

Evaluator(s): ______

	1st try EVALUATOR	1st try STUDENT	2nd try EVALUATOR	2nd try STUDENT
Tie & Rig a Tensionless Hitch on Anchor Point.				
Tie & Rig Figure 8 on a Bight on Anchor Point.				
Tie & Rig Figure 8 Follow Through on Anchor Point.				
Tie & Rig a Water Knot in Webbing on Anchor Point.				
Tie & Rig Load Sharing Anchor on Two Anchor Points.				
Rig Self-Equalizing Anchor on Two Anchor Points.				
Rig Self-Equalizing Anchor on Three or More Anchor Points.				

High Angle Skills Checklist

BELAYING

Instructions

As each student completes a skill check, the evaluator notes the date in the evaluator column. If the student demonstrates the prescribed level of competence for the skill, the instructor signs his initials; if the student fails, the instructor places a 'U' (Unsatisfactory). If successful, the student signs his initials in the adjacent column to the right.

Student:

Evaluator(s): ______________________

____

____

	1st try EVALUATOR	1st try STUDENT	2nd try EVALUATOR	2nd try STUDENT
Repeat from memory and in correct sequence the belay voice communications.				
On level ground, use a Munter Hitch to belay a person moving away from the belayer.				
On level ground, use a Munter Hitch to belay a person moving toward the belayer.				
Using a belay practice system, use a Munter Hitch to catch a dropped weight as it is being lowered.				
On level ground, use a belay plate to belay a person moving away from the belayer.				
On level ground, use a belay plate to belay a person moving toward the belayer.				
Using a belay practice system, use a belay plate to catch a dropped weight as it is being raised.				
Using a belay practice system, use a belay plate to catch a dropped weight as it is being lowered.				

High Angle Skills Checklist

RAPPELING

Instructions

As each student completes a skill check, the evaluator notes the date in the evaluator column. If the student demonstrates the prescribed level of competence for the skill, the instructor signs his initials; if the student fails, the instructor places a 'U' (Unsatisfactory). If successful, the student signs his initials in the adjacent column to the right.

Student:

Evaluator(s): ______________________

____

____

	1st try EVALUATOR	1st try STUDENT	2nd try EVALUATOR	2nd try STUDENT
Rappel Using a Figure 8 with Ears, and:				
1. Attach it to harness and rope correctly.				
2. Maintain control during the entire rappel.				
3. Partway down, come to a complete stop.				
4. Lock off the Figure 8 descender securely.				
5. Unlock the descender and complete the rappel.				
Rappel Using a Brake Bar Rack, and:				
1. Attach it to harness and rope correctly.				
2. Maintain control during the entire rappel.				
3. Partway down, come to a complete stop.				
4. Lock off the Brake Bar Rack securely.				
5. Unlock the descender and complete the rappel.				

High Angle Skills Checklist

ASCENDING

Instructions

As each student completes a skill check, the evaluator notes the date in the evaluator column. If the student demonstrates the prescribed level of competence for the skill, the instructor signs his initials; if the student fails, the instructor places a 'U' (Unsatisfactory). If successful, the student signs his initials in the adjacent column to the right.

Student: ____________________

Evaluator(s): ____________________

____

____

	1st try		2nd try	
	EVALUATOR	STUDENT	EVALUATOR	STUDENT
Correctly tie a Prusik Knot onto a fixed rope.				
Ascend a rope safely and efficiently with an ascending system using three ascenders.				
Tie off short while ascending a fixed rope.				
Safely and efficiently change over from ascending to rappeling.				
Safely and efficiently change over from rappeling to ascending.				
Extricate oneself from a simulated jammed rappel device using ascenders.				

High Angle Skills Checklist

ONE PERSON RESCUE

Instructions

As each student completes a skill check, the evaluator notes the date in the evaluator column. If the student demonstrates the prescribed level of competence for the skill, the instructor signs his initials; if the student fails, the instructor places a 'U' (Unsatisfactory). If successful, the student signs his initials in the adjacent column to the right.

Student:

Evaluator(s): ______________________

____

____

	1st try		2nd try	
	EVALUATOR	STUDENT	EVALUATOR	STUDENT
Safely and efficiently perform a one-person rescue of a person wearing a seat harness.				
Tie onto a subject either a hasty seat harness, a hasty seat harness with a chest harness, or a hasty body harness.				
Safely and efficiently perform a one-person rescue of a person wearing a seat harness.				

High Angle Skills Checklist

SLOPE EVACUATION

Instructions

As each student completes a skill check, the evaluator notes the date in the evaluator column. If the student demonstrates the prescribed level of competence for the skill, the instructor signs his initials; if the student fails, the instructor places a 'U' (Unsatisfactory). If successful, the student signs his initials in the adjacent column to the right.

Student:

Evaluator(s): ______________________

____

____

	1st try		2nd try	
	EVALUATOR	STUDENT	EVALUATOR	STUDENT
Repeat from memory the voice communications used in slope evacuation.				
Correctly rig a litter for slope evacuation.				
Package a rescue subject for slope evacuation.				
Correctly rig and anchor a brake system for slope evacuation using a Figure 8 with ears.				
Correctly rig and anchor a brake system for slope evacuation using a Brake Rack.				
Correctly rig a tree wrap brake system.				
Correctly rig and anchor a belay system for slope evacuation.				
Correctly rig and anchor a 1:1 (counterbalance) hauling system for slope evacuation.				

High Angle Skills Checklist

HIGH ANGLE LOWERING

Instructions

As each student completes a skill check, the evaluator notes the date in the evaluator column. If the student demonstrates the prescribed level of competence for the skill, the instructor signs his initials; if the student fails, the instructor places a 'U' (Unsatisfactory). If successful, the student signs his initials in the adjacent column to the right.

Student:

Evaluator(s): ______________________

____

____

	1st try EVALUATOR	1st try STUDENT	2nd try EVALUATOR	2nd try STUDENT
Repeat from memory the voice communications used in high angle lowering.				
Correctly rig a litter for high angle lowering.				
Package a rescue subject for high angle lowering.				
Using a litter rigged and with a simulated rescue load:				
1. Correctly anchor a brake bar rack for lowering.				
2. Correctly attach the litter to lowering rope.				
3. Correctly rig and anchor a belay for litter.				
4. Correctly lace lowering rope to brake bar rack.				
5. Correctly lower litter with simulated load, stop lowering, bring rack to full stop, tie off rack, then unlock rack to continue lower.				
6. Correctly belay a simulated rescue load during a lowering.				
7. Perform a knot pass during a simulated lowering.				

High Angle Skills Checklist

HAULING

Instructions

As each student completes a skill check, the evaluator notes the date in the evaluator column. If the student demonstrates the prescribed level of competence for the skill, the instructor signs his initials; if the student fails, the instructor places a 'U' (Unsatisfactory). If successful, the student signs his initials in the adjacent column to the right.

Student:

Evaluator(s): ______________________

____

____

	1st try		2nd try	
	EVALUATOR	STUDENT	EVALUATOR	STUDENT
Repeat from memory the voice communications used in rescue hauling.				
Using a simulated rescue load, rig the following with appropriate anchors, pulley placement, haul cams, and safety cams: 1. A 1:1 haul system.				
2. A 2:1 haul system.				
3. A 3:1 haul system ("Z-Rig").				
4. A 4:1 haul system ("Piggy-Back Rig").				

High Angle Skills Checklist

HIGHLINES

Instructions

As each student completes a skill check, the evaluator notes the date in the evaluator column. If the student demonstrates the prescribed level of competence for the skill, the instructor signs his initials; if the student fails, the instructor places a 'U' (Unsatisfactory). If successful, the student signs his initials in the adjacent column to the right.

Student:

Evaluator(s): ______________________

____

____

	1st try		2nd try	
	EVALUATOR	STUDENT	EVALUATOR	STUDENT
Estimate the amount of visible sag needed in a highline under the following conditions:				
1. 200 pound load on 100 foot span.				
2. 200 pound load on 200 foot span.				
3. 400 pound load on a 100 foot span.				
4. 100 pound load on a 200 foot span.				
Correctly rig a litter for a highline with the following elements:				
1. Spider.				
2. Litter Tender attachments and safety.				
3. Attachments to main line.				
4. Lowering/belay line and tag line.				

APPENDIX III

STANDARDS SETTING ORGANIZATIONS & FURTHER READING

Standards Setting Organizations

ANSI
American National Standards Institute, Inc.
1430 Broadway
New York, N.Y. 10018
ANSI is concerned with standards for safety belts, harnesses, lanyards, lifelines, and drop lines for construction and industrial use.

ASTM
1916 Race Street
Philadelphia, Pennsylvania 19103
ASTM is the facilitator for standards created through the full consensus method. Standards for equipment and techniques used in search and rescue are currently being developed through the ASTM process.

NFPA
National Fire Protection Association
Batterymarch Park
Quincy, Massachusetts 02269
NFPA is concerned with rescue equipment used by firefighters on the fireground.

UIAA
c/o American Alpine Assn.
113 East 90th Street
New York, NY 10128-1589
The UIAA sets standards for rope, helmets, hardware, and other equipment used in recreational climbing.

OSHA
U.S. Occupational Health and Safety Administration
Department of Labor
Washington, D.C. 20210
OSHA is concerned with equipment used in the work place.

Further Reading

Frank, James A. and Smith, Jerrold B. 1987. *Rope Rescue Manual.* Santa Barbara: California Mountain Company. 138 pp.

Leonard, R. and A. Wexler. 1956. *Belaying the Leader, An Omnibus on Climbing Safety.* San Francisco: Sierra Club. 85 pp.

MacInnes, H. 1972. *International Mountain Rescue Handbook.* New York: Charles ScribnerUs Sons. 218 pp.

Montgomery, N.R. 1977. *Single Rope Techniques: A Guide for Vertical Cavers.* Sydney: Sydney Speleological Society. 122 pp.

National Cave Rescue Commission and National Speleological Society. 1988. *Manual of U.S. Cave Rescue Techniques.* Huntsville, Alabama: National Speleological Society. 260 pp.

Padgett, A. and Smith, B. *1987. On Rope: North American Vertical Rope Techniques for Caving, Search and Rescue, and Mountaineering.* Huntsville, Alabama: National Speleological Society. 341 pp.

Peters E., ed. 1987. *Mountaineering: The Freedom of the Hills,* Fourth edition, Seattle: The Mountaineers. 560 pp.

Paulcke W. and Dumler H. 1973. *Hazards in Mountaineering.* New York: Oxford University Press. 161 pp.

Robbins, R. 1971. *Basic Rockcraft.* Glendale, California: La Siesta Press. 70 pp.

Setnicka, T.J. 1980. *Wilderness Search and Rescue.* Boston: Appalachian Mountain Club. 640 pp.

Wheelock, W. 1967. *Ropes, Knots, and Slings for Climbers.* Glendale, California: La Siesta Press. 36 pp.

INDEX

NOTES

NOTES

NOTES

NOTES

NOTES

NOTES

NOTES